ARTHUR P. NORTON, 1876–1955

TWENTIETH EDITION

Norton's Star Atlas

and Reference Handbook

EPOCH 2000.0

EDITED BY

Ian Ridpath

Pi Press

NEW YORK

PI PRESS

An Imprint of Pearson Education, Inc.
1185 Avenue of the Americas, New York, New York 10036

Pi Press offers discounts for bulk purchases. For information contact U.S. Corporate
and Government Sales, 1-800-382-3419, corpsales@pearsontechgroup.com
For sales outside the U.S.A., please contact International Sales, 1-317-581-3793,
international@pearsontechgroup.com

Company and product names mentioned herein are the trademarks
or registered trademarks of their respective owners.

Images of Mars provided courtesy of Malin Space Science Systems.
For more information, contact Malin Space Science Systems at www.msss.com

Printed in the United States of America

Second Printing

Library of Congress Cataloging-in-Publication Data
A CIP catalog record for this book can be obtained from the Library of Congress.

Pi Press books are listed at www.pipress.net

ISBN 0-13-145164-2

Pearson Education Ltd.
Pearson Education Australia Pty., Limited
Pearson Education Singapore, Pte. Ltd.
Pearson Education North Asia Ltd.
Pearson Education Canada, Ltd.
Pearson Educación de Mexico, S.A. de C.V.
Pearson Education — Japan
Pearson Education Malaysia, Pte. Ltd.

Contents

vii   *Foreword by Leif J. Robinson*

ix   *Preface*

1   Chapter One: Position and Time

1   The Heavens Above

3   Position

7   Date and Time

23   Chapter Two: Practical Astronomy

23   Observing

28   Astronomical Instruments

40   Astronomical Imaging

49   Chapter Three: The Solar System

49   The Sun

54   The Moon

65   The Planets and Their Satellites

71   Mercury

73   Venus

74   Mars

78   Jupiter

82   Saturn

85   Uranus, Neptune, and Pluto

86   Minor Planets

88   Comets

91   Meteors

94   Eclipses

97   Occultations

98   Aurorae, Noctilucent Clouds and the Zodiacal Light

101   Artificial Satellites

103   Chapter Four: Stars, Nebulae and Galaxies

103   The Stars – Constellations and Nomenclature

106   Radiation, Magnitude and Luminosity

110   Distances, Motions and Physical Parameters

115   Spectral Classification

117   Stellar Evolution

119   Double Stars

122   Variable Stars

130   Clusters, Nebulae and Galaxies

139   Chapter Five: Star Charts

179   Appendix

179   *Units and Notation*

181   *Astronomical Constants*

181   *Symbols and Abbreviations*

182   *Useful Addresses*

185   Glossary

192   Index

Foreword

Once in a blue moon a book appears that dramatically and forever changes its subject. Equally rare is a book that remains in print for a century. *Norton's Star Atlas* is such a work, since 1910 being the quintessential guide to bright stars and fuzzy things that dot our nighttime sky. *Norton's* did for the heavens what Roger Tory Peterson's field guides did for nature study on Earth. Both works not only made it easy and fun to recognize new sights, they also prompted learning.

In 1956 the Los Angeles Astronomical Society held a prize draw among its junior members. The prize was, by the standards of this 17-year-old, a very nice mounting for a 6-inch telescope. Since only a winner would tell this story, I'll go on to say that my parents added a first-class mirror and a couple of eyepieces. Thus my career as a telescopic observer was launched.

What happened next, of course, was purchasing a copy of *Norton's* – already famous and then in its 12th edition. It was not a big book, but it sure had a big impact on this budding amateur. Since then, I've referred to *Norton's* thousands of times as I sought sights in the sky or checked facts at my editor's desk.

As I scan the maps in that ancient edition, I'm amazed at the history recorded upon them. There's a pencil line documenting the path of a fireball I saw in 1961 – one so bright that it made my dark-adapted eyes see a *negative* image of my surroundings. And there is a little 'X' marking the position of what is now known as TT Coronae Borealis. As a youngster I proved this star varies in brightness, so I felt very paternal toward it. On Chart 9 another 'X' identifies 3C 273. Undoubtedly I made that mark in 1963, the year quasars first caused astronomers to realize that many entities in the Universe were gravity-powered, bizarre, and violent.

There is a clutter of other marks on my charts. Some mean nothing to me now, but their subjects must once have seemed very important. Oh yes…that dot must be Nova Cygni 1975, and the one below it Nova Delphini 1967 – now *that* was an interesting star, a so-called slow nova that I observed every possible night for months. My old *Norton's* is truly an astronomical diary. Yours will surely turn into one too.

Looking back, *Norton's* had a twofold influence. One, of course, was its collection of star charts, a pioneering presentation of six shield-shaped gores and two round polar plots. All of them cover huge sweeps of sky, which greatly helped me recognize constellations and relate one to another. Yet, at the same time, the charts had enough detail to encourage me to seek countless new acquaintances.

Norton's second influence was its text. It took some time for me to appreciate this handbook section – after all, observing is much more fun than reading! But once I began to explore it, I found a gold mine of observing tips, explanations of technical matters and snippets of history. The text also raised many 'how to' and 'what if' questions. These prompted more reading and led me to an ever greater appreciation of astronomy. I can trace a lot of what I know to inspiration from *Norton's*. It's no exaggeration to say that it started me on a life-long love affair with astronomy. Heaven knows how many others would say the same – there must be a lot of us!

This latest edition of *Norton's* continues a tradition of upgrades, to make this resource even more indispensable and user friendly. Since 1989, when I wrote my first foreword to *Norton's*, amateur astronomy has undergone an unprecedented revolution. This is especially evident in the proliferation of inexpensive robotic telescopes and increasingly sophisticated software. When the two are linked, one has an enormously powerful system that can automatically point to any celestial object at just the touch of a button.

Thus some may wonder about the utility of a paper atlas. I'll simply point out that a book doesn't need electric power. But don't forget to put batteries in your flashlight!

– LEIF J. ROBINSON
EDITOR EMERITUS, *Sky & Telescope* MAGAZINE

Preface

This, the 20th edition of *Norton's Star Atlas,* has been completely redesigned and reset to give it a fresh, modern appearance for a new century. The star charts have been relabelled to make them more legible and attractive. Among the updates and improvements to the reference handbook and its data tables, we have introduced new sections on computer-controlled telescopes and CCD imaging, both of which have changed the face of amateur astronomy in the five years since the 19th edition appeared. We have also enlarged the section on observing deep-sky objects, which are popular targets for amateurs. Other major improvements include completely new maps of the Moon and Mars. Throughout, the needs of the active observer have been kept uppermost in mind.

In its previous editions, *Norton's* earned the reputation of being the most famous and most widely used star atlas in the world, and its reference handbook has become an indispensable companion for observers of all standards. We believe that this latest edition will carry the tradition of *Norton's Star Atlas* well into the twenty-first century.

History

Norton's Star Atlas first appeared in 1910. It achieved immediate success, due largely to its uniquely convenient arrangement of charts in slices, or 'gores', each covering approximately one-fifth of the sky, and its inclusion of stars down to sixth magnitude, the naked-eye limit. The Atlas was intended for owners of small telescopes, particularly those who wanted to find the objects of interest that were listed in two famous observing guides by nineteenth-century amateur astronomers: *Celestial Objects for Common Telescopes* by the Rev. T. W. Webb, and *Cycle of Celestial Objects* by W. H. Smyth. Over the years *Norton's* established an international reputation, becoming a standard reference work for amateur and professional astronomers alike.

The author of the Atlas, Arthur Philip Norton (1876–1955), was an amateur astronomer; his full-time occupation was as a schoolmaster. Had it not been for his Atlas he would have remained almost unknown in the world of astronomy.

Norton was born in Cardiff, Wales, the son of a clergyman. His interest in astronomy started when, as a schoolboy, he was given a telescope that had belonged to his great-grandfather. After receiving his BA degree from Trinity College, Dublin, he taught at various schools in England. For 22 years Norton taught geography and mathematics at the Judd School, Tonbridge, Kent, retiring in 1936. Norton published no books other than the Atlas on which his fame rests (and a simplified version, the *Popular Star Atlas,* which appeared in 1949), but during his lifetime it went through numerous editions, and he updated the star charts twice.

Back in 1910, when *Norton's Star Atlas* first appeared, there were no officially recognized boundaries to the constellations – a deficiency that the International Astronomical Union rectified in 1930. For the 5th edition of his Atlas, published in 1933, Arthur Norton redrew the charts to incorporate the newly defined IAU constellation boundaries, and he set the magnitude limit of the stars at 6.2, based on the *Harvard Revised Photometry* catalogue (the magnitude limit of the 1st edition had not been precisely defined). Norton by now had to cope with the fact that the sight in his left eye was badly blurred as a result of a blood clot behind the retina, but it did not affect the quality of his charts.

Celestial cartographers are faced with a problem that does not afflict their terrestrial counterparts: the coordinates of all stars are gradually changing with time, because of an effect called precession. This means that all star charts are bound to become progressively out of date. The epoch (i.e. the reference date for the star positions) of the original *Norton's* was 1920. For the 9th edition, published in 1943, Norton redrew his charts again, this time for the standard epoch of 1950.0, and further extended the magnitude limit of the stars to 6.35. That version of the charts remained in print long after his death.

Inevitably, with the passage of time, another change of epoch became necessary. For the 18th edition of *Norton's,* published in 1989, the charts were redrawn to the standard epoch of 2000.0, using technology that Norton could hardly have dreamed of. And for the first time *Norton's* contained nothing by Arthur Norton himself, although his influence lived on in more than just the title.

The Charts

An early decision in preparing the star charts for the new epoch was to retain their existing arrangement, which had stood the test of time. A subtle difference was that both the polar charts and the equatorial gores now used the same projection, known as Lambert's azimuthal equidistant projection, which allows large areas of sky to be represented with little distortion. Norton never stated the projections that he used; the gores were apparently plotted on a modified globular projection of his own devising.

In the new charts, the plane of the projection surface touches the celestial sphere at the poles for the polar charts, and at the celestial equator for the gores. To minimize distortion, each gore has been projected from its central meridian at the equator. All the projections were generated by computer for maximum accuracy.

The projection software was written at the cartographic company of John Bartholomew & Son in Edinburgh. The outlines of the Milky Way, the Magellanic Clouds, the galactic equator and the ecliptic were added by our cartographic consultant, Mike Swan. In addition to being a former professional cartographer with the Ordnance Survey, he is a deep-sky observer with the Webb Society. His expertise in both astronomy and cartography was a vital ingredient in the project.

With all the data converted into machine-readable form, the charts were generated on film at Bartholomews. These films then went to Mike Swan for hand-labelling and final checking. Films combining the star charts and labelling were output at Bartholomews, from which printing plates were produced. For the 20th edition of Norton's, the charts were redrawn and relabelled by the book's designer, Charles Nix and his associate Gary Robbins.

Data

For information on positions and magnitudes of stars we adopted the Yale *Bright Star Catalogue* (BS) and its Supplement. The BS contains the same stars as the *Harvard Revised Photometry* catalogue that Arthur Norton used for his charts, but with considerably improved magnitude measurements. We chose a magnitude limit of 6.49 (i.e. encompassing all stars of 6th magnitude and brighter), against the 6.35 used for the 1950.0 maps.

Dorrit Hoffleit of Yale University Observatory, senior author of the BS, and Wayne Warren of the National Space Science Data Center in Greenbelt, Maryland, supplied magnetic tapes of the 5th edition of the *Bright Star Catalogue*. They also supplied tapes of the 1983 Supplement to the BS, from which we extracted stars brighter than mag. 6.50 that were not included in the main BS, having been missed by the original *Harvard Revised Photometry*.

Data required for the maps were extracted from the BS tapes by the Royal Observatory, Edinburgh, and were supplied to

Bartholomew with the constellation boundaries added. Data for the galactic charts were also supplied by the ROE.

Even in a computerized operation such as this, considerable manual intervention was still necessary. Since the BS does not include deep-sky objects, lists of star clusters, nebulae and galaxies were drawn up by Mike Swan for addition to the stellar database. He also spent many hours identifying variable stars, and stars that are both variable and double, for depiction by special symbols on the charts. Ordinary double stars were identified directly from the BS tapes.

The charts for the current edition of *Norton's Star Atlas* show over 8800 stars. The star symbols are graduated in whole-magnitude steps for ease of identification. The few stars of magnitude 0 and −1 are given the same size symbol as stars of magnitude +1. The percentage of stars in each magnitude range is as follows:

MAGNITUDE RANGE	PERCENTAGE OF STARS IN *Norton's Star Atlas*
−1.50 to +1.49	0.25
+1.50 to +2.49	0.9
+2.50 to +3.49	2.5
+3.50 to +4.49	7.2
+4.50 to +5.49	22.6
+5.50 to +6.49	66.5

Double and Multiple Stars

Stars that are listed in the BS as double or multiple are identified on the charts by a special symbol (a line bisecting the star dot) if their separation is at least 0.1 arcsec. The exceptions to this system are stars whose components are wide enough to be plotted separately; these do not carry the double-star symbol unless they have other, closer companions. Spectroscopic binaries and other exceptionally close doubles (e.g. those found by occultation studies or speckle interferometry) are not denoted by the double-star symbol on the maps.

In the list of interesting objects that accompanies each chart in *Norton's Star Atlas*, the double stars are restricted to those with a combined magnitude brighter than 6.5. For the 20th edition, position, separation and magnitude data for these stars were checked against the latest edition of the Washington Double Star Catalogue. All the double stars in these lists are labelled on the charts.

Variable Stars

Those variable stars with a range of at least 0.1 mag. and a maximum magnitude brighter than 6.5, as listed in the BS and other sources consulted by us, are identified by a variable-star symbol. This symbol consists of a ring surrounding a solid dot; the size of the outer ring indicates the maximum magnitude of the star. Those variables, including novae, whose minimum brightness takes them below our chart limit of mag. 6.49 are denoted by an open circle only. More than 500 variable stars are identified on the charts, including over 40 that are not in the BS or its Supplement but for which we found evidence of maxima above mag. 6.5 (certain classes of variable, particularly those of long period, have ranges of variation that are not precisely bounded). A combined symbol is used to identify nearly 150 stars that are both variable and double.

The lists of variable stars that accompany the charts are believed to include all variables that have an amplitude of at least 0.4 mag. and a maximum brighter than approximately mag. 6.5. All the variables contained in these lists are labelled on the charts.

Deep-sky Objects

Separate symbols are used to distinguish each class of deep-sky object: open star clusters, globular star clusters, diffuse nebulae, planetary nebulae and galaxies. Additionally, the charts depict the true shape and extent of nebulae and galaxies that are larger than about 0°.5 in apparent diameter. In all, over 600 deep-sky objects are shown. The most interesting of them are briefly described in the notes accompanying each chart.

Reference Handbook

Over the years, the Reference Handbook section has become as valuable a part of *Norton's* as the Atlas itself. In the 1st edition the text amounted to only 18 pages, mostly written by James Gall Inglis (1865–1939). By the 5th edition in 1933 the text had grown to 51 pages, and by the 17th edition in 1978 it covered 116 pages. For the 18th edition the text was rewritten almost entirely, while attempting to retain the essential character of *Norton's*. The emphasis remained on reference information and practical observing advice that is often difficult to obtain elsewhere, a philosophy that has been maintained in subsequent editions.

As in previous editions we have decided against giving a bibliography, but mention must be made of *Burnham's Celestial Handbook* (in three volumes) by Robert Burnham Jr. (Dover Publications). Although now dated, this is an invaluable companion to *Norton's* and remains a classic guide for observers. A more compact handbook, with individual constellation charts and notes on objects of interest, is *Guide to Stars and Planets* by Ian Ridpath and Wil Tirion (HarperCollins U.K., Princeton University Press U.S.).

Acknowledgements

The editor of *Norton's Star Atlas* acknowledges with gratitude the efforts of the following contributors who have reviewed and, where necessary, revised the text and provided new material for the current edition: Steve Bell and Catherine Hohenkerk of H.M. Nautical Almanac Office (position and time), Robin Scagell (instruments and observing), Nik Szymanek (astronomical imaging), Geoff Elston (the Sun), Peter Grego (the Moon), Robert Steele (Mercury and Venus), Richard McKim (Mars), Ian Phelps (Jupiter and Saturn), Andrew Hollis (Uranus, Neptune, Pluto and minor planets), Tom McEwan (aurorae, noctilucent clouds and the zodiacal light), Roger Griffin (stars), Storm Dunlop (variable stars) and Darren Bushnall (observing deep-sky objects). Additional information was supplied by Jean Meeus, Jon Harper, Jonathan McDowell and Jonathan Shanklin. We are grateful to Leif Robinson of *Sky & Telescope* for his foreword.

The map of the Moon combines a shaded airbrush map produced by the U.S. Geological Survey (USGS) and a mosaic of images from the Clementine probe; these were supplied by Mark Rosiek of the USGS, Flagstaff, Arizona. The map of Mars was produced from images taken by the Mars Orbiter Camera on the Mars Global Surveyor spacecraft and was supplied by Michael Caplinger of Malin Space Science Systems, San Diego, California.

This edition was designed and typeset by Charles Nix, who also produced the line diagrams. John Woodruff provided valuable assistance with the proofs and made many useful suggestions. In particular I would like to thank Stephen Morrow at Pi Press for his enthusiastic support of *Norton's Star Atlas* which made this edition possible. Any errors or omissions are the responsibility of the editor who will, as ever, be pleased to hear from users of the book with suggestions for improvements.

– Ian Ridpath

CHAPTER ONE

Position and Time

The Heavens Above

The Celestial Sphere

All astronomical objects can be considered as lying on an imaginary sphere surrounding the Earth, called the celestial sphere (Figure 1). Like any sphere, the celestial sphere has two poles and an equator. The *celestial poles* lie directly above the poles of the Earth, while the *celestial equator* lies directly above the Earth's equator. The celestial sphere appears to rotate once a day around the celestial poles, actually as a result of the rotation of the Earth on its axis.

An observer standing on the surface of the Earth sees only half the celestial sphere at any one time. The visible half of the celestial sphere is bounded by the observer's *horizon*, a plane that cuts the celestial sphere 90° from the observer's *zenith*. The zenith is the point on the celestial sphere directly above the observer. Directly beneath the observer is the point called the *nadir*.

For the coordinate systems used to measure the positions of objects on the celestial sphere, see the section on Position (p. 3).

Daily Rotation

Every day the celestial sphere appears to turn as the Earth rotates, causing the daily rising and setting of the Sun, stars and other celestial bodies. As seen from the equator, all stars rise at right angles to the horizon and remain above the horizon for 12 hours. But as seen from the poles, stars move in circles parallel to the horizon and remain permanently above the horizon, never rising or setting.

At intermediate latitudes, the apparent motion of the stars lies between these two extremes. Some stars rise and set, but others circle around the pole without setting; these are known as circumpolar stars (see below). At any latitude, the stars that rise and set always do so at the same points on the horizon. This is not the case for the Sun, Moon and planets, which move against the celestial sphere and hence rise and set at different points from day to day.

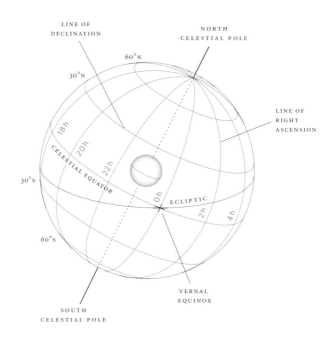

FIGURE 1. *The celestial sphere, with the Earth at its centre, showing lines of right ascension and declination, and the ecliptic.*

The length of time an object spends above the horizon as seen from a particular location depends on its declination (angular distance north or south of the celestial equator). Stars on the celestial equator rise due east and set due west, and are above the horizon for 12 hours as seen from anywhere on Earth (except at the poles). Stars nearer the visible pole are above the horizon for longer than 12 hours, whereas stars more than 90° from the pole set in less than 12 hours.

Table 1 gives the length of time that objects at various declinations take to reach the meridian (see p. 2) after rising; the total time for which the object is above the horizon from rising to setting is twice the time given in this table.

CIRCUMPOLAR STARS are stars that never set as seen from a given location. The area of sky that is circumpolar depends on the observer's latitude on Earth, since the altitude of the celestial

TABLE 1. *Semi-diurnal arcs. For stars of a particular declination, the time interval between rising or setting of the star and its transit over the observer's meridian (i.e. its culmination) is given for different latitudes. The total time for which the star is visible, from rising to setting, is twice this interval. The figures are calculated from the formula:*

$$\cos(\text{semi-diurnal arc}) = -\tan(\text{dec.}) \times \tan(\text{lat.})$$

To convert degrees to hours, divide by 15. Atmospheric refraction is ignored.

LATITUDE OF OBSERVER °	DECLINATION OF STAR (north declination for northern observers, south declination for southern observers)							DECLINATION OF STAR (south declination for southern observers, north declination for northern observers)					
	30° h m	25° h m	20° h m	15° h m	10° h m	05° h m	00° h m	05° h m	10° h m	15° h m	20° h m	25° h m	30° h m
5	06 12	06 09	06 07	06 05	06 04	06 02	06 00	05 58	05 56	05 55	05 53	05 51	05 48
10	06 23	06 19	06 15	06 11	06 07	06 04	06 00	05 56	05 53	05 49	05 45	05 41	05 37
15	06 36	06 29	06 22	06 16	06 11	06 05	06 00	05 55	05 49	05 44	05 38	05 31	05 24
20	06 49	06 39	06 30	06 22	06 15	06 07	06 00	05 53	05 45	05 38	05 30	05 21	05 11
25	07 02	06 50	06 39	06 29	06 19	06 09	06 00	05 51	05 41	05 31	05 21	05 10	04 58
30	07 18	07 02	06 49	06 36	06 23	06 12	06 00	05 48	05 37	05 24	05 11	04 58	04 42
35	07 35	07 16	06 59	06 43	06 28	06 14	06 00	05 46	05 32	05 17	05 01	04 44	04 25
40	07 56	07 32	07 11	06 52	06 34	06 17	06 00	05 43	05 26	05 08	04 49	04 28	04 04
45	08 21	07 51	07 25	07 02	06 41	06 20	06 00	05 40	05 19	04 58	04 35	04 09	03 39
50	08 54	08 15	07 43	07 14	06 49	06 24	06 00	05 36	05 11	04 46	04 17	03 45	03 06
55	09 42	08 47	08 05	07 30	06 58	06 29	06 00	05 31	05 02	04 30	03 55	03 13	02 18
60	—	09 35	08 36	07 51	07 11	06 35	06 00	05 25	04 49	04 09	03 24	02 25	—

pole equals the observer's latitude. For example, from latitude 40° the celestial pole has an altitude above the horizon of 40°, and all objects within 40° of the pole are circumpolar. Equally, there is a corresponding area within 40° of the opposite pole that is never visible. At the poles, all stars visible are circumpolar. From the equator, none are circumpolar.

The Meridian

An observer's meridian is the imaginary line that runs from north to south in the sky, passing through both celestial poles and through the zenith. When an object lies on the meridian it is said to be *at culmination* or *in transit* (both phrases mean the same thing).

A circumpolar object culminates twice. *Upper culmination* refers to its highest altitude as it passes from east to west across the meridian. *Lower culmination* refers to its lowest altitude as it passes across the meridian between the pole and the horizon. When used without qualification, 'culmination' refers to the moment when a celestial body reaches its greatest altitude above the horizon, i.e. upper culmination.

The Ecliptic

During the year the Sun appears to move once around the celestial sphere, as a result of the Earth's orbital motion. The Sun's yearly path on the celestial sphere is called the ecliptic. It is inclined to the celestial equator by about 23½°, an angle known as the *obliquity of the ecliptic* (see p. 3), which results from the axial tilt of the Earth. During the year the Sun's declination ranges from 23½° north to 23½° south, reaching its greatest values north and south at the summer and winter solstices respectively. If the Earth's axis were not tilted, the Sun would appear to move along the celestial equator and there would be no seasons on Earth.

When the Sun crosses the celestial equator at the vernal equinox, around March 21 every year, its position is right ascension 0h, declination 0°. Six months later, around September 23 (the autumnal equinox), it is at right ascension 12h, declination 0°. Between these dates it has been north of the celestial equator; for the following six months of the year, between September 23 and March 21, the Sun lies south of the celestial equator. Table 7 on page 17 gives the right ascension and declination of the Sun at various times of the year.

The *ecliptic poles* lie 90° north and south of the ecliptic. They are at right ascension 18h, declination 66½° north; and right ascension 6h, declination 66½° south.

OBLIQUITY OF THE ECLIPTIC (SYMBOL ε) is the angle at which the celestial equator is tilted with respect to the ecliptic; it is equal to the tilt of the Earth's axis from the perpendicular to the plane of its orbit. The obliquity of the ecliptic varies slightly with time due to the effects of nutation and the gravitational pulls of the planets on the Earth. Nutation causes a variation of up to 9″.2 from the mean value every 18.6 years, while planetary precession is currently causing the mean value of the obliquity to decrease by 0″.47 per year. On 2000 January 1 the obliquity was 23° 26′ 21″, and on 2050 January 1 it will be 23° 25′ 58″.

The Equinoxes

These two points mark where the ecliptic intersects the celestial equator. The Sun reaches these points around March 21 and September 23 each year. At an equinox the centre of the Sun lies exactly on the celestial equator but, although the word 'equinox' means 'equal night', day and night are not exactly equal at the equinoxes. There are two reasons for this. Firstly, sunrise and sunset are calculated for the upper limb of the Sun, not its centre; secondly, the effect of refraction in the Earth's atmosphere lifts the image of the Sun by about half a degree at the horizon. These both have the effect of slightly lengthening the day.

VERNAL EQUINOX (SYMBOL ♈) This is the point where the Sun crosses the celestial equator moving northwards, about March 21 each year, and is hence the ascending node of the ecliptic. The vernal (or spring) equinox is also known as the *first point of Aries*, since it lay in the constellation Aries 2000 years ago when its position was determined by the Greeks. Since then it has moved into the neighbouring constellation of Pisces as a result of the effect of precession (p. 5).

The vernal equinox is the zero point for the measurement of right ascension. Precession moves the vernal equinox westwards on the celestial sphere by about 0.14 arcsec per day.

AUTUMNAL EQUINOX (SYMBOL ♎) This is the point where the Sun crosses the celestial equator moving southwards, about September 23 each year, and is hence the descending node of the ecliptic. It is also known as the *first point of Libra*, although precession has moved it into Virgo.

COLURES The *equinoctial colure* is the hour circle (see p.4) that passes through the vernal and autumnal equinoxes, and is therefore the hour circle of right ascension 0h and 12h. The *solstitial colure* is the hour circle that passes through the summer and winter solstices, i.e. the hour circle of right ascension 6h and 18h.

The Solstices

The two points on the ecliptic at which the Sun reaches its greatest declination of 23½° north or south of the celestial equator, around June 21 (the summer solstice) and December 22 (the winter solstice) each year. At the *summer solstice* the Sun lies directly overhead at noon on the Tropic of Cancer, and the northern hemisphere then has its longest day and shortest night (vice versa in the southern hemisphere). At the *winter solstice* the Sun lies overhead on the Tropic of Capricorn so that the northern hemisphere has its shortest day and longest night (vice versa in the southern hemisphere).

Paradoxically, the dates of latest sunrise and earliest sunset do not coincide with the shortest day; neither do the dates of earliest sunrise and latest sunset coincide with the longest day. The reason is the changing value of the equation of time (p.17), as a result of which sunrise and sunset are both getting later each day at the solstices, although by different amounts. The net result is that the latest sunrise actually occurs after the shortest day, and the earliest sunset before it, while the earliest sunrise occurs before the longest day, and the latest sunset after it. In each case the exact date depends on the observer's latitude, but the offset from the solstices is greater at lower latitudes. The effect is more pronounced at the December solstice than at the June solstice because the day-to-day increase in the equation of time is greater in December.

Position

Celestial Coordinate Systems

Astronomical positions can be measured on the celestial sphere by using one of five different systems of coordinates, each of which suits a particular requirement. Four give positions as seen from the Earth; each uses a different reference plane (see Table 2). A fifth system, heliocentric coordinates, gives positions as seen from the Sun.

In catalogues, equatorial and ecliptic coordinates are listed for *geocentric* positions, i.e. as the object would be seen from the centre of the Earth. With distant objects such as stars and galaxies the observer's specific location on the Earth makes no difference to the object's observed position, but for objects in the Solar System slight corrections are needed to give the *topocentric* coordinates, i.e. as observed from a given place on the surface of the Earth.

EQUATORIAL COORDINATES are the ones most usually encountered in astronomical work. Their reference plane is the celestial equator, and the coordinates used are right ascension and declination, although hour angle and polar distance are sometimes used instead.

Right ascension (RA, symbol α) and *declination* (dec., symbol δ) are the celestial equivalents of longitude and latitude on the Earth. Declination is measured in degrees, from 0° at the celestial equator to 90° at the celestial poles.

The zero line of right ascension is the equivalent of the Greenwich meridian on the Earth. It passes through the point where the Sun crosses the celestial equator from the southern into the northern hemisphere each year; this happens around March 21, at the vernal equinox. Right ascension is measured eastwards from the vernal equinox along the celestial equator in hours, minutes and seconds, from 0 to 24 hours, although sometimes degrees are used. Each hour of right ascension is equivalent to 15° (and thus 1° is equivalent to 4 minutes).

An *hour circle* is the circle passing through a body on the celestial sphere and through the celestial poles; it is perpendicular to the celestial equator. *Hour angle* is the angle between the meridian and the hour circle through a celestial body, measured westwards along the celestial equator. An object on the meridian has an hour angle of 0h; an object that crossed the meridian one hour ago has an hour angle of 1h, and so on. Hence the hour angle is the time that has elapsed since the object last crossed the meridian.

Polar distance is the angular distance of an object from the celestial pole, measured along the hour circle. It is equal to 90° minus the object's declination.

HORIZONTAL COORDINATES, also called *azimuthal coordinates*, are the simplest positional system. Their reference plane is the observer's horizon, and the object's position is given in terms of altitude and azimuth (or zenith distance).

Altitude (symbol *h* or *a*) is the angle of elevation of a celestial object above the observer's horizon, measured perpendicular to the horizon. When the object is on the horizon, it has an altitude of 0°; an object at the zenith has an altitude of 90°. The *zenith distance* (symbol *z*) is sometimes used in place of altitude. This is the angular distance of an object from the zenith, and is equal to 90° minus the altitude. *Azimuth* (symbol *A*) is the angular distance measured clockwise around the horizon from due north, through east, to the point where a circle through the zenith and an object intersects the horizon. An object due north has an azimuth of 0°, an object due east has an azimuth of 90°, and so on.

The altitude and azimuth of an object at a particular time depend on the point on the Earth's surface from which they are measured, i.e. they are purely topocentric. Hence horizontal coordinates must be worked out for the observer's location and the required time from the right ascension and declination of the object.

ECLIPTIC COORDINATES take the ecliptic as their reference plane. Although they are less commonly met with than equatorial coordinates, they are sometimes used to give the positions of bodies

TABLE 2. *Planes of reference for coordinate systems, and points of origin.*

PLANE OF REFERENCE	COORDINATES
Celestial equator	Declination; right ascension from the vernal equinox
Ecliptic (from Earth's centre)	Celestial (geocentric) latitude; celestial (geocentric) longitude from the vernal equinox
Ecliptic (from Sun's centre)	Heliocentric latitude; heliocentric longitude from the vernal equinox
Horizon of the observer	Altitude; azimuth from north
Meridian	Declination from the celestial equator; hour angle from the meridian
Galactic plane	Galactic latitude; galactic longitude from the centre of the Galaxy
Sun's equator	Heliographic latitude; heliographic longitude from arbitrary zero
Equator of the Moon or a planet	Planetographic latitude and longitude from an agreed prime meridian
Limb of the Sun, Moon or a planet	Distance from either the north point or the vertex (the uppermost point)

in the Solar System, as seen from the Earth's centre. Their coordinates are celestial longitude and celestial latitude.

Celestial (or *ecliptic*) *longitude* (symbol λ) is measured along the ecliptic from 0° to 360° east of the vernal equinox. *Celestial* (or *ecliptic*) *latitude* (symbol β) is measured from 0° to 90° at right angles to the ecliptic, along a circle passing through the object and the ecliptic poles.

GALACTIC COORDINATES are used for studies of the positions of objects within our Galaxy. The reference plane is the galactic equator, which is inclined at about 63° to the celestial equator. The coordinates are *galactic latitude* (symbol *b*), which is measured perpendicular to the galactic equator from 0° to 90°, and *galactic longitude* (symbol *l*), which is measured in degrees eastwards along the galactic equator. The zero point of galactic longitude lies in the direction of the galactic centre; its position as adopted by the International Astronomical Union in 1959 is RA 17h 45.6m, dec. −28° 56′.2 (2000.0 coordinates). The north galactic pole, which lies in Coma Berenices, is at RA 12h 51.4m, dec. 27° 7′.7 (2000.0 coordinates); the south galactic pole lies in Sculptor, at RA 00h 51.4m, dec. −27° 7′.7.

HELIOCENTRIC COORDINATES give the positions of objects as viewed from the centre of the Sun, and are used in particular for bodies in the Solar System. The reference plane of the heliocentric system is the ecliptic, and the coordinates are *heliocentric latitude* (symbol *b*) and *heliocentric longitude* (symbol *l*); the zero point of heliocentric longitude is the vernal equinox. The Earth's heliocentric longitude at a given instant is equal to the Sun's geocentric longitude plus 180°. More rigorous would be to use *barycentric* positions. These refer to the centre of mass of the Solar System, which is offset from the centre of the Sun because of the presence of the planets, in particular Jupiter, the most massive.

Planes of Reference

The planes of reference used for the various position systems, with the coordinates and their points of origin, are summarized in Table 2.

The *invariable plane* of the Solar System is defined by its total angular momentum (i.e. by the rotational spins and orbital motions of all the planets and moons), and passes through the centre of mass of the Solar System. It forms an unvarying reference plane since its position in space is not changed by planetary perturbations, unlike the ecliptic. It is inclined at 1°.58 to the ecliptic.

Star Places

The positions of the 'fixed' stars are usually given in terms of their right ascension and declination. But precession (see below) is constantly changing the positions of the equator and equinoxes on the celestial sphere, with consequent changes in the right ascensions and declinations of the stars. In addition, there are other effects (described below), such as nutation, aberration, parallax, proper motion and refraction, that affect the observed place of a star. In precise work it is therefore necessary to apply corrections for these various effects.

Three forms of a star's place are commonly used: the mean place, the true place and the apparent place.

The *mean place* of a star is its heliocentric (barycentric) position on the celestial sphere with the effects of refraction, parallax and aberration removed, and is reduced to the mean equator and equinox for a stated epoch. The mean place varies with time, but only as a result of proper motion and precession, and is the position given in star catalogues.

The *true place* of a star is given by the heliocentric (barycentric) coordinates of the star on the celestial sphere at the instant of observation, i.e. with the effects of annual parallax and aberration resulting from the movement of the Earth removed, and is referred to the true equator and equinox at that instant.

The *apparent place* of a star is its position on the celestial sphere as would actually be observed from the centre of the Earth (i.e. in geocentric coordinates) at a given time. It is referred to the true equator and equinox at the instant of observation.

TRUE EQUATOR AND EQUINOX   These are the actual positions of the equator and equinox as observed at any given time. They are constantly shifting because of the progressive effect of precession and the cyclical effect of nutation.

MEAN EQUATOR AND EQUINOX   These are the positions of the equator and equinox, corrected for the effect of nutation, and hence affected only by precession. Positions in star catalogues are referred to a stated mean equator and equinox, usually for the start or the middle of a given year.

STANDARD EPOCH is a set date and time used for comparing star coordinates and other data. Since 1984, the standard epoch for coordinates has been 2000 January 1.5, denoted by J2000.0. The prefix J signifies the Julian epoch, which is based on the Julian year of exactly 365.25 days. The current standard epoch is exactly one Julian century (36 525 days) removed from the standard epoch of 1900 January 0.5. It is usual for a standard epoch to be retained for half a century or so. Future epochs will be based on multiples of the Julian year.

PRECESSION is a westward movement of the equinoxes on the celestial sphere. Its main component, caused by the gravitational pulls of the Sun and Moon on the Earth's equatorial bulge, is known as *lunisolar precession*, and there is a small additional effect due to the pulls of the planets called *planetary precession*. The sum of these two components, called *general precession*, amounts to about 50″.3 per year, or 1° every 71.6 years. Precession causes all stars to move parallel to the ecliptic, so that the stars visible from a given place or at a given time of night gradually change throughout one cycle of precession.

Because of precession the right ascension and declination of stars are continually changing. Catalogues and atlases such as this one give the star positions for a standard epoch, which by international agreement is currently the start of the year 2000. To find the approximate position of a star for another date, Table 3 can be used.

One complete cycle of precession lasts 25 800 years, during which time the Earth's axis, inclined at 23½° to the perpendicular of its orbit, traces out a circle on the celestial sphere with a radius of 23½°. Hence Polaris (Alpha (α) Ursae Minoris) is only temporarily the closest bright star to the north celestial pole. Around 4800 years ago Thuban (Alpha (α) Draconis) lay a mere 0°.1 from the pole. Deneb (Alpha (α) Cygni) will be the brightest star near the pole in about 8000 years' time, at a distance of 7°.5, and in 11 300 years' time Vega (Alpha (α) Lyrae) will be at its closest to the pole, 5°.7 away.

NUTATION   The circular path of precession that the celestial pole traces out on the celestial sphere is not perfectly smooth, but slightly wavy. This irregularity is called nutation, the result of a

TABLE 3. *Precession.*

Precession in RA over 10.0 years

DEC.	HOURS OF RA FOR OBJECTS WITH NORTHERLY DECLINATION												
	0, 12	1, 11	2, 10	3, 9	4, 8	5, 7	6	18	19, 17	20, 16	21, 15	22, 14	23, 13
	HOURS OF RA FOR OBJECTS WITH SOUTHERLY DECLINATION												
	0, 12	23, 13	22, 14	21, 15	20, 16	19, 17	18	6	5, 7	4, 8	3, 9	2, 10	1, 11
°	m	m	m	m	m	m	m	m	m	m	m	m	m
80	+0.51	+0.84	+1.15	+1.41	+1.61	+1.73	+1.78	−0.75	−0.71	−0.58	−0.38	−0.12	+0.19
70	+0.51	+0.67	+0.82	+0.95	+1.04	+1.10	+1.12	−0.10	−0.08	−0.02	+0.08	+0.21	+0.35
60	+0.51	+0.61	+0.71	+0.79	+0.85	+0.89	+0.90	+0.13	+0.14	+0.18	+0.24	+0.32	+0.41
50	+0.51	+0.58	+0.65	+0.70	+0.74	+0.77	+0.78	+0.25	+0.26	+0.28	+0.32	+0.38	+0.44
40	+0.51	+0.56	+0.61	+0.64	+0.67	+0.69	+0.70	+0.33	0.33	+0.35	+0.38	+0.42	+0.46
30	+0.51	+0.55	+0.58	+0.60	+0.62	+0.64	+0.64	+0.38	+0.39	+0.40	+0.42	+0.45	+0.48
20	+0.51	+0.53	+0.55	+0.57	+0.58	+0.59	+0.59	+0.43	+0.43	+0.44	+0.46	+0.47	+0.49
10	+0.51	+0.52	+0.53	+0.54	+0.55	+0.55	+0.55	+0.47	+0.47	+0.48	+0.48	+0.49	+0.50
0	+0.51	+0.51	+0.51	+0.51	+0.51	+0.51	+0.51	+0.51	+0.51	+0.51	+0.51	+0.51	+0.51

Precession in declination over 10.0 years

HOURS OF RA FOR NORTHERN AND SOUTHERN OBJECTS												
0	1, 23	2, 22	3, 21	4, 20	5, 19	6, 18	7, 17	8, 16	9, 15	10, 14	11, 13	12
′	′	′	′	′	′	′	′	′	′	′	′	′
+3.3	+3.2	+2.9	+2.4	+1.7	+0.9	0.0	−0.9	−1.7	−2.4	−2.9	−3.2	−3.3

regular 'nodding' of the Earth's poles towards and away from the ecliptic poles; it perceptibly modifies the changes in star positions caused by precession. As a result of the varying distances and relative positions of the Moon and Sun, their gravitational pulls on the Earth vary in both strength and direction. The net effect is a combination of three components: *lunar nutation* (which causes the pole to wander from its mean position by $\pm9''$ in a period of 18.6 y), *solar nutation* ($\pm1''.2$ in 0.5 y) and *fortnightly nutation* ($\pm0''.1$ in 15 d). As the 18.6-year lunar nutation has the greatest effect, the Earth's axis passes the mean position about 2750 times during one precessional cycle of 25 800 years. Nutation causes a slight variation in the obliquity of the ecliptic, and this component is called *nutation in obliquity*. The component of nutation measured along the ecliptic is termed *nutation in longitude*, or the *equation of the equinoxes*.

ABERRATION The finite velocity of light combines with the Earth's orbital velocity of about $30\,\mathrm{km\,s^{-1}}$ to produce a small displacement of celestial objects from their true positions known as *annual aberration*. During the year a star seems to move in a small ellipse around its true position, the eccentricity of the ellipse varying from a circle for a star at the ecliptic pole to a straight line for a star on the ecliptic. The maximum amount by which a star is displaced from its true position is $20''.5$, known as the *constant of aberration*; it is the angle whose tangent is obtained by dividing the Earth's mean orbital speed by the speed of light. There is also a small additional effect due to the Earth's speed of rotation, called *diurnal aberration*, which amounts to $0''.32$ at most.

HOW TO USE TABLE 3

To estimate precession in RA, look at the column for the RA hour nearest to that being precessed. Note whether declination is north (positive) or south (negative) and choose the correct column. Read off the tabulated value against the declination; interpolate as necessary for greater accuracy. The table gives the correction for 10 years of precession, so multiply the figure as necessary for the required interval. Then apply the resulting correction to the object's RA.

For declination, read off the correction, interpolating as necessary, and multiply the figure for the required time interval. Apply the resulting correction to the object's declination.

Example: *The 1950 coordinates of the star Beta (β) Capricorni are RA 20h 18.2m, dec. $-14°\ 56'.5$. What is the star's approximate position in 2000 coordinates?*

From the table, the correction for 10 years is +0.56m. Multiply by 5 to obtain the correction for 50 years, and add to the 1950 RA, thus:

RA 1950	20h 18.2m
5 × (+0.56m)	+2.8m
RA 2000	20h 21.0m

The approximate correction for 10 years in declination is $+1'.9$, so the change over 50 years is

dec. 1950	$-14°\ 56'.5$
5 × (+1'.9)	+9'.5
dec. 2000	$-14°\ 47'.0$

Take care to add the correction algebraically: for like signs, add; for unlike signs (as here), subtract.

The actual 2000 coordinates for Beta (β) Capricorni are RA 20h 21.0m, dec. $-14°46'.9$. Inaccuracies will occur when estimating precession over long periods of time for stars with large proper motions.

Coordinates may be estimated for a previous epoch by reversing the sign of each correction factor.

ANNUAL PARALLAX (SYMBOL π) is the difference between the geocentric and heliocentric positions of a star. The apparent position of nearby stars varies slightly throughout the year because the Earth is constantly changing its position as it orbits the Sun. Annual parallax is defined as the angle subtended at the star by the semi-major axis of the Earth's orbit. Measurement of parallax is the only direct way of determining the distances of individual stars.

PROPER MOTION (SYMBOL μ) is the motion of a star relative to the Sun as projected onto the celestial sphere. Proper motion is tabulated in star catalogues as changes in right ascension (μ_{α}) and declination (μ_{δ}) per year or per century. The largest proper motion known is for Barnard's Star, 10″.4 per year.

REFRACTION Light paths are bent as they pass through the Earth's atmosphere, so the observed altitude of an object is greater than its true altitude. The amount of refraction ranges from just over half a degree at the horizon to zero at the zenith, and it changes with atmospheric conditions. Table 4 gives refraction for selected altitudes.

LATITUDE VARIATION (POLAR MOTION) The measured declinations of stars show minute, irregular cyclic changes up to a maximum value of 0″.04, as a result of the Earth's poles of rotation wandering around a mean position in a counter-clockwise direction. The variation consists of two components. The principal component arises from the Earth's axis of rotation being slightly inclined to its axis of symmetry; this causes a polar wandering of maximum amplitude ±0″.3 in latitude (equivalent to a circle with a radius of about 9 metres on the ground) in a period of 428 days. The origin of the second component is seasonal movements of air masses, and it has an amplitude of ±0″.18 (±5 metres) in a period of one year.

TABLE 4. *Refraction. For altitudes of 15° or more, the refraction R, in degrees, can be calculated from the following formula:*

$$R = \frac{0.00452p}{(273 + T)\tan a}$$

where p is the atmospheric pressure in millibars, T is the temperature in degrees Celsius, and a is the altitude in degrees. For altitudes below 15° this simple formula becomes increasingly inaccurate, and the following more precise formula must be used:

$$R = \frac{p(0.1594 + 0.0196a + 0.00002a^2)}{(273 + T)(1 + 0.505a + 0.0845a^2)}$$

The values given below for the amount of refraction (in arc minutes) for various altitudes have been calculated from the above formulae for a pressure of 1013.25 millibar (1 atmosphere) and a temperature of 10°C.

ALTITUDE (a) °	REFRACTION (R) ′	ALTITUDE (a) °	REFRACTION (R) ′
90	0.0	14	3.8
80	0.2	13	4.1
70	0.4	12	4.4
65	0.5	11	4.8
60	0.6	10	5.3
55	0.7	9	5.9
50	0.8	8	6.5
45	1.0	7	7.4
40	1.2	6	8.4
35	1.4	5	9.8
30	1.7	4	11.7
25	2.1	3	14.3
20	2.7	2	18.2
18	3.0	1	24.2
16	3.4	0.5	28.5
15	3.6	0	34.2

the planets, and consequently there are many variants of the day and year. The most important of these variants are defined below, but few of the numerical values quoted are truly constant.

Date and Time

The main time-scales used in astronomy are based on natural divisions: the rotation of the Earth (the day) and its orbital motion around the Sun (the year). The day is subdivided into artificial units of hours, minutes and seconds; the second, originally defined as a fraction of the day, is now defined in terms of atomic properties (specifically, the duration of 9 192 631 770 cycles of radiation corresponding to the transition between two hyperfine levels of the ground state of the caesium-133 atom).

In practice, the Earth's rotation is affected by precession, nutation and tidal friction, and also by the slight, unpredictable effects of winds, ocean currents and motions of material inside the planet. The Earth's orbital motion also departs from the perfect ellipse of Kepler's laws because of the gravitational effects of the Moon and

Writing the Date

In astronomical work, the date is usually written in order of successively smaller units of increasing precision, i.e. year, month, day and either fractions of a day or hours, minutes and seconds, thus:

2001 January 1d 2h 34m 4.8s

or

2001 January 1.107

JULIAN DATE (JD) Astronomers use a system of dates that counts the number of days that have elapsed since a given starting date; it is useful for comparing dates separated by long intervals since it is unaffected by changes in the calendar. The Julian day is reckoned from Greenwich noon and is given in decimal form, not hours and minutes. For example, 2000 January 1 at Greenwich noon is JD 245 1545.0. The starting point for the Julian day system is Greenwich noon on 4713 BC January 1, sufficiently long ago for past astronomical events which we might wish to consider to have positive Julian day numbers. Any time less than 12 h (0.5 d) belongs to the Julian day preceding the civil date. The Julian date at midnight may be calculated for the years from 1901 to 2099 as follows:

The Gregorian calendar date is Y, M, D, where Y is the year, M the month and D the day.

If $M > 2$, set $y = Y$ and $m = M - 3$;
otherwise set $y = Y - 1$ and $m = M + 9$.

Then

$$JD = 172 1103.5 + INT(365.25y) + INT(30.6m + 0.5) + D$$

where INT means 'take the whole-number part inside the brackets'.

To find the Julian date at H hours past Greenwich midnight, add $H/24$ to the JD already calculated for 0h UT.

Example: Find the Julian date of 1989 June 7 at 18h UT.

$Y = 1989, M = 6, D = 7, H = 18$

Since $M > 2$, $y = 1989, m = 3$

JD at 0h UT $= 172 1103.5 + INT(726482.5) + INT(92.3) + 7$
$\qquad = 172 1103.5 + 726482 + 92 + 7$
JD at 18h UT $= 2447684.5 + 18/24 = 244 7685.25$

MODIFIED JULIAN DATE (MJD) When dealing with dates close to the present, the modified Julian date is more convenient to use: it is JD $- 2400000.5$. Hence, for example, the MJD of 1989 June 7.75 is 47684.75. Note that the MJD begins at Greenwich midnight.

The Day

The unit of time called the day is based on the rotation of the Earth upon its axis. The day relative to the Sun (the solar day) is about 4 minutes longer than the day relative to the stars (the sidereal day) because of the Earth's orbital motion around the Sun, which causes the apparent position of the Sun against the star background to change each day.

There are two types of solar day: the apparent solar day and the mean solar day.

APPARENT SOLAR DAY The interval between two successive meridian transits of the centre of the Sun. This time interval is not constant, since the Earth's orbit around the Sun is not circular but elliptical, and the Sun moves along the ecliptic, not the celestial equator.

MEAN SUN An imaginary body that moves along the celestial equator at a constant speed, on which a uniform time-scale can be based.

MEAN SOLAR DAY The interval between two successive meridian transits of the mean Sun, equal to the mean value of the apparent solar day. The mean solar day is the one used for civil time-keeping purposes.

SIDEREAL DAY The time interval between two successive transits of the vernal equinox. The sidereal day is 3m 55.91s shorter than the mean solar day.

SUNRISE AND SUNSET are the times at which the upper limb of the Sun lies on the horizon, the effect of refraction by the Earth's atmosphere being taken into account. The times of sunrise and sunset would differ noticeably from the observed times if calculated for the centre of the Sun. At sunrise and sunset the Sun's centre is 50′ below the horizon, based on adopted values of 34′ for refraction at the horizon and 16′ for the semi-diameter of the Sun. Table 5 gives the approximate times of sunrise and sunset at various latitudes on the Earth in any year.

TWILIGHT is the period in the evening after sunset or in the morning before sunrise when the sky is not completely dark because of scattering of sunlight in the atmosphere. Three types of twilight are defined:

Civil twilight begins and ends when the centre of the Sun is 6° below the horizon.

Nautical twilight begins and ends when the centre of the Sun is 12° below the horizon. During nautical twilight the brightest stars are visible and the sea horizon can be seen.

Astronomical twilight begins and ends when the centre of the Sun is 18° below the horizon, so that in a clear sky 6th-magnitude stars are just visible at the zenith.

Twilight lengthens with the distance of the observer from the equator. Table 6 (pp. 13–16) gives times for the beginning and ending of astronomical twilight at various latitudes on the Earth throughout the year. Note that at high latitudes astronomical twilight is perpetual during the summer.

The Measurement of Time

APPARENT SOLAR TIME is the time shown on a sundial, which records the motion of the real Sun across the sky. But apparent solar time is not uniform because the motion of the real Sun varies throughout the year (see Apparent solar day, this page). For more uniform time-keeping, mean solar time is used.

(text continued on p.17)

TABLE 5. *Times of sunrise and sunset at various latitudes.*

JANUARY — SUNRISE

	50°S	45°S	40°S	30°S	20°S	10°S	0	10°N	20°N	30°N	40°N	45°N	50°N	55°N	60°N	65°N	70°N
	h	h	h	h	h	h	h	h	h	h	h	h	h	h	h	h	h
1	3.9	4.3	4.6	5.0	5.4	5.7	6.0	6.3	6.6	6.9	7.4	7.6	8.0	8.4	9.0	10.1	—
6	4.0	4.4	4.7	5.1	5.5	5.8	6.0	6.3	6.6	6.9	7.4	7.6	8.0	8.4	9.0	10.0	—
11	4.1	4.5	4.7	5.2	5.5	5.8	6.1	6.3	6.6	7.0	7.4	7.6	7.9	8.3	8.9	9.8	—
16	4.2	4.6	4.8	5.2	5.6	5.8	6.1	6.4	6.6	6.9	7.3	7.6	7.9	8.2	8.8	9.6	—
21	4.4	4.7	4.9	5.3	5.6	5.9	6.1	6.4	6.6	6.9	7.3	7.5	7.8	8.1	8.6	9.4	11.0
26	4.5	4.8	5.0	5.4	5.7	5.9	6.1	6.4	6.6	6.9	7.2	7.4	7.7	8.0	8.5	9.1	10.4
31	4.7	4.9	5.1	5.5	5.7	6.0	6.2	6.4	6.6	6.9	7.2	7.4	7.6	7.9	8.3	8.9	9.9

SUNSET

	50°S	45°S	40°S	30°S	20°S	10°S	0	10°N	20°N	30°N	40°N	45°N	50°N	55°N	60°N	65°N	70°N
	h	h	h	h	h	h	h	h	h	h	h	h	h	h	h	h	h
1	20.2	19.8	19.5	19.1	18.7	18.4	18.1	17.8	17.5	17.2	16.8	16.5	16.1	15.7	15.1	14.0	—
6	20.2	19.8	19.5	19.1	18.7	18.4	18.2	17.9	17.6	17.2	16.8	16.6	16.2	15.8	15.2	14.2	—
11	20.1	19.8	19.5	19.1	18.8	18.5	18.2	17.9	17.6	17.3	16.9	16.7	16.3	15.9	15.4	14.5	—
16	20.1	19.7	19.5	19.1	18.8	18.5	18.2	18.0	17.7	17.4	17.0	16.8	16.5	16.1	15.6	14.7	—
21	20.0	19.7	19.4	19.1	18.8	18.5	18.2	18.0	17.8	17.5	17.1	16.9	16.6	16.2	15.8	15.0	13.4
26	19.9	19.6	19.4	19.0	18.7	18.5	18.3	18.0	17.8	17.5	17.2	17.0	16.7	16.4	16.0	15.3	14.1
31	19.8	19.5	19.3	19.0	18.7	18.5	18.3	18.1	17.9	17.6	17.3	17.1	16.9	16.6	16.2	15.6	14.6

FEBRUARY — SUNRISE

	50°S	45°S	40°S	30°S	20°S	10°S	0	10°N	20°N	30°N	40°N	45°N	50°N	55°N	60°N	65°N	70°N
	h	h	h	h	h	h	h	h	h	h	h	h	h	h	h	h	h
1	4.7	4.9	5.1	5.5	5.7	6.0	6.2	6.4	6.6	6.8	7.1	7.3	7.6	7.9	8.2	8.8	9.8
6	4.8	5.1	5.2	5.5	5.8	6.0	6.2	6.4	6.6	6.8	7.1	7.2	7.4	7.7	8.0	8.5	9.3
11	5.0	5.2	5.4	5.6	5.8	6.0	6.2	6.3	6.5	6.7	7.0	7.1	7.3	7.5	7.8	8.2	8.9
16	5.1	5.3	5.5	5.7	5.9	6.0	6.2	6.3	6.5	6.7	6.9	7.0	7.2	7.3	7.6	8.0	8.5
21	5.3	5.4	5.5	5.7	5.9	6.0	6.2	6.3	6.4	6.6	6.8	6.9	7.0	7.2	7.4	7.7	8.1
26	5.4	5.5	5.6	5.8	5.9	6.1	6.2	6.3	6.4	6.5	6.6	6.7	6.8	7.0	7.1	7.4	7.7

SUNSET

	50°S	45°S	40°S	30°S	20°S	10°S	0	10°N	20°N	30°N	40°N	45°N	50°N	55°N	60°N	65°N	70°N
	h	h	h	h	h	h	h	h	h	h	h	h	h	h	h	h	h
1	19.7	19.5	19.3	19.0	18.7	18.5	18.3	18.1	17.9	17.6	17.3	17.1	16.9	16.6	16.2	15.7	14.7
6	19.6	19.4	19.2	18.9	18.7	18.5	18.3	18.1	17.9	17.7	17.4	17.2	17.0	16.8	16.4	16.0	15.2
11	19.5	19.3	19.1	18.9	18.6	18.5	18.3	18.1	18.0	17.8	17.5	17.4	17.2	17.0	16.7	16.3	15.6
16	19.3	19.1	19.0	18.8	18.6	18.4	18.3	18.1	18.0	17.8	17.6	17.5	17.3	17.1	16.9	16.5	16.0
21	19.2	19.0	18.9	18.7	18.5	18.4	18.3	18.2	18.0	17.9	17.7	17.6	17.5	17.3	17.1	16.8	16.4
26	19.0	18.9	18.8	18.6	18.5	18.4	18.3	18.2	18.1	17.9	17.8	17.7	17.6	17.5	17.3	17.1	16.8

MARCH — SUNRISE

	50°S	45°S	40°S	30°S	20°S	10°S	0	10°N	20°N	30°N	40°N	45°N	50°N	55°N	60°N	65°N	70°N
	h	h	h	h	h	h	h	h	h	h	h	h	h	h	h	h	h
1	5.5	5.6	5.7	5.8	6.0	6.1	6.2	6.2	6.3	6.4	6.6	6.6	6.7	6.8	7.0	7.2	7.5
6	5.6	5.7	5.8	5.9	6.0	6.1	6.1	6.2	6.3	6.3	6.4	6.5	6.6	6.6	6.7	6.9	7.1
11	5.8	5.8	5.9	6.0	6.0	6.1	6.1	6.2	6.2	6.2	6.3	6.3	6.4	6.4	6.5	6.6	6.7
16	5.9	5.9	6.0	6.0	6.0	6.1	6.1	6.1	6.1	6.2	6.2	6.2	6.2	6.2	6.2	6.3	6.3
21	6.0	6.1	6.1	6.1	6.1	6.1	6.1	6.1	6.1	6.1	6.0	6.0	6.0	6.0	6.0	6.0	5.9
26	6.2	6.2	6.1	6.1	6.1	6.1	6.0	6.0	6.0	5.9	5.9	5.9	5.8	5.8	5.7	5.7	5.5
31	6.3	6.3	6.2	6.2	6.1	6.1	6.0	6.0	5.9	5.8	5.8	5.7	5.7	5.6	5.5	5.3	5.2

SUNSET

	50°S	45°S	40°S	30°S	20°S	10°S	0	10°N	20°N	30°N	40°N	45°N	50°N	55°N	60°N	65°N	70°N
	h	h	h	h	h	h	h	h	h	h	h	h	h	h	h	h	h
1	18.9	18.8	18.7	18.6	18.4	18.4	18.3	18.2	18.1	18.0	17.9	17.8	17.7	17.6	17.4	17.3	17.0
6	18.7	18.6	18.6	18.5	18.4	18.3	18.2	18.2	18.1	18.0	18.0	17.9	17.8	17.8	17.7	17.5	17.3
11	18.5	18.5	18.4	18.4	18.3	18.3	18.2	18.2	18.1	18.1	18.0	18.0	18.0	17.9	17.9	17.8	17.7
16	18.4	18.3	18.3	18.3	18.2	18.2	18.2	18.2	18.2	18.1	18.1	18.1	18.1	18.1	18.1	18.0	18.0
21	18.2	18.2	18.2	18.2	18.2	18.2	18.2	18.2	18.2	18.2	18.2	18.2	18.2	18.3	18.3	18.3	18.4
26	18.0	18.0	18.0	18.1	18.1	18.1	18.1	18.2	18.2	18.2	18.3	18.3	18.4	18.4	18.5	18.6	18.7
31	17.8	17.9	17.9	18.0	18.0	18.1	18.1	18.2	18.2	18.3	18.4	18.4	18.5	18.6	18.7	18.8	19.0

TABLE 5 (continued). Times of sunrise and sunset at various latitudes.

APRIL

SUNRISE

	50°S	45°S	40°S	30°S	20°S	10°S	0	10°N	20°N	30°N	40°N	45°N	50°N	55°N	60°N	65°N	70°N
	h	h	h	h	h	h	h	h	h	h	h	h	h	h	h	h	h
1	6.3	6.3	6.2	6.2	6.1	6.1	6.0	6.0	5.9	5.8	5.7	5.7	5.6	5.5	5.4	5.3	5.1
6	6.5	6.4	6.3	6.2	6.1	6.1	6.0	5.9	5.8	5.7	5.6	5.5	5.4	5.3	5.2	5.0	4.7
11	6.6	6.5	6.4	6.3	6.2	6.1	6.0	5.9	5.8	5.6	5.5	5.4	5.3	5.1	4.9	4.7	4.3
16	6.7	6.6	6.5	6.3	6.2	6.1	5.9	5.8	5.7	5.5	5.4	5.2	5.1	4.9	4.7	4.4	3.9
21	6.9	6.7	6.6	6.4	6.2	6.1	5.9	5.8	5.6	5.5	5.2	5.1	4.9	4.7	4.4	4.1	3.4
26	7.0	6.8	6.7	6.4	6.2	6.1	5.9	5.7	5.6	5.4	5.1	5.0	4.8	4.5	4.2	3.7	3.0

SUNSET

	50°S	45°S	40°S	30°S	20°S	10°S	0	10°N	20°N	30°N	40°N	45°N	50°N	55°N	60°N	65°N	70°N
	h	h	h	h	h	h	h	h	h	h	h	h	h	h	h	h	h
1	17.8	17.8	17.9	17.9	18.0	18.1	18.1	18.2	18.2	18.3	18.4	18.5	18.5	18.6	18.7	18.9	19.1
6	17.6	17.7	17.7	17.8	17.9	18.0	18.1	18.2	18.3	18.4	18.5	18.6	18.7	18.8	18.9	19.1	19.5
11	17.4	17.5	17.6	17.8	17.9	18.0	18.1	18.2	18.3	18.4	18.6	18.7	18.8	18.9	19.1	19.4	19.8
16	17.3	17.4	17.5	17.7	17.8	17.9	18.1	18.2	18.3	18.5	18.7	18.8	18.9	19.1	19.3	19.7	20.2
21	17.1	17.2	17.4	17.6	17.7	17.9	18.0	18.2	18.3	18.5	18.7	18.9	19.0	19.3	19.5	19.9	20.6
26	16.9	17.1	17.3	17.5	17.7	17.9	18.0	18.2	18.4	18.6	18.8	19.0	19.2	19.4	19.7	20.2	21.0

MAY

SUNRISE

	50°S	45°S	40°S	30°S	20°S	10°S	0	10°N	20°N	30°N	40°N	45°N	50°N	55°N	60°N	65°N	70°N
	h	h	h	h	h	h	h	h	h	h	h	h	h	h	h	h	h
1	7.1	6.9	6.7	6.5	6.3	6.1	5.9	5.7	5.5	5.3	5.0	4.8	4.6	4.3	4.0	3.4	2.5
6	7.2	7.0	6.8	6.5	6.3	6.1	5.9	5.7	5.5	5.2	4.9	4.7	4.5	4.2	3.8	3.1	2.0
11	7.3	7.1	6.9	6.6	6.3	6.1	5.9	5.7	5.4	5.2	4.8	4.6	4.3	4.0	3.5	2.8	1.4
16	7.5	7.2	7.0	6.6	6.4	6.1	5.9	5.6	5.4	5.1	4.7	4.5	4.2	3.9	3.3	2.5	0.4
21	7.6	7.3	7.1	6.7	6.4	6.1	5.9	5.6	5.4	5.1	4.7	4.4	4.1	3.7	3.2	2.3	—
26	7.7	7.4	7.1	6.7	6.4	6.1	5.9	5.6	5.3	5.0	4.6	4.3	4.0	3.6	3.0	2.0	—
31	7.8	7.4	7.2	6.8	6.5	6.2	5.9	5.6	5.3	5.0	4.6	4.3	3.9	3.5	2.9	1.7	—

SUNSET

	50°S	45°S	40°S	30°S	20°S	10°S	0	10°N	20°N	30°N	40°N	45°N	50°N	55°N	60°N	65°N	70°N
	h	h	h	h	h	h	h	h	h	h	h	h	h	h	h	h	h
1	16.8	17.0	17.2	17.4	17.6	17.8	18.0	18.2	18.4	18.6	18.9	19.1	19.3	19.6	20.0	20.5	21.4
6	16.6	16.9	17.1	17.3	17.6	17.8	18.0	18.2	18.4	18.7	19.0	19.2	19.4	19.7	20.2	20.8	22.0
11	16.5	16.8	17.0	17.3	17.5	17.8	18.0	18.2	18.5	18.7	19.1	19.3	19.6	19.9	20.4	21.1	22.6
16	16.4	16.7	16.9	17.2	17.5	17.8	18.0	18.2	18.5	18.8	19.2	19.4	19.7	20.0	20.6	21.4	—
21	16.3	16.6	16.8	17.2	17.5	17.8	18.0	18.3	18.5	18.8	19.2	19.5	19.8	20.2	20.8	21.7	—
26	16.2	16.5	16.8	17.2	17.5	17.7	18.0	18.3	18.6	18.9	19.3	19.6	19.9	20.3	20.9	22.0	—
31	16.2	16.5	16.7	17.1	17.5	17.8	18.0	18.3	18.6	18.9	19.4	19.6	20.0	20.4	21.1	22.3	—

JUNE

SUNRISE

	50°S	45°S	40°S	30°S	20°S	10°S	0	10°N	20°N	30°N	40°N	45°N	50°N	55°N	60°N	65°N	70°N
	h	h	h	h	h	h	h	h	h	h	h	h	h	h	h	h	h
1	7.8	7.5	7.2	6.8	6.5	6.2	5.9	5.6	5.3	5.0	4.6	4.3	3.9	3.5	2.8	1.6	—
6	7.9	7.5	7.3	6.8	6.5	6.2	5.9	5.6	5.3	5.0	4.5	4.2	3.9	3.4	2.7	1.4	—
11	7.9	7.6	7.3	6.9	6.5	6.2	5.9	5.6	5.3	5.0	4.5	4.2	3.8	3.4	2.6	1.2	—
16	8.0	7.6	7.3	6.9	6.5	6.2	5.9	5.7	5.3	5.0	4.5	4.2	3.8	3.3	2.6	1.1	—
21	8.0	7.6	7.4	6.9	6.6	6.3	6.0	5.7	5.4	5.0	4.5	4.2	3.8	3.3	2.6	1.0	—
26	8.0	7.7	7.4	6.9	6.6	6.3	6.0	5.7	5.4	5.0	4.5	4.2	3.9	3.4	2.6	1.1	—

SUNSET

	50°S	45°S	40°S	30°S	20°S	10°S	0	10°N	20°N	30°N	40°N	45°N	50°N	55°N	60°N	65°N	70°N
	h	h	h	h	h	h	h	h	h	h	h	h	h	h	h	h	h
1	16.1	16.5	16.7	17.1	17.5	17.8	18.0	18.3	18.6	18.9	19.4	19.7	20.0	20.5	21.1	22.3	—
6	16.1	16.4	16.7	17.1	17.5	17.8	18.0	18.3	18.6	19.0	19.4	19.7	20.1	20.6	21.3	22.6	—
11	16.1	16.4	16.7	17.1	17.5	17.8	18.1	18.3	18.7	19.0	19.5	19.8	20.1	20.6	21.4	22.8	—
16	16.1	16.4	16.7	17.1	17.5	17.8	18.1	18.4	18.7	19.0	19.5	19.8	20.2	20.7	21.4	23.0	—
21	16.1	16.4	16.7	17.1	17.5	17.8	18.1	18.4	18.7	19.1	19.5	19.8	20.2	20.7	21.5	23.1	—
26	16.1	16.4	16.7	17.2	17.5	17.8	18.1	18.4	18.7	19.1	19.5	19.8	20.2	20.7	21.5	23.0	—

JULY — SUNRISE

	50°S	45°S	40°S	30°S	20°S	10°S	0	10°N	20°N	30°N	40°N	45°N	50°N	55°N	60°N	65°N	70°N
	h	h	h	h	h	h	h	h	h	h	h	h	h	h	h	h	h
1	8.0	7.7	7.4	6.9	6.6	6.3	6.0	5.7	5.4	5.0	4.6	4.3	3.9	3.4	2.7	1.2	—
6	8.0	7.6	7.4	6.9	6.6	6.3	6.0	5.7	5.4	5.1	4.6	4.3	4.0	3.5	2.8	1.5	—
11	7.9	7.6	7.3	6.9	6.6	6.3	6.0	5.8	5.5	5.1	4.7	4.4	4.0	3.6	2.9	1.7	—
16	7.8	7.5	7.3	6.9	6.6	6.3	6.0	5.8	5.5	5.2	4.7	4.5	4.1	3.7	3.1	2.0	—
21	7.8	7.5	7.2	6.9	6.6	6.3	6.0	5.8	5.5	5.2	4.8	4.6	4.2	3.8	3.3	2.3	—
26	7.7	7.4	7.2	6.8	6.5	6.3	6.0	5.8	5.6	5.3	4.9	4.6	4.3	4.0	3.4	2.6	—
31	7.6	7.3	7.1	6.8	6.5	6.3	6.0	5.8	5.6	5.3	5.0	4.7	4.5	4.1	3.6	2.9	1.2

JULY — SUNSET

	50°S	45°S	40°S	30°S	20°S	10°S	0	10°N	20°N	30°N	40°N	45°N	50°N	55°N	60°N	65°N	70°N
	h	h	h	h	h	h	h	h	h	h	h	h	h	h	h	h	h
1	16.1	16.5	16.8	17.2	17.5	17.8	18.1	18.4	18.7	19.1	19.5	19.8	20.2	20.7	21.4	22.9	—
6	16.2	16.5	16.8	17.2	17.6	17.9	18.1	18.4	18.7	19.1	19.5	19.8	20.2	20.6	21.3	22.7	—
11	16.3	16.6	16.8	17.3	17.6	17.9	18.2	18.4	18.7	19.1	19.5	19.8	20.1	20.6	21.2	22.4	—
16	16.4	16.7	16.9	17.3	17.6	17.9	18.2	18.4	18.7	19.0	19.5	19.7	20.1	20.5	21.1	22.1	—
21	16.5	16.7	17.0	17.3	17.7	17.9	18.2	18.4	18.7	19.0	19.4	19.7	20.0	20.4	20.9	21.9	—
26	16.6	16.8	17.0	17.4	17.7	17.9	18.2	18.4	18.7	19.0	19.3	19.6	19.9	20.2	20.8	21.6	—
31	16.7	16.9	17.1	17.4	17.7	17.9	18.2	18.4	18.6	18.9	19.3	19.5	19.7	20.1	20.6	21.3	22.8

AUGUST — SUNRISE

	50°S	45°S	40°S	30°S	20°S	10°S	0	10°N	20°N	30°N	40°N	45°N	50°N	55°N	60°N	65°N	70°N
	h	h	h	h	h	h	h	h	h	h	h	h	h	h	h	h	h
1	7.5	7.3	7.1	6.8	6.5	6.3	6.0	5.8	5.6	5.3	5.0	4.7	4.5	4.1	3.7	2.9	1.4
6	7.4	7.2	7.0	6.7	6.5	6.2	6.0	5.8	5.6	5.4	5.0	4.8	4.6	4.3	3.9	3.2	2.0
11	7.3	7.1	6.9	6.6	6.4	6.2	6.0	5.8	5.6	5.4	5.1	4.9	4.7	4.4	4.1	3.5	2.6
16	7.1	6.9	6.8	6.5	6.4	6.2	6.0	5.8	5.7	5.5	5.2	5.0	4.8	4.6	4.3	3.8	3.0
21	7.0	6.8	6.7	6.5	6.3	6.1	6.0	5.9	5.7	5.5	5.3	5.1	5.0	4.8	4.5	4.1	3.4
26	6.8	6.7	6.6	6.4	6.2	6.1	6.0	5.9	5.7	5.6	5.4	5.2	5.1	4.9	4.7	4.3	3.8
31	6.6	6.5	6.4	6.3	6.2	6.1	6.0	5.8	5.7	5.6	5.4	5.3	5.2	5.1	4.9	4.6	4.2

AUGUST — SUNSET

	50°S	45°S	40°S	30°S	20°S	10°S	0	10°N	20°N	30°N	40°N	45°N	50°N	55°N	60°N	65°N	70°N
	h	h	h	h	h	h	h	h	h	h	h	h	h	h	h	h	h
1	16.7	16.9	17.1	17.5	17.7	17.9	18.2	18.4	18.6	18.9	19.2	19.4	19.7	20.1	20.5	21.2	22.7
6	16.8	17.0	17.2	17.5	17.7	18.0	18.2	18.4	18.6	18.8	19.1	19.3	19.6	19.9	20.3	20.9	22.1
11	16.9	17.1	17.3	17.6	17.8	18.0	18.1	18.3	18.5	18.8	19.0	19.2	19.4	19.7	20.1	20.6	21.5
16	17.0	17.2	17.4	17.6	17.8	18.0	18.1	18.3	18.5	18.7	18.9	19.1	19.3	19.5	19.8	20.3	21.1
21	17.2	17.3	17.4	17.6	17.8	18.0	18.1	18.3	18.4	18.6	18.8	19.0	19.1	19.3	19.6	20.0	20.6
26	17.3	17.4	17.5	17.7	17.8	18.0	18.1	18.2	18.3	18.5	18.7	18.8	19.0	19.1	19.4	19.7	20.2
31	17.4	17.5	17.6	17.7	17.9	18.0	18.1	18.2	18.3	18.4	18.6	18.7	18.8	18.9	19.1	19.4	19.8

SEPTEMBER — SUNRISE

	50°S	45°S	40°S	30°S	20°S	10°S	0	10°N	20°N	30°N	40°N	45°N	50°N	55°N	60°N	65°N	70°N
	h	h	h	h	h	h	h	h	h	h	h	h	h	h	h	h	h
1	6.6	6.5	6.4	6.3	6.1	6.0	5.9	5.8	5.7	5.6	5.5	5.4	5.2	5.1	4.9	4.6	4.2
6	6.4	6.3	6.3	6.2	6.1	6.0	5.9	5.8	5.8	5.7	5.5	5.5	5.4	5.3	5.1	4.9	4.6
11	6.2	6.2	6.1	6.1	6.0	5.9	5.9	5.8	5.8	5.7	5.6	5.6	5.5	5.4	5.3	5.1	4.9
16	6.0	6.0	6.0	6.0	5.9	5.9	5.9	5.8	5.8	5.7	5.7	5.7	5.6	5.6	5.5	5.4	5.3
21	5.9	5.9	5.9	5.9	5.8	5.8	5.8	5.8	5.8	5.8	5.8	5.8	5.7	5.7	5.7	5.6	5.6
26	5.7	5.7	5.7	5.7	5.8	5.8	5.8	5.8	5.8	5.8	5.9	5.9	5.9	5.9	5.9	5.9	5.9

SEPTEMBER — SUNSET

	50°S	45°S	40°S	30°S	20°S	10°S	0	10°N	20°N	30°N	40°N	45°N	50°N	55°N	60°N	65°N	70°N
	h	h	h	h	h	h	h	h	h	h	h	h	h	h	h	h	h
1	17.4	17.5	17.6	17.7	17.9	18.0	18.1	18.2	18.3	18.4	18.5	18.6	18.7	18.9	19.1	19.3	19.7
6	17.6	17.6	17.7	17.8	17.9	18.0	18.0	18.1	18.2	18.3	18.4	18.5	18.6	18.7	18.8	19.0	19.3
11	17.7	17.7	17.8	17.8	17.9	17.9	18.0	18.1	18.1	18.2	18.3	18.3	18.4	18.5	18.6	18.7	18.9
16	17.8	17.8	17.8	17.9	17.9	17.9	18.0	18.0	18.0	18.1	18.1	18.2	18.2	18.2	18.3	18.4	18.5
21	17.9	17.9	17.9	17.9	17.9	17.9	17.9	17.9	18.0	18.0	18.0	18.0	18.0	18.0	18.1	18.1	18.1
26	18.1	18.0	18.0	18.0	17.9	17.9	17.9	17.9	17.9	17.9	17.9	17.8	17.8	17.8	17.8	17.8	17.8

TABLE 5 *(continued)*. *Times of sunrise and sunset at various latitudes.*

OCTOBER — SUNRISE

	50°S	45°S	40°S	30°S	20°S	10°S	0	10°N	20°N	30°N	40°N	45°N	50°N	55°N	60°N	65°N	70°N
	h	h	h	h	h	h	h	h	h	h	h	h	h	h	h	h	h
1	5.5	5.5	5.6	5.6	5.7	5.7	5.8	5.8	5.8	5.9	5.9	6.0	6.0	6.0	6.1	6.1	6.2
6	5.3	5.4	5.4	5.5	5.6	5.7	5.7	5.8	5.9	5.9	6.0	6.1	6.1	6.2	6.3	6.4	6.6
11	5.1	5.2	5.3	5.4	5.6	5.6	5.7	5.8	5.9	6.0	6.1	6.2	6.3	6.4	6.5	6.7	6.9
16	5.0	5.1	5.2	5.4	5.5	5.6	5.7	5.8	5.9	6.0	6.2	6.3	6.4	6.4	6.7	6.7	7.3
21	4.8	4.9	5.1	5.3	5.4	5.6	5.7	5.8	5.9	6.1	6.3	6.4	6.5	6.7	6.9	7.2	7.6
26	4.6	4.8	5.0	5.2	5.4	5.5	5.7	5.8	6.0	6.2	6.4	6.5	6.7	6.9	7.1	7.5	8.0
31	4.5	4.7	4.8	5.1	5.3	5.5	5.7	5.8	6.0	6.2	6.5	6.6	6.8	7.0	7.3	7.7	8.4

OCTOBER — SUNSET

	50°S	45°S	40°S	30°S	20°S	10°S	0	10°N	20°N	30°N	40°N	45°N	50°N	55°N	60°N	65°N	70°N
	h	h	lı	h	h	h	h	h	h	h	h	h	h	h	h	h	h
1	18.2	18.1	18.1	18.0	18.0	17.9	17.9	17.8	17.8	17.8	17.7	17.7	17.6	17.6	17.6	17.5	17.4
6	18.3	18.2	18.2	18.1	18.0	17.9	17.9	17.8	17.7	17.7	17.6	17.5	17.5	17.4	17.3	17.2	17.0
11	18.4	18.3	18.3	18.1	18.0	17.9	17.8	17.8	17.7	17.6	17.4	17.4	17.3	17.2	17.1	16.9	16.6
16	18.6	18.4	18.3	18.2	18.0	17.9	17.8	17.7	17.6	17.5	17.3	17.2	17.1	17.0	16.8	16.6	16.2
21	18.7	18.6	18.4	18.2	18.1	17.9	17.8	17.7	17.5	17.4	17.2	17.1	17.0	16.8	16.6	16.3	15.8
26	18.9	18.7	18.5	18.3	18.1	17.9	17.8	17.6	17.5	17.3	17.1	17.0	16.8	16.6	16.3	16.0	15.4
31	19.0	18.8	18.6	18.4	18.1	18.0	17.8	17.6	17.4	17.2	17.0	16.8	16.6	16.4	16.1	15.7	15.0

NOVEMBER — SUNRISE

	50°S	45°S	40°S	30°S	20°S	10°S	0	10°N	20°N	30°N	40°N	45°N	50°N	55°N	60°N	65°N	70°N
	h	h	h	h	h	h	h	h	h	h	h	h	h	h	h	h	h
1	4.4	4.7	4.8	5.1	5.3	5.5	5.7	5.8	6.0	6.2	6.5	6.6	6.8	7.1	7.4	7.8	8.5
6	4.3	4.5	4.7	5.0	5.3	5.5	5.7	5.9	6.1	6.3	6.6	6.7	7.0	7.2	7.6	8.1	8.9
11	4.2	4.4	4.6	5.0	5.2	5.5	5.7	5.9	6.1	6.4	6.7	6.9	7.1	7.4	7.8	8.4	9.4
16	4.1	4.3	4.6	4.9	5.2	5.5	5.7	5.9	6.2	6.4	6.8	7.0	7.2	7.6	8.0	8.7	9.9
21	4.0	4.3	4.5	4.9	5.2	5.5	5.7	5.9	6.2	6.5	6.9	7.1	7.4	7.7	8.2	8.9	10.6
26	3.9	4.2	4.5	4.9	5.2	5.5	5.7	6.0	6.3	6.6	7.0	7.2	7.5	7.9	8.4	9.2	—

NOVEMBER — SUNSET

	50°S	45°S	40°S	30°S	20°S	10°S	0	10°N	20°N	30°N	40°N	45°N	50°N	55°N	60°N	65°N	70°N
	h	h	h	h	h	h	h	h	h	h	h	h	h	h	h	h	h
1	19.0	18.8	18.6	18.4	18.1	18.0	17.8	17.6	17.4	17.2	17.0	16.8	16.6	16.4	16.1	15.6	14.9
6	19.2	18.9	18.7	18.4	18.2	18.0	17.8	17.6	17.4	17.2	16.9	16.7	16.5	16.2	15.9	15.4	14.5
11	19.3	19.0	18.8	18.5	18.2	18.0	17.8	17.6	17.4	17.1	16.8	16.6	16.4	16.1	15.7	15.1	14.1
16	19.4	19.2	18.9	18.6	18.3	18.0	17.8	17.6	17.3	17.1	16.7	16.5	16.2	15.9	15.5	14.8	13.6
21	19.6	19.3	19.0	18.6	18.3	18.1	17.8	17.6	17.3	17.0	16.7	16.4	16.2	15.8	15.3	14.6	13.0
26	19.7	19.4	19.1	18.7	18.4	18.1	17.8	17.6	17.3	17.0	16.6	16.4	16.1	15.7	15.2	14.3	—

DECEMBER — SUNRISE

	50°S	45°S	40°S	30°S	20°S	10°S	0	10°N	20°N	30°N	40°N	45°N	50°N	55°N	60°N	65°N	70°N
	h	h	h	h	h	h	h	h	h	h	h	h	h	h	h	h	h
1	3.8	4.2	4.4	4.9	5.2	5.5	5.8	6.0	6.3	6.6	7.0	7.3	7.6	8.0	8.6	9.5	—
6	3.8	4.1	4.4	4.9	5.2	5.5	5.8	6.1	6.4	6.7	7.1	7.4	7.7	8.1	8.7	9.7	—
11	3.8	4.1	4.4	4.9	5.2	5.5	5.8	6.1	6.4	6.8	7.2	7.5	7.8	8.2	8.9	9.9	—
16	3.8	4.1	4.4	4.9	5.3	5.6	5.9	6.2	6.5	6.8	7.3	7.5	7.9	8.3	9.0	10.1	—
21	3.8	4.2	4.5	4.9	5.3	5.6	5.9	6.2	6.5	6.9	7.3	7.6	7.9	8.4	9.0	10.2	—
26	3.8	4.2	4.5	5.0	5.3	5.7	5.9	6.2	6.5	6.9	7.3	7.6	8.0	8.4	9.1	10.2	—
31	3.9	4.3	4.6	5.0	5.4	5.7	6.0	6.3	6.6	6.9	7.4	7.6	8.0	8.4	9.0	10.1	—

DECEMBER — SUNSET

	50°S	45°S	40°S	30°S	20°S	10°S	0	10°N	20°N	30°N	40°N	45°N	50°N	55°N	60°N	65°N	70°N
	h	h	h	h	h	h	h	h	h	h	h	h	h	h	h	h	h
1	19.8	19.5	19.2	18.8	18.4	18.1	17.9	17.6	17.3	17.0	16.6	16.3	16.0	15.6	15.0	14.1	—
6	19.9	19.6	19.3	18.8	18.5	18.2	17.9	17.6	17.3	17.0	16.6	16.3	16.0	15.6	15.0	14.0	—
11	20.0	19.7	19.4	18.9	18.5	18.2	17.9	17.7	17.4	17.0	16.6	16.3	16.0	15.5	14.9	13.8	—
16	20.1	19.7	19.4	19.0	18.6	18.3	18.0	17.7	17.4	17.0	16.6	16.3	16.0	15.5	14.9	13.8	—
21	20.2	19.8	19.5	19.0	18.6	18.3	18.0	17.7	17.4	17.1	16.6	16.4	16.0	15.6	14.9	13.8	—
26	20.2	19.8	19.5	19.0	18.7	18.4	18.1	17.8	17.5	17.1	16.7	16.4	16.1	15.6	15.0	13.8	—
31	20.2	19.8	19.5	19.1	18.7	18.4	18.1	17.8	17.5	17.2	16.7	16.5	16.1	15.7	15.1	14.0	—

TABLE 6. *Times at which astronomical twilight begins and ends at various latitudes.*

JANUARY — BEGINNING

	50°S	45°S	40°S	30°S	20°S	10°S	0	10°N	20°N	30°N	40°N	45°N	50°N	55°N	60°N	65°N	70°N
	h	h	h	h	h	h	h	h	h	h	h	h	h	h	h	h	h
1	—	1.8	2.5	3.4	4.0	4.4	4.7	5.0	5.3	5.5	5.7	5.9	6.0	6.1	6.3	6.5	6.8
6	—	1.9	2.6	3.5	4.1	4.5	4.8	5.1	5.3	5.5	5.8	5.9	6.0	6.1	6.3	6.5	6.7
11	0.5	2.1	2.8	3.6	4.1	4.5	4.8	5.1	5.3	5.5	5.8	5.9	6.0	6.1	6.2	6.4	6.6
16	1.1	2.3	2.9	3.7	4.2	4.6	4.9	5.1	5.3	5.5	5.7	5.8	5.9	6.0	6.2	6.3	6.5
21	1.5	2.5	3.0	3.8	4.3	4.6	4.9	5.1	5.3	5.5	5.7	5.8	5.9	6.0	6.1	6.2	6.3
26	1.8	2.7	3.2	3.9	4.3	4.7	4.9	5.2	5.3	5.5	5.7	5.7	5.8	5.9	6.0	6.0	6.1
31	2.1	2.9	3.3	4.0	4.4	4.7	5.0	5.2	5.3	5.5	5.6	5.7	5.7	5.8	5.8	5.9	5.9

ENDING

	50°S	45°S	40°S	30°S	20°S	10°S	0	10°N	20°N	30°N	40°N	45°N	50°N	55°N	60°N	65°N	70°N
	h	h	h	h	h	h	h	h	h	h	h	h	h	h	h	h	h
1	—	22.3	21.6	20.7	20.1	19.7	19.4	19.1	18.8	18.6	18.4	18.3	18.1	18.0	17.8	17.6	17.4
6	—	22.3	21.5	20.7	20.1	19.7	19.4	19.1	18.9	18.7	18.4	18.3	18.2	18.1	17.9	17.7	17.5
11	23.6	22.2	21.5	20.7	20.1	19.7	19.4	19.2	18.9	18.7	18.5	18.4	18.3	18.2	18.0	17.9	17.7
16	23.2	22.0	21.4	20.6	20.1	19.7	19.5	19.2	19.0	18.8	18.6	18.5	18.4	18.3	18.2	18.0	17.9
21	22.9	21.9	21.3	20.6	20.1	19.7	19.5	19.2	19.0	18.8	18.7	18.6	18.5	18.4	18.3	18.2	18.1
26	22.6	21.7	21.2	20.5	20.1	19.7	19.5	19.3	19.1	18.9	18.8	18.7	18.6	18.5	18.5	18.4	18.3
31	22.3	21.6	21.1	20.5	20.0	19.7	19.5	19.3	19.1	19.0	18.8	18.8	18.7	18.7	18.6	18.6	18.6

FEBRUARY — BEGINNING

	50°S	45°S	40°S	30°S	20°S	10°S	0	10°N	20°N	30°N	40°N	45°N	50°N	55°N	60°N	65°N	70°N
	h	h	h	h	h	h	h	h	h	h	h	h	h	h	h	h	h
1	2.2	2.9	3.4	4.0	4.4	4.7	5.0	5.2	5.3	5.5	5.6	5.7	5.7	5.8	5.8	5.8	5.9
6	2.5	3.1	3.5	4.1	4.5	4.8	5.0	5.2	5.3	5.4	5.5	5.6	5.6	5.6	5.6	5.6	5.6
11	2.7	3.3	3.7	4.2	4.5	4.8	5.0	5.2	5.3	5.4	5.4	5.5	5.5	5.5	5.5	5.4	5.3
16	3.0	3.4	3.8	4.3	4.6	4.8	5.0	5.1	5.2	5.3	5.4	5.4	5.3	5.3	5.3	5.2	5.0
21	3.2	3.6	3.9	4.4	4.6	4.9	5.0	5.1	5.2	5.2	5.3	5.2	5.2	5.1	5.0	4.9	4.6
26	3.4	3.8	4.0	4.4	4.7	4.9	5.0	5.1	5.1	5.2	5.1	5.1	5.0	5.0	4.8	4.6	4.3

ENDING

	50°S	45°S	40°S	30°S	20°S	10°S	0	10°N	20°N	30°N	40°N	45°N	50°N	55°N	60°N	65°N	70°N
	h	h	h	h	h	h	h	h	h	h	h	h	h	h	h	h	h
1	22.2	21.5	21.1	20.4	20.0	19.7	19.5	19.3	19.1	19.0	18.9	18.8	18.8	18.7	18.7	18.6	18.6
6	22.0	21.4	20.9	20.4	20.0	19.7	19.5	19.3	19.2	19.0	18.9	18.9	18.9	18.9	18.9	18.9	18.9
11	21.7	21.2	20.8	20.3	19.9	19.7	19.5	19.3	19.2	19.1	19.0	19.0	19.0	19.0	19.0	19.1	19.2
16	21.5	21.0	20.7	20.2	19.9	19.6	19.5	19.3	19.2	19.2	19.1	19.1	19.1	19.2	19.2	19.3	19.5
21	21.2	20.8	20.5	20.1	19.8	19.6	19.4	19.3	19.3	19.2	19.2	19.2	19.3	19.3	19.4	19.6	19.9
26	21.0	20.6	20.4	20.0	19.7	19.6	19.4	19.3	19.3	19.3	19.3	19.3	19.4	19.5	19.6	19.9	20.2

MARCH — BEGINNING

	50°S	45°S	40°S	30°S	20°S	10°S	0	10°N	20°N	30°N	40°N	45°N	50°N	55°N	60°N	65°N	70°N
	h	h	h	h	h	h	h	h	h	h	h	h	h	h	h	h	h
1	3.5	3.9	4.1	4.5	4.7	4.9	5.0	5.1	5.1	5.1	5.1	5.0	4.9	4.8	4.7	4.4	4.0
6	3.7	4.0	4.2	4.5	4.8	4.9	5.0	5.0	5.0	5.0	4.9	4.9	4.8	4.6	4.4	4.1	3.6
11	3.9	4.1	4.3	4.6	4.8	4.9	5.0	5.0	5.0	4.9	4.8	4.7	4.6	4.4	4.1	3.8	3.1
16	4.1	4.3	4.4	4.7	4.8	4.9	4.9	4.9	4.9	4.8	4.7	4.6	4.4	4.2	3.9	3.4	2.5
21	4.2	4.4	4.5	4.7	4.8	4.9	4.9	4.9	4.9	4.7	4.5	4.4	4.2	3.9	3.6	3.0	1.8
26	4.4	4.5	4.6	4.8	4.9	4.9	4.9	4.9	4.8	4.6	4.4	4.2	4.0	3.7	3.2	2.5	—
31	4.5	4.6	4.7	4.8	4.9	4.9	4.9	4.8	4.7	4.5	4.2	4.0	3.8	3.4	2.9	1.9	—

ENDING

	50°S	45°S	40°S	30°S	20°S	10°S	0	10°N	20°N	30°N	40°N	45°N	50°N	55°N	60°N	65°N	70°N
	h	h	h	h	h	h	h	h	h	h	h	h	h	h	h	h	h
1	20.9	20.5	20.3	19.9	19.7	19.5	19.4	19.3	19.3	19.3	19.4	19.4	19.5	19.6	19.8	20.0	20.4
6	20.6	20.3	20.1	19.8	19.6	19.5	19.4	19.3	19.3	19.4	19.4	19.5	19.6	19.8	20.0	20.3	20.9
11	20.4	20.2	20.0	19.7	19.5	19.4	19.4	19.3	19.4	19.4	19.5	19.6	19.8	20.0	20.2	20.6	21.3
16	20.2	20.0	19.8	19.6	19.5	19.4	19.3	19.3	19.4	19.5	19.6	19.8	19.9	20.1	20.5	21.0	21.9
21	20.0	19.8	19.7	19.5	19.4	19.3	19.3	19.3	19.4	19.5	19.7	19.9	20.1	20.3	20.7	21.3	22.6
26	19.8	19.6	19.5	19.4	19.3	19.3	19.3	19.3	19.4	19.6	19.8	20.0	20.2	20.5	21.0	21.8	—
31	19.6	19.5	19.4	19.3	19.2	19.2	19.3	19.3	19.5	19.6	19.9	20.1	20.4	20.8	21.3	22.3	—

TABLE 6 (continued). *Times at which astronomical twilight begins and ends at various latitudes.*

APRIL — BEGINNING

	50°S	45°S	40°S	30°S	20°S	10°S	0	10°N	20°N	30°N	40°N	45°N	50°N	55°N	60°N	65°N	70°N
	h	h	h	h	h	h	h	h	h	h	h	h	h	h	h	h	h
1	4.5	4.7	4.7	4.9	4.9	4.9	4.9	4.8	4.7	4.5	4.2	4.0	3.7	3.4	2.8	1.8	—
6	4.7	4.8	4.8	4.9	4.9	4.9	4.8	4.7	4.6	4.4	4.0	3.8	3.5	3.1	2.4	0.9	—
11	4.8	4.9	4.9	4.9	4.9	4.9	4.8	4.7	4.5	4.3	3.9	3.6	3.3	2.8	2.0	—	—
16	4.9	5.0	5.0	5.0	5.0	4.9	4.8	4.6	4.4	4.2	3.7	3.4	3.0	2.5	1.4	—	—
21	5.0	5.1	5.1	5.0	5.0	4.9	4.8	4.6	4.4	4.0	3.6	3.3	2.8	2.1	0.5	—	—
26	5.2	5.2	5.1	5.1	5.0	4.9	4.7	4.5	4.3	3.9	3.4	3.1	2.6	1.7	—	—	—

ENDING

	50°S	45°S	40°S	30°S	20°S	10°S	0	10°N	20°N	30°N	40°N	45°N	50°N	55°N	60°N	65°N	70°N
	h	h	h	h	h	h	h	h	h	h	h	h	h	h	h	h	h
1	19.6	19.5	19.4	19.3	19.2	19.2	19.3	19.3	19.5	19.7	19.9	20.2	20.4	20.8	21.4	22.4	—
6	19.4	19.3	19.2	19.2	19.2	19.2	19.2	19.3	19.5	19.7	20.1	20.3	20.6	21.0	21.7	23.4	—
11	19.2	19.2	19.1	19.1	19.1	19.1	19.2	19.4	19.5	19.8	20.2	20.4	20.8	21.3	22.2	—	—
16	19.1	19.0	19.0	19.0	19.0	19.1	19.2	19.4	19.6	19.9	20.3	20.6	21.0	21.6	22.7	—	—
21	18.9	18.9	18.9	18.9	19.0	19.1	19.2	19.4	19.6	19.9	20.4	20.7	21.2	21.9	—	—	—
26	18.8	18.8	18.8	18.8	18.9	19.0	19.2	19.4	19.6	20.0	20.5	20.9	21.4	22.3	—	—	—

MAY — BEGINNING

	50°S	45°S	40°S	30°S	20°S	10°S	0	10°N	20°N	30°N	40°N	45°N	50°N	55°N	60°N	65°N	70°N
	h	h	h	h	h	h	h	h	h	h	h	h	h	h	h	h	h
1	5.3	5.2	5.2	5.1	5.0	4.9	4.7	4.5	4.2	3.8	3.3	2.9	2.3	1.3	—	—	—
6	5.4	5.3	5.3	5.2	5.0	4.9	4.7	4.5	4.2	3.8	3.1	2.7	2.0	0.6	—	—	—
11	5.5	5.4	5.3	5.2	5.1	4.9	4.7	4.4	4.1	3.7	3.0	2.5	1.8	—	—	—	—
16	5.6	5.5	5.4	5.3	5.1	4.9	4.7	4.4	4.1	3.6	2.9	2.4	1.5	—	—	—	—
21	5.7	5.6	5.5	5.3	5.1	4.9	4.7	4.4	4.0	3.5	2.8	2.2	1.2	—	—	—	—
26	5.7	5.6	5.5	5.3	5.1	4.9	4.7	4.4	4.0	3.5	2.7	2.0	0.8	—	—	—	—
31	5.8	5.7	5.6	5.4	5.2	4.9	4.7	4.4	4.0	3.4	2.6	1.9	0.3	—	—	—	—

ENDING

	50°S	45°S	40°S	30°S	20°S	10°S	0	10°N	20°N	30°N	40°N	45°N	50°N	55°N	60°N	65°N	70°N
	h	h	h	h	h	h	h	h	h	h	h	h	h	h	h	h	h
1	18.6	18.6	18.7	18.8	18.9	19.0	19.2	19.4	19.7	20.1	20.6	21.0	21.6	22.7	—	—	—
6	18.5	18.5	18.6	18.7	18.8	19.0	19.2	19.4	19.7	20.1	20.8	21.2	21.9	23.4	—	—	—
11	18.4	18.5	18.5	18.7	18.8	19.0	19.2	19.5	19.8	20.2	20.9	21.4	22.2	—	—	—	—
16	18.3	18.4	18.5	18.6	18.8	19.0	19.2	19.5	19.8	20.3	21.0	21.6	22.4	—	—	—	—
21	18.2	18.3	18.4	18.6	18.8	19.0	19.2	19.5	19.9	20.4	21.1	21.7	22.8	—	—	—	—
26	18.2	18.3	18.4	18.6	18.8	19.0	19.2	19.5	19.9	20.4	21.2	21.9	23.2	—	—	—	—
31	18.1	18.2	18.3	18.5	18.8	19.0	19.3	19.6	20.0	20.5	21.3	22.0	—	—	—	—	—

JUNE — BEGINNING

	50°S	45°S	40°S	30°S	20°S	10°S	0	10°N	20°N	30°N	40°N	45°N	50°N	55°N	60°N	65°N	70°N
	h	h	h	h	h	h	h	h	h	h	h	h	h	h	h	h	h
1	5.8	5.7	5.6	5.4	5.2	4.9	4.7	4.3	4.0	3.4	2.6	1.9	—	—	—	—	—
6	5.9	5.8	5.6	5.4	5.2	4.9	4.7	4.3	3.9	3.4	2.5	1.8	—	—	—	—	—
11	5.9	5.8	5.7	5.4	5.2	5.0	4.7	4.4	3.9	3.4	2.5	1.7	—	—	—	—	—
16	6.0	5.8	5.7	5.5	5.2	5.0	4.7	4.4	3.9	3.4	2.5	1.7	—	—	—	—	—
21	6.0	5.9	5.7	5.5	5.3	5.0	4.7	4.4	4.0	3.4	2.5	1.7	—	—	—	—	—
26	6.0	5.9	5.8	5.5	5.3	5.0	4.7	4.4	4.0	3.4	2.5	1.7	—	—	—	—	—

ENDING

	50°S	45°S	40°S	30°S	20°S	10°S	0	10°N	20°N	30°N	40°N	45°N	50°N	55°N	60°N	65°N	70°N
	h	h	h	h	h	h	h	h	h	h	h	h	h	h	h	h	h
1	18.1	18.2	18.3	18.5	18.8	19.0	19.3	19.6	20.0	20.5	21.4	22.1	—	—	—	—	—
6	18.1	18.2	18.3	18.5	18.8	19.0	19.3	19.6	20.0	20.6	21.4	22.2	—	—	—	—	—
11	18.0	18.2	18.3	18.5	18.8	19.0	19.3	19.6	20.1	20.6	21.5	22.3	—	—	—	—	—
16	18.0	18.2	18.3	18.5	18.8	19.0	19.3	19.7	20.1	20.7	21.6	22.4	—	—	—	—	—
21	18.1	18.2	18.3	18.6	18.8	19.1	19.3	19.7	20.1	20.7	21.6	22.4	—	—	—	—	—
26	18.1	18.2	18.3	18.6	18.8	19.1	19.4	19.7	20.1	20.7	21.6	22.4	—	—	—	—	—

JULY — BEGINNING

	50°S	45°S	40°S	30°S	20°S	10°S	0	10°N	20°N	30°N	40°N	45°N	50°N	55°N	60°N	65°N	70°N
	h	h	h	h	h	h	h	h	h	h	h	h	h	h	h	h	h
1	6.0	5.9	5.8	5.5	5.3	5.0	4.8	4.4	4.0	3.4	2.5	1.8	—	—	—	—	—
6	6.0	5.9	5.7	5.5	5.3	5.0	4.8	4.4	4.0	3.5	2.6	1.9	—	—	—	—	—
11	6.0	5.8	5.7	5.5	5.3	5.1	4.8	4.5	4.1	3.5	2.7	2.0	—	—	—	—	—
16	5.9	5.8	5.7	5.5	5.3	5.1	4.8	4.5	4.1	3.6	2.8	2.1	0.8	—	—	—	—
21	5.8	5.8	5.7	5.5	5.3	5.1	4.8	4.5	4.2	3.7	2.9	2.3	1.2	—	—	—	—
26	5.8	5.7	5.6	5.4	5.3	5.1	4.8	4.6	4.2	3.7	3.0	2.5	1.5	—	—	—	—
31	5.7	5.6	5.5	5.4	5.2	5.1	4.8	4.6	4.3	3.8	3.1	2.6	1.8	—	—	—	—

JULY — ENDING

	50°S	45°S	40°S	30°S	20°S	10°S	0	10°N	20°N	30°N	40°N	45°N	50°N	55°N	60°N	65°N	70°N
	h	h	h	h	h	h	h	h	h	h	h	h	h	h	h	h	h
1	18.1	18.3	18.4	18.6	18.8	19.1	19.4	19.7	20.1	20.7	21.6	22.4	—	—	—	—	—
6	18.2	18.3	18.4	18.6	18.9	19.1	19.4	19.7	20.1	20.7	21.5	22.3	—	—	—	—	—
11	18.2	18.3	18.5	18.7	18.9	19.1	19.4	19.7	20.1	20.6	21.5	22.2	—	—	—	—	—
16	18.3	18.4	18.5	18.7	18.9	19.1	19.4	19.7	20.1	20.6	21.4	22.0	23.3	—	—	—	—
21	18.4	18.5	18.6	18.7	18.9	19.1	19.4	19.7	20.0	20.5	21.3	21.9	23.0	—	—	—	—
26	18.5	18.5	18.6	18.8	19.0	19.2	19.4	19.7	20.0	20.5	21.2	21.7	22.6	—	—	—	—
31	18.5	18.6	18.7	18.8	19.0	19.2	19.4	19.6	20.0	20.4	21.1	21.6	22.3	—	—	—	—

AUGUST — BEGINNING

	50°S	45°S	40°S	30°S	20°S	10°S	0	10°N	20°N	30°N	40°N	45°N	50°N	55°N	60°N	65°N	70°N
	h	h	h	h	h	h	h	h	h	h	h	h	h	h	h	h	h
1	5.7	5.6	5.5	5.4	5.2	5.1	4.8	4.6	4.3	3.8	3.2	2.6	1.9	—	—	—	—
6	5.5	5.5	5.5	5.3	5.2	5.0	4.8	4.6	4.3	3.9	3.3	2.8	2.1	0.5	—	—	—
11	5.4	5.4	5.4	5.3	5.2	5.0	4.8	4.6	4.3	4.0	3.4	3.0	2.4	1.3	—	—	—
16	5.3	5.3	5.3	5.2	5.1	5.0	4.8	4.6	4.4	4.0	3.5	3.1	2.6	1.7	—	—	—
21	5.1	5.2	5.2	5.1	5.1	5.0	4.8	4.7	4.4	4.1	3.6	3.3	2.8	2.1	—	—	—
26	5.0	5.0	5.0	5.0	5.0	4.9	4.8	4.7	4.5	4.2	3.7	3.4	3.0	2.4	1.2	—	—
31	4.8	4.9	4.9	5.0	4.9	4.9	4.8	4.7	4.5	4.2	3.8	3.6	3.2	2.7	1.8	—	—

AUGUST — ENDING

	50°S	45°S	40°S	30°S	20°S	10°S	0	10°N	20°N	30°N	40°N	45°N	50°N	55°N	60°N	65°N	70°N
	h	h	h	h	h	h	h	h	h	h	h	h	h	h	h	h	h
1	18.6	18.6	18.7	18.8	19.0	19.2	19.4	19.6	19.9	20.4	21.0	21.5	22.3	—	—	—	—
6	18.7	18.7	18.8	18.9	19.0	19.2	19.4	19.6	19.9	20.3	20.9	21.4	22.0	23.5	—	—	—
11	18.8	18.8	18.8	18.9	19.0	19.2	19.3	19.5	19.8	20.2	20.8	21.2	21.8	22.8	—	—	—
16	18.9	18.9	18.9	18.9	19.0	19.2	19.3	19.5	19.8	20.1	20.6	21.0	21.5	22.3	—	—	—
21	19.0	19.0	19.0	19.0	19.1	19.1	19.3	19.5	19.7	20.0	20.5	20.8	21.3	22.0	23.8	—	—
26	19.1	19.0	19.0	19.0	19.1	19.1	19.2	19.4	19.6	19.9	20.3	20.6	21.0	21.6	22.7	—	—
31	19.2	19.1	19.1	19.1	19.1	19.1	19.2	19.3	19.5	19.8	20.2	20.4	20.8	21.3	22.1	—	—

SEPTEMBER — BEGINNING

	50°S	45°S	40°S	30°S	20°S	10°S	0	10°N	20°N	30°N	40°N	45°N	50°N	55°N	60°N	65°N	70°N
	h	h	h	h	h	h	h	h	h	h	h	h	h	h	h	h	h
1	4.8	4.9	4.9	4.9	4.9	4.9	4.8	4.7	4.5	4.2	3.9	3.6	3.2	2.7	1.9	—	—
6	4.6	4.7	4.8	4.8	4.9	4.8	4.8	4.7	4.5	4.3	4.0	3.7	3.4	3.0	2.3	0.7	—
11	4.4	4.6	4.6	4.7	4.8	4.8	4.7	4.7	4.5	4.4	4.1	3.9	3.6	3.2	2.6	1.6	—
16	4.2	4.4	4.5	4.6	4.7	4.7	4.7	4.7	4.6	4.4	4.2	4.0	3.7	3.4	2.9	2.1	—
21	4.0	4.2	4.3	4.5	4.6	4.7	4.7	4.7	4.6	4.5	4.3	4.1	3.9	3.6	3.2	2.6	1.2
26	3.8	4.0	4.2	4.4	4.5	4.6	4.7	4.7	4.6	4.5	4.3	4.2	4.0	3.8	3.5	3.0	2.0

SEPTEMBER — ENDING

	50°S	45°S	40°S	30°S	20°S	10°S	0	10°N	20°N	30°N	40°N	45°N	50°N	55°N	60°N	65°N	70°N
	h	h	h	h	h	h	h	h	h	h	h	h	h	h	h	h	h
1	19.2	19.2	19.1	19.1	19.1	19.1	19.2	19.3	19.5	19.8	20.1	20.4	20.7	21.2	22.0	—	—
6	19.3	19.3	19.2	19.1	19.1	19.1	19.2	19.3	19.4	19.6	20.0	20.2	20.5	20.9	21.6	23.0	—
11	19.5	19.4	19.3	19.2	19.1	19.1	19.1	19.2	19.3	19.5	19.8	20.0	20.3	20.6	21.2	22.2	—
16	19.6	19.5	19.4	19.2	19.1	19.1	19.1	19.2	19.3	19.4	19.7	19.8	20.1	20.4	20.8	21.6	—
21	19.8	19.6	19.4	19.3	19.1	19.1	19.1	19.1	19.2	19.3	19.5	19.7	19.8	20.1	20.5	21.1	22.4
26	19.9	19.7	19.5	19.3	19.2	19.1	19.1	19.1	19.1	19.2	19.4	19.5	19.6	19.9	20.2	20.7	21.6

OCTOBER BEGINNING

	50°S	45°S	40°S	30°S	20°S	10°S	0	10°N	20°N	30°N	40°N	45°N	50°N	55°N	60°N	65°N	70°N
	h	h	h	h	h	h	h	h	h	h	h	h	h	h	h	h	h
1	3.6	3.9	4.0	4.3	4.5	4.6	4.6	4.6	4.6	4.6	4.4	4.3	4.2	4.0	3.7	3.3	2.5
6	3.4	3.7	3.9	4.2	4.4	4.5	4.6	4.6	4.6	4.6	4.5	4.4	4.3	4.2	3.9	3.6	3.0
11	3.2	3.5	3.7	4.1	4.3	4.5	4.6	4.6	4.7	4.7	4.6	4.5	4.5	4.3	4.2	3.9	3.4
16	3.0	3.3	3.6	4.0	4.2	4.4	4.5	4.6	4.7	4.7	4.7	4.6	4.6	4.5	4.4	4.2	3.8
21	2.7	3.1	3.4	3.9	4.2	4.4	4.5	4.6	4.7	4.8	4.8	4.8	4.7	4.7	4.6	4.4	4.2
26	2.5	2.9	3.3	3.8	4.1	4.3	4.5	4.6	4.7	4.8	4.9	4.9	4.8	4.8	4.8	4.7	4.5
31	2.2	2.8	3.1	3.7	4.0	4.3	4.5	4.6	4.8	4.9	4.9	5.0	5.0	5.0	4.9	4.9	4.8

ENDING

	50°S	45°S	40°S	30°S	20°S	10°S	0	10°N	20°N	30°N	40°N	45°N	50°N	55°N	60°N	65°N	70°N
	h	h	h	h	h	h	h	h	h	h	h	h	h	h	h	h	h
1	20.1	19.8	19.6	19.4	19.2	19.1	19.0	19.0	19.0	19.1	19.2	19.3	19.4	19.6	19.9	20.3	21.0
6	20.2	19.9	19.7	19.4	19.2	19.1	19.0	19.0	19.0	19.0	19.1	19.2	19.3	19.4	19.6	20.0	20.5
11	20.4	20.1	19.8	19.5	19.3	19.1	19.0	18.9	18.9	18.9	18.9	19.0	19.1	19.2	19.4	19.6	20.1
16	20.6	20.2	19.9	19.6	19.3	19.1	19.0	18.9	18.8	18.8	18.8	18.9	18.9	19.0	19.1	19.3	19.7
21	20.8	20.4	20.1	19.6	19.3	19.1	19.0	18.9	18.8	18.7	18.7	18.7	18.8	18.8	18.9	19.0	19.3
26	21.0	20.5	20.2	19.7	19.4	19.1	19.0	18.8	18.7	18.6	18.6	18.6	18.6	18.6	18.7	18.8	18.9
31	21.3	20.7	20.3	19.8	19.4	19.2	19.0	18.8	18.7	18.6	18.5	18.5	18.5	18.5	18.5	18.5	18.6

NOVEMBER BEGINNING

	50°S	45°S	40°S	30°S	20°S	10°S	0	10°N	20°N	30°N	40°N	45°N	50°N	55°N	60°N	65°N	70°N
	h	h	h	h	h	h	h	h	h	h	h	h	h	h	h	h	h
1	2.2	2.7	3.1	3.7	4.0	4.3	4.5	4.6	4.8	4.9	5.0	5.0	5.0	5.0	5.0	4.9	4.8
6	1.9	2.5	3.0	3.6	4.0	4.3	4.5	4.7	4.8	4.9	5.0	5.1	5.1	5.1	5.2	5.2	5.1
11	1.7	2.4	2.9	3.5	3.9	4.2	4.5	4.7	4.8	5.0	5.1	5.2	5.2	5.3	5.3	5.4	5.4
16	1.4	2.2	2.7	3.4	3.9	4.2	4.5	4.7	4.9	5.0	5.2	5.3	5.3	5.4	5.5	5.6	5.7
21	1.1	2.1	2.6	3.4	3.8	4.2	4.5	4.7	4.9	5.1	5.3	5.4	5.5	5.5	5.6	5.8	5.9
26	0.7	1.9	2.5	3.3	3.8	4.2	4.5	4.7	5.0	5.2	5.4	5.5	5.6	5.7	5.8	5.9	6.1

ENDING

	50°S	45°S	40°S	30°S	20°S	10°S	0	10°N	20°N	30°N	40°N	45°N	50°N	55°N	60°N	65°N	70°N
	h	h	h	h	h	h	h	h	h	h	h	h	h	h	h	h	h
1	21.3	20.8	20.4	19.8	19.4	19.2	19.0	18.8	18.7	18.6	18.5	18.5	18.4	18.4	18.5	18.5	18.6
6	21.6	20.9	20.5	19.9	19.5	19.2	19.0	18.8	18.6	18.5	18.4	18.4	18.3	18.3	18.3	18.3	18.3
11	21.9	21.1	20.6	20.0	19.6	19.2	19.0	18.8	18.6	18.5	18.3	18.3	18.2	18.2	18.1	18.1	18.0
16	22.2	21.3	20.8	20.1	19.6	19.3	19.0	18.8	18.6	18.4	18.3	18.2	18.1	18.1	18.0	17.9	17.8
21	22.5	21.5	20.9	20.2	19.7	19.3	19.0	18.8	18.6	18.4	18.2	18.2	18.1	18.0	17.9	17.8	17.6
26	22.9	21.7	21.1	20.3	19.8	19.4	19.1	18.8	18.6	18.4	18.2	18.1	18.0	17.9	17.8	17.6	17.5

DECEMBER BEGINNING

	50°S	45°S	40°S	30°S	20°S	10°S	0	10°N	20°N	30°N	40°N	45°N	50°N	55°N	60°N	65°N	70°N
	h	h	h	h	h	h	h	h	h	h	h	h	h	h	h	h	h
1	0.3	1.8	2.5	3.3	3.8	4.2	4.5	4.8	5.0	5.2	5.4	5.5	5.7	5.8	5.9	6.1	6.3
6	—	1.7	2.4	3.3	3.8	4.2	4.5	4.8	5.1	5.3	5.5	5.6	5.7	5.9	6.0	6.2	6.5
11	—	1.6	2.4	3.3	3.8	4.2	4.6	4.9	5.1	5.3	5.6	5.7	5.8	6.0	6.1	6.3	6.6
16	—	1.6	2.4	3.3	3.9	4.3	4.6	4.9	5.1	5.4	5.6	5.8	5.9	6.0	6.2	6.4	6.7
21	—	1.6	2.4	3.3	3.9	4.3	4.7	4.9	5.2	5.4	5.7	5.8	5.9	6.1	6.3	6.5	6.8
26	—	1.7	2.4	3.4	3.9	4.4	4.7	5.0	5.2	5.5	5.7	5.8	6.0	6.1	6.3	6.5	6.8
31	—	1.7	2.5	3.4	4.0	4.4	4.7	5.0	5.3	5.5	5.7	5.9	6.0	6.1	6.3	6.5	6.8

ENDING

	50°S	45°S	40°S	30°S	20°S	10°S	0	10°N	20°N	30°N	40°N	45°N	50°N	55°N	60°N	65°N	70°N
	h	h	h	h	h	h	h	h	h	h	h	h	h	h	h	h	h
1	23.5	21.9	21.2	20.4	19.8	19.4	19.1	18.9	18.6	18.4	18.2	18.1	18.0	17.8	17.7	17.5	17.3
6	—	22.0	21.3	20.4	19.9	19.5	19.2	18.9	18.6	18.4	18.2	18.1	18.0	17.8	17.7	17.5	17.2
11	—	22.2	21.4	20.5	19.9	19.5	19.2	18.9	18.7	18.4	18.2	18.1	17.9	17.8	17.6	17.4	17.2
16	—	22.3	21.5	20.6	20.0	19.6	19.2	19.0	18.7	18.5	18.2	18.1	18.0	17.8	17.6	17.4	17.2
21	—	22.3	21.5	20.6	20.0	19.6	19.3	19.0	18.7	18.5	18.3	18.1	18.0	17.8	17.7	17.5	17.2
26	—	22.4	21.6	20.7	20.1	19.7	19.3	19.0	18.8	18.5	18.3	18.2	18.0	17.9	17.7	17.5	17.2
31	—	22.3	21.6	20.7	20.1	19.7	19.4	19.1	18.8	18.6	18.4	18.2	18.1	18.0	17.8	17.6	17.3

TABLE 7. *The equation of time and the Sun's longitude, RA and declination at 0h UT for various dates throughout the year.*

The longitude of the Sun is given to the nearest degree, and the equation of time (EQN) to the nearest minute; the RA is given in whole hours at the nearest date; and the declination is given to the nearest degree. For the RA on intermediate dates, add 4 minutes per day, and for longitude add 1° per day. The equation of time and declination can be interpolated by inspection. The dates have been chosen so that the apparent longitude of the Sun is an exact multiple of 5°.

The intervals between dates in the table vary between 4 and 6 days because the longitude has been rounded to the nearest degree, and also because the elliptical orbit of the Earth means that the change of the Sun's longitude during the year is not uniform.

DATE	LONG. °	EQN m	RA h	DEC. °	DATE	LONG. °	EQN m	RA h	DEC. °
Dec. 27	275	−1		−23	June 27	95	−3		+23
Jan. 1	280	−3		−23	July 2	100	−4		+23
6	285	−6	19	−23	7	105	−5	7	+23
10	290	−7		−22	13	110	−6		+22
15	295	−9		−21	18	115	−6		+21
20	300	−11	20	−20	23	120	−6	8	+20
25	305	−12		−19	28	125	−6		+19
30	310	−13		−18	Aug. 2	130	−6		+18
Feb. 4	315	−14	21	−16	8	135	−6	9	+16
9	320	−14		−15	13	140	−5		+15
14	325	−14		−13	18	145	−4		+13
19	330	−14	22	−11	23	150	−3	10	+12
24	335	−13		−10	28	155	−1		+10
Mar. 1	340	−12		−8	Sept. 3	160	0		+8
6	345	−11	23	−6	8	165	+2	11	+6
11	350	−10		−4	13	170	+4		+4
16	355	−9		−2	18	175	+6		+2
21	0	−7	0	0	23	180	+7	12	0
26	5	−6		+2	28	185	+9		−2
31	10	−4		+4	Oct. 3	190	+11		−4
Apr. 5	15	−3	1	+6	8	195	+12	13	−6
10	20	−1		+8	14	200	+14		−8
15	25	0		+10	19	205	+15		−10
20	30	+1	2	+11	24	210	+16	14	−12
25	35	+2		+13	29	215	+16		−13
May 1	40	+3		+15	Nov. 3	220	+16		−15
6	45	+3	3	+16	8	225	+16	15	−16
11	50	+4		+18	13	230	+16		−18
16	55	+4		+19	18	235	+15		−19
21	60	+3	4	+20	22	240	+14	16	−20
26	65	+3		+21	27	245	+13		−21
June 1	70	+2		+22	Dec. 2	250	+11		−22
6	75	+1	5	+23	7	255	+9	17	−23
11	80	+1		+23	12	260	+7		−23
16	85	−1		+23	17	265	+4		−23
22	90	−2	6	+23.4	22	270	+2	18	−23.4

MEAN SOLAR TIME is the time as shown on a clock, with the irregularities of apparent solar time smoothed out. Mean solar time is based on the movement of the imaginary mean Sun (see p. 8), although it is still affected by slight variations in the rotation of the Earth. The difference between mean solar time and apparent solar time is given by the equation of time.

GREENWICH MEAN TIME (GMT) is the mean solar time at the longitude of Greenwich, counted from midnight. In 1928, on the recommendation of the International Astronomical Union, GMT became known as Universal Time (UT). Before 1925, astron-omers reckoned GMT from Greenwich noon, which avoided the need to change the date during a night's observing; this is now called Greenwich Mean Astronomical Time (GMAT).

EQUATION OF TIME The correction to be applied to apparent solar time to obtain mean solar time:

$$\text{mean solar time} = \text{apparent solar time} - \text{equation of time}$$

The equation of time is greatest in early November, when apparent solar time is over 16 minutes ahead of mean solar time, which is shown by giving it a positive value. In mid-February apparent solar time is over 14 minutes behind mean solar time, shown by a negative value. The difference is zero four times a year: on April 15, June 14, September 1 and December 25. Table 7 gives the values of the equation of time to the nearest minute for various dates throughout the year.

SIDEREAL TIME is the time that has elapsed since the vernal equinox last crossed the meridian, i.e. it is the local hour angle of the vernal equinox. At any place, the sidereal time is equal to the right ascension of a star that is on the meridian.

GREENWICH SIDEREAL TIME (GST) is the time that has elapsed since the vernal equinox last crossed the meridian at Greenwich. Table 8 gives approximate values of Greenwich sidereal time throughout the year. This table is intended for use with the star charts to find which part of the sky is near the meridian at a given date and time. Precise values for Greenwich sidereal time at 0h UT for every date are given in *The Astronomical Almanac* and, to a lower precision suitable for most purposes, in the *Handbook* of the British Astronomical Association.

LOCAL SIDEREAL TIME (LST) To obtain local sidereal time, the longitude of the observer relative to Greenwich must be taken into account. For each degree of longitude east of Greenwich, add 4 minutes to GST to obtain LST (1h for every 15°). For each degree west of Greenwich, subtract 4 minutes from GST.

GREENWICH HOUR ANGLE (GHA) To find the Greenwich hour angle of a star, subtract the star's right ascension from the Greenwich sidereal time.

LOCAL HOUR ANGLE (LHA) To find the local hour angle of a star, subtract the star's right ascension from the local sidereal time; alternatively, add the longitude east to the GHA.

TIME ZONES Figure 2 shows how the Earth is divided into 24 time zones, each 15° broad, with the prime zone centred on the Greenwich meridian. Countries in each zone keep time that is usually an exact number of hours (or in some cases half-hours) different from UT. Time in the zones to the east of Greenwich is ahead of UT, while times to the west of Greenwich are behind UT. The date changes at the International Date Line, on the opposite side of the Earth from the Greenwich meridian.

UNIVERSAL TIME (UT) is the name by which Greenwich Mean Time became known for scientific purposes in 1928. However, precise observations have shown that the daily rotation of the Earth, on which UT is based, has various irregularities and hence can no longer be used as the basis of a uniform system of time. Several versions of UT are now defined:

UT0 is mean solar time determined directly from observations of the stars; because of the motion of the Earth's poles its value depends on the location of the observatory.

UT1 is UT0 corrected for the slight wandering of the Earth's geographical poles. UT1 is the time-scale used by astronomers and navigators, and is what is usually meant when the term UT is used without further qualification.

UT2 is UT1 corrected for seasonal variations in the Earth's rate of spin. It was used as the basis of time signals between 1956 and 1972, but its role as a quickly established, nearly uniform time-scale is now filled by International Atomic Time (see below).

UTC, Coordinated Universal Time, is the time given by broadcast time signals since 1972. It is derived from atomic clocks, so that one second of UTC is exactly the same length as one second of International Atomic Time. UTC is kept within 0.9 second of UT1 by introducing or deleting one second, known as a leap second, at the end of December, June, March or September, as necessary. This time-scale is widely known as GMT, although that term is no longer used in astronomy.

INTERNATIONAL ATOMIC TIME (TAI) is the time given by atomic clocks that maintain a continuous count of seconds. TAI differs from UTC by an exact number of seconds, as a result of the introduction of leap seconds in the UTC time-scale to take account of changes in the rotation rate of the Earth. TAI and UTC are the recommended time-scales for the precise dating of observations.

BARYCENTRIC DYNAMICAL TIME (TDB) is the time system used in describing the motions of bodies with respect to the barycentre of the Solar System. It is used for ephemerides of Solar System objects such as those produced by the Jet Propulsion Laboratory. However, for practical purposes ephemerides are required for the positions of Solar System bodies as seen from the centre of the Earth. Clocks run at different rates depending on the strength of the gravitational field in which they are operating, as predicted by the special theory of relativity. Since the Earth moves in an elliptical orbit around the Sun, a hypothetical clock at the centre of the Earth would not keep TDB. Instead, it defines a time-scale known as Terrestrial Time (TT), which is used for geocentric ephemerides. TDB does not differ from TT by more than 1.6 ms, which can be neglected for all but the most critical applications. Before 1991, TT was known as Terrestrial Dynamical Time (TDT). On 1977 January 1 at 0h TAI, TT was defined by the IAU as 1977 January 1 0h 0m 32.184s.

Irregularities in the rotation of the Earth and the general slowing of the Earth's rotation by tidal friction and other forces mean that Universal Time (UT1) is falling behind TT at the rate of about

TABLE 8. *Greenwich sidereal time. The RA on the Greenwich meridian is given for the dates and times indicated. For times after midnight, add 1 day to the date at the side.*

Intermediate dates: *Add to the RA for the previous date the number of minutes from the 7- or 8-day interval table below.*

7-day interval:	1d	2d	3d	4d	5d	6d	8 day interval:	1d	2d	3d	4d	5d	6d	7d
Add minutes:	4m	9m	13m	17m	21m	26m	Add minutes:	4m	8m	12m	15m	19m	23m	26m

Intermediate minutes of mean time: *Add the same number of RA minutes to the previous RA hour. Thus Apr. 6 at 1709h = RA 6h 09m.*

							UT							
	1700	1800	1900	2000	2100	2200	2300	0000	0100	0200	0300	0400	0500	0600
DATE	h	h	h	h	h	h	h	h	h	h	h	h	h	h
Jan. 5	0	1	2	3	4	5	6	7	8	9	10	11	12	13
13	0½	1½	2½	3½	4½	5½	6½	7½	8½	9½	10½	11½	12½	13½
21	1	2	3	4	5	6	7	8	9	10	11	12	13	14
28	1½	2½	3½	4½	5½	6½	7½	8½	9½	10½	11½	12½	13½	14½
Feb. 5	2	3	4	5	6	7	8	9	10	11	12	13	14	15
13	2½	3½	4½	5½	6½	7½	8½	9½	10½	11½	12½	13½	14½	15½
20	3	4	5	6	7	8	9	10	11	12	13	14	15	16
28	3½	4½	5½	6½	7½	8½	9½	10½	11½	12½	13½	14½	15½	16½
Mar. 7	4	5	6	7	8	9	10	11	12	13	14	15	16	17
15	4½	5½	6½	7½	8½	9½	10½	11½	12½	13½	14½	15½	16½	17½
22	5	6	7	8	9	10	11	12	13	14	15	16	17	18
29	5½	6½	7½	8½	9½	10½	11½	12½	13½	14½	15½	16½	17½	18½
Apr. 6	6	7	8	9	10	11	12	13	14	15	16	17	18	19
14	6½	7½	8½	9½	10½	11½	12½	13½	14½	15½	16½	17½	18½	19½
22	7	8	9	10	11	12	13	14	15	16	17	18	19	20
29	7½	8½	9½	10½	11½	12½	13½	14½	15½	16½	17½	18½	19½	20½
May 7	8	9	10	11	12	13	14	15	16	17	18	19	20	21
15	8½	9½	10½	11½	12½	13½	14½	15½	16½	17½	18½	19½	20½	21½
22	9	10	11	12	13	14	15	16	17	18	19	20	21	22
30	9½	10½	11½	12½	13½	14½	15½	16½	17½	18½	19½	20½	21½	22½
June 6	10	11	12	13	14	15	16	17	18	19	20	21	22	23
14	10½	11½	12½	13½	14½	15½	16½	17½	18½	19½	20½	21½	22½	23½
22	11	12	13	14	15	16	17	18	19	20	21	22	23	0
29	11½	12½	13½	14½	15½	16½	17½	18½	19½	20½	21½	22½	23½	0½
July 7	12	13	14	15	16	17	18	19	20	21	22	23	0	1
15	12½	13½	14½	15½	16½	17½	18½	19½	20½	21½	22½	23½	0½	1½
22	13	14	15	16	17	18	19	20	21	22	23	0	1	2
30	13½	14½	15½	16½	17½	18½	19½	20½	21½	22½	23½	0½	1½	2½
Aug. 6	14	15	16	17	18	19	20	21	22	23	0	1	2	3
14	14½	15½	16½	17½	18½	19½	20½	21½	22½	23½	0½	1½	2½	3½
22	15	16	17	18	19	20	21	22	23	0	1	2	3	4
29	15½	16½	17½	18½	19½	20½	21½	22½	23½	0½	1½	2½	3½	4½
Sept. 6	16	17	18	19	20	21	22	23	0	1	2	3	4	5
13	16½	17½	18½	19½	20½	21½	22½	23½	0½	1½	2½	3½	4½	5½
21	17	18	19	20	21	22	23	0	1	2	3	4	5	6
29	17½	18½	19½	20½	21½	22½	23½	0½	1½	2½	3½	4½	5½	6½
Oct. 6	18	19	20	21	22	23	0	1	2	3	4	5	6	7
14	18½	19½	20½	21½	22½	23½	0½	1½	2½	3½	4½	5½	6½	7½
21	19	20	21	22	23	0	1	2	3	4	5	6	7	8
29	19½	20½	21½	22½	23½	0½	1½	2½	3½	4½	5½	6½	7½	8½
Nov. 6	20	21	22	23	0	1	2	3	4	5	6	7	8	9
13	20½	21½	22½	23½	0½	1½	2½	3½	4½	5½	6½	7½	8½	9½
21	21	22	23	0	1	2	3	4	5	6	7	8	9	10
28	21½	22½	23½	0½	1½	2½	3½	4½	5½	6½	7½	8½	9½	10½
Dec. 6	22	23	0	1	2	3	4	5	6	7	8	9	10	11
14	22½	23½	0½	1½	2½	3½	4½	5½	6½	7½	8½	9½	10½	11½
21	23	0	1	2	3	4	5	6	7	8	9	10	11	12
29	23½	0½	1½	2½	3½	4½	5½	6½	7½	8½	9½	10½	11½	12½

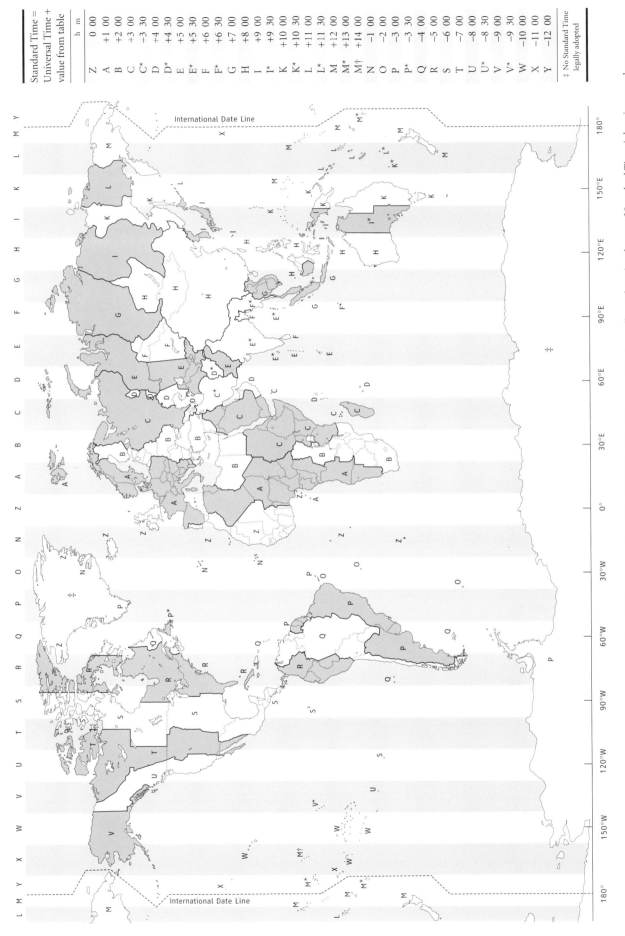

Standard Time = Universal Time + value from table		
	h	m
Z	0	00
A	+1	00
B	+2	00
C	+3	00
C*	+3	30
D	+4	00
D*	+4	30
E	+5	00
E*	+5	30
F	+6	00
F*	+6	30
G	+7	00
H	+8	00
I	+9	00
I*	+9	30
K	+10	00
K*	+10	30
L	+11	00
L*	+11	30
M	+12	00
M*	+13	00
M†	+14	00
N	−1	00
O	−2	00
P	−3	00
P*	−3	30
Q	−4	00
R	−5	00
S	−6	00
T	−7	00
U	−8	00
U*	−8	30
V	−9	00
V*	−9	30
W	−10	00
X	−11	00
Y	−12	00

‡ No Standard Time legally adopted

FIGURE 2. *World map of time zones, corrected to June 2003. Zone boundaries are approximate. Daylight Saving Time (Summer Time), usually one hour in advance of Standard Time, is kept in some places. Based on a map compiled by HM Nautical Almanac Office © Copyright Council for the Central Laboratory of the Research Councils. Reproduced with permission of HM Nautical Almanac Office. Map outline © Mountain High Maps.*

TABLE 9. *Values of ΔT, the amount by which Universal Time (UT1) lags behind Terrestrial Time (TT), from 1900 to 2020. Values for 2000.5 onwards are predicted using a rate of about 1 s per year, but this rate may change.*

YEAR	ΔT (s)	YEAR	ΔT (s)	YEAR	ΔT (s)
1900.5	−2	1960.5	+33	1995.5	+61
1910.5	+11	1965.5	+36	2000.5	+64
1920.5	+22	1970.5	+41	2005.5	+66
1930.5	+24	1975.5	+46	2010.5	+67
1940.5	+25	1980.5	+51	2015.5	+70
1950.5	+29	1985.5	+55	2020.5	+73
1955.5	+31	1990.5	+57	2025.5	+77

Source: Dennis D. McCarthy, U.S. Naval Observatory.

1 s every 500 days. The difference between TT and UT1 is known as ΔT. Table 9 gives the observed and predicted values of ΔT. This quantity can be determined only by observation, so that predictions of ΔT cannot be made with accuracy very far into the future

The Year

TROPICAL YEAR The time taken by the Sun to complete one circuit of the celestial sphere from one vernal equinox to the next. Because of precession, the vernal equinox has an annual retrograde motion of 50″.29 across the celestial sphere, so during the tropical year the Sun covers an angular distance of 360° − 50″.29. The tropical year lasts 365.242 19 d, and is the year on which the calendar is based, since the declination of the Sun determines the occurrence of the seasons.

SIDEREAL YEAR The time taken by the Earth to orbit the Sun once with respect to the celestial sphere. In a sidereal year the Sun covers an angular distance of exactly 360° relative to the star background, taking 365.256 36 d.

ANOMALISTIC YEAR The interval between one perihelion of the Earth and the next. The anomalistic year is slightly longer than the sidereal year because of a gradual advance of the Earth's perihelion caused by the gravitational pulls of the planets. During an anomalistic year the Sun covers an angular distance of 360° + 11″.64, taking 365.259 64 d.

ECLIPSE YEAR The time taken for the Sun to return to the same node of the Moon's orbit; it lasts 346.620 08 d. This is considerably shorter than a sidereal year, because the nodes of the Moon's orbit regress by about 19° per year. The eclipse year is responsible for the regular recurrence of both solar and lunar eclipses, which can take place only when these bodies are within a small distance of the node. Nineteen eclipse years are 6585.78 d, almost exactly the same duration as the Saros cycle of 223 synodic months, or 6585.32 d.

The Month

SYNODIC MONTH The interval between successive new moons. It is also known as a *lunation*. Its mean length is 29.530 59 d, but the actual value can range between about 29.27 and 29.83 d. The shortest synodic months occur in July, when the Earth is near aphelion, and the longest in January, when the Earth is near perihelion.

SIDEREAL MONTH The period taken by the Moon to make one complete circuit of the celestial sphere as seen from the Earth. Its mean value is 27.321 66 d.

TROPICAL MONTH The time taken for the Moon to orbit the Earth once with respect to the vernal equinox. Its mean value is 27.321 58 d.

ANOMALISTIC MONTH The interval between successive perigees of the Moon. Its mean value is 27.554 55 d.

DRACONIC MONTH The interval between successive passages of the Moon through its ascending node. Its mean value is 27. 212 22 d.

CHAPTER TWO

Practical Astronomy

Observing

THE HUMAN EYE

The eye is the astronomer's most basic observational instrument. It consists of a hollow sphere filled with a transparent substance through which light rays are focused by a crystalline lens. The image is formed on a screen (the *retina*) which is covered with many nerve endings. Some of these nerves (the *cones*) give the sensation of colour, while the *rods*, which respond to much lower levels of illumination, interpret the image as shades of grey.

DEFECTS OF THE EYE include an inability to focus on nearby or distant objects – *hypermetropia* (long sight) and *myopia* (short sight), respectively – and *astigmatism*, which results from the lens being distorted. An astigmatic eye focuses a point object such as a star into an oval or a short line. All these defects may be compensated for by wearing suitable spectacles or contact lenses.

Long or short sight is of no importance in telescopic astronomy since the instrument may be focused to suit the individual. Astigmatism is more of a problem, particularly at low magnifications, as the telescope's *exit pupil* is then large, and most or all of the defective lens is being used. At high magnifications, when the exit pupil is narrow, only the centre of the eye's lens is used, and the effect of any astigmatism is minimized. In particular, astigmatic users of binoculars may be forced to wear their spectacles while observing, which has the result of forcing the eyes back from the eyepieces, with the consequent loss of part of the field of view (except in binoculars with extra-large eye relief). The advantage of wearing contact lenses to counteract serious astigmatism is obvious.

The eye is not perfectly *achromatic* (free from false colour), but this is rarely if ever noticeable in practice. The defect of *spherical aberration* (in which the edge and the centre of the eye's lens bring light to different foci, producing a blurred image) is, like astigmatism, most noticeable at very low magnifications when the whole of the lens is being used. Since fine image detail is not being sought at these times, the defect is of little practical importance.

With age, the eye loses its ability to focus on objects over a wide range of distance. More importantly, the lens may become opaque. This can be corrected by a cataract operation, which has the effect of improving the eye's sensitivity to violet and near-ultraviolet wavelengths, which are absorbed both by glass in the telescope and by the eye's own lens.

OBSERVING TECHNIQUES

The observing eye should always be in its most relaxed state – in other words, focused at infinity. When straining to make out elusive details, it is easy for the eye to change its focus involuntarily. At intervals, therefore, particularly when observing planetary surfaces, it pays to pause, relax, and then refocus carefully on a nearby star. This chapter deals with general aspects of observation; notes on the observation of specific objects are included in Chapters 3 and 4.

RESOLVING POWER The resolving power of the normal eye on naked-eye stars is about 4 arc minutes – in other words, two stars separated by this amount can be made out individually (although some observers can do better). It follows that a magnification of about ×240 will permit two stars 1 arcsec apart to be distinguished. Practical experience shows that this is the limit of resolution, or resolving power, of a telescope with an aperture of 110 mm, from which it may be deduced that a magnification of about $2.2D$ (where D is the aperture in millimetres) will permit all the detail in any telescope's image to be made out (although particular circumstances may modify this rule). In any case, sheer resolution is not always the most important factor when selecting the best magnification.

OBSERVING FAINT OBJECTS The retina achieves maximum sensitivity through the secretion of a hormone (rhodopsin) which stimulates the rods. This happens only under conditions of poor illumination, and sensitivity can continue to improve noticeably for several minutes, and detectably for perhaps half an hour. Such *dark-adaptation* is rapidly destroyed upon exposure to bright light.

The dark-adapted eye loses sensitivity at the long-wavelength

(text continued on p.26)

North Polar Limiting Magnitude Chart

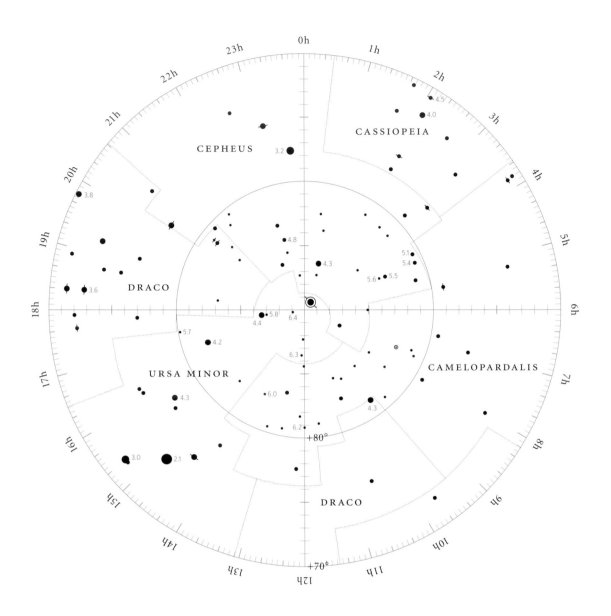

FIGURE 3(a). *North polar magnitudes. This chart can be used to ascertain the faintest stars visible to the naked eye on any given night, as a guide to atmospheric transparency. It shows stars within 20° of the north celestial pole. The magnitudes and positions of all stars down to mag. 5.5 are shown in the outer 10° circle; in the inner circle all stars plotted in the Atlas are shown, with sample magnitudes down to 6.5.*

South Polar Limiting Magnitude Chart

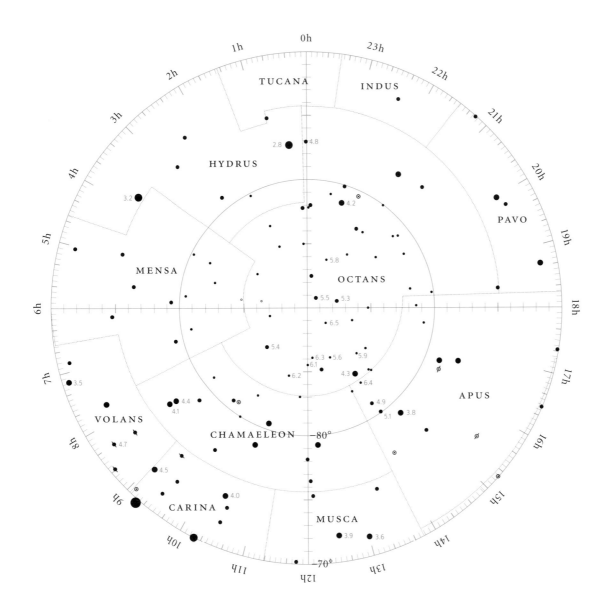

FIGURE 3(b). *South polar magnitudes. This chart can be used to ascertain the faintest stars visible to the naked eye on any given night, as a guide to atmospheric transparency. It shows stars within 20° of the south celestial pole. The magnitudes and positions of all stars down to mag. 5.5 are shown in the outer 10° circle; in the inner circle all stars plotted in the Atlas are shown, with sample magnitudes down to 6.5.*

(red) end of the visible spectrum. In daylight, a normal eye has a visual range of approximately 400–750 nm, with a peak sensitivity in the yellow-green region of the spectrum, around 555 nm. For a dark-adapted eye the visual range is approximately 400–620 nm, and the peak sensitivity moves into the green region of the spectrum, around 510 nm. This shift of sensitivity is known as the *Purkinje effect*, and explains why moonlight appears bluer than direct sunlight.

Since the most sensitive part of the retina is an annular region around its centre, very faint objects may be detectable only by *averted vision*, in which the observer looks slightly to one side of the object being observed.

ATMOSPHERIC CONDITIONS

A celestial object is best observed when it is as far as possible above the haze and mist of the horizon; *culmination*, or maximum altitude, occurs when it crosses the meridian.

Nights when the air is most transparent, and the stars appear most brilliant, are not necessarily ideal for astronomical work since such conditions are often accompanied by turbulent air currents, which cause the stars to twinkle to the naked eye and 'boil' when viewed through a telescope. Very transparent nights may, however, be ideal for observing faint, extended objects such as comets and large nebulae.

Slight haze is often a sign of steady air, and good views of the Moon, planets and bright stars may be obtained under such conditions.

TRANSPARENCY is a measure of the clarity of the atmosphere. It always depends on altitude: even under perfect conditions a star will appear about three magnitudes fainter just above the horizon than at the zenith, and in practice horizon haze will usually increase this difference considerably. However, by selecting stars at a constant altitude (i.e. around the celestial pole) it is possible to classify the transparency of any night by noting the faintest star visible with the naked eye. Figure 3, on pp. 24–5, indicates stars of suitable magnitudes for this purpose.

In practice, transparency as measured in this way is not always determined simply by the absorption of light by water or dirt in the atmosphere. Unless the observer is situated far from any built-up area, sky-brightening caused by artificial lighting reflected off airborne particles can also make the stars appear dim by contrast. Faint auroral glows in the upper atmosphere can have a similar effect.

SEEING is a term used to indicate the steadiness of the air, as judged by the appearance of the telescopic image. The two are connected by the fact that air currents are caused by masses of air at different temperatures, and the refractive index of air changes with temperature: therefore the currents cause the image to flicker.

There are two basic components, often called 'high' and 'low'

seeing. High seeing is affected by currents at altitudes of between about a thousand metres and several kilometres; the quality of low seeing, over which the observer has some control, depends on conditions near the ground and even inside the telescope itself. For example, after a summer's day the night air in contact with the ground around the telescope is warmed and rises, creating turbulence; similarly, warm air trapped inside an observatory, or in a telescope tube, can cause unsteadiness. Bad high-level seeing causes a star's image to move irregularly, or sharp undulations to cross a planet's disk; bad low-level seeing takes the form of sudden loss of focus. Seeing conditions also affect telescopes of different aperture in different ways. At first sight Jupiter may appear less distinct in a 300-mm telescope than in a 100-mm, since the larger aperture is affected by turbulent air across nine times the cross-sectional area of atmosphere.

Since seeing conditions can have a considerable influence on what is visible, they must always be recorded. The classification of seeing is likely to be more subjective than that of transparency, and expressions such as 'good' or 'poor' have little general meaning. More precise descriptions, such as 'boiling with steady moments' or 'image unsteady and rather diffuse', should be used. Several numerical scales have also been devised. For lunar and planetary work the *Antoniadi scale* is widely used:

  I   perfect seeing, without a quiver

  II slight undulations, with moments of calm lasting
      several seconds

  III moderate seeing, with larger air tremors

  IV poor seeing, with constant troublesome undulations

  V   very bad seeing, scarcely allowing a rough sketch
      to be made.

RECORDING OBSERVATIONS

All observations should be written down at the time they are made. The notes should be clearly worded, and should have entered on them the year, month, day, hour and minute (UT) of the observation, together with the aperture and magnification used and the seeing conditions.

OBSERVATIONAL RECORDS should be copied up from the notes made at the telescope as soon as possible, while they are still fresh in the mind. Many observers keep a separate book for each type of object being observed. Each observation should be given a serial number to permit later indexing.

ILLUMINATION of the notebook, star atlas or reference work used at the telescope is often provided by an ordinary torch (flashlight) with a red filter over the front or, more effectively, by a red light-emitting diode (LED) instead of the bulb and filter. It is important to use only the minimum light necessary, so that dark-adaptation is affected as little as possible.

It is inconvenient to hold a torch as well as manipulate eye-pieces, refer to sources, adjust the telescope, and hold the notebook and write in it. The observing light is best secured to the telescope tube or stand, with a readily accessible switch.

TIMING OBSERVATIONS

With the wide availability of reliable and accurate watches, the need for a specialized observatory clock has diminished. A digital watch checked against a radio time-signal or the telephone time service earlier in the evening will remain accurate to within a fraction of a second throughout the night, which is sufficient for general observational purposes.

For higher precision, clocks and watches with built-in radio receivers are now available. They tune in periodically, usually every hour, to specialized radio transmitters which broadcast accurate time-signals so that they can adjust themselves accordingly. Global Positioning System (GPS) receivers are another source of exceptionally accurate time signals. Home computer programs are also available which link via the Internet to numerous atomic clocks in different parts of the world and so maintain very precise time.

For setting an equatorial telescope with circles, a clock showing sidereal time is necessary. A mechanical (wind-up) alarm clock, adjusted to gain at a rate of 4 minutes a day, may be set to the sidereal time as derived from Table 8 (p.19) before the observing session begins. Alternatively, electronic sidereal clocks are available.

PRECISE TIMING of astronomical phenomena (such as eclipses of Jupiter's satellites, occultations of stars by the Moon or fixes on artificial satellites) requires a stopwatch. The stopwatch may either be started against a time-signal and stopped at the observed phenomenon, or vice versa. If a series of phenomena are to be observed, a split-action stopwatch will allow successive timings to be made and noted.

It is a good idea to test the accuracy of the watch from time to time against a reliable time source and then make any necessary allowances in timings. Some observers obtain precise timings by recording the event on video, incorporating an accurately measured electronic time-stamp onto the sequence of pictures.

DIRECTIONS IN AN INVERTING TELESCOPE

The north–south and east–west orientations of the telescopic field of view depend on its position in the sky. North can be at the top, bottom, right or left of the field, or any position in between;

the same is true of the other cardinal points. The fact that most astronomical telescopes invert the image adds further to the initial confusion.

It is convenient to use the drift of a celestial object as a reference. As the Earth spins, the Sun, Moon, planets and stars all appear to move through the field of view of a stationary telescope from east (*following*, or f) to west (*preceding*, or p). Having used the drift of objects through the field of view to establish the p and f points, the observer may determine the north and south points by remembering 'PSFN' (or 'poisonous snakes feel nice') as the clockwise order of the four directions in the field. The field of view is often divided into four quadrants: north–following, south–following, south–preceding and north–preceding. Accurate *position angle* (PA) measurements of one object with respect to another use a 0–360° scale (north 0°, east or following 90°, south 180°, and west or preceding 270°).

Figure 4 shows the approximate orientation of the field of view of an inverting telescope, in either the northern or the southern hemisphere.

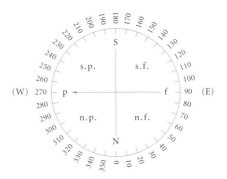

FIGURE 4. *Inverted field of view, and position angle. The inverted telescopic field of view, and the direction of travel of a celestial object across it, as seen by an observer in the northern hemisphere looking due south or – if turned through 180° (the relative positions remain unchanged) – by an observer in the southern hemisphere looking due north. The position angle (PA) scale is shown around the outside.*

ANGULAR DISTANCES ON THE CELESTIAL SPHERE

The distance from thumb to little finger of an outstretched hand at arm's length is about 20°. The following approximate separations between pairs of stars may also be found useful in determining distances on the celestial sphere; others can easily be measured from the star charts. The separations are measured along a great circle passing through the two stars.

½° = the angular diameter of the Sun or Moon

1¼° = δ to ε Orionis, or β to λ Crucis

2° = α to γ Aquilae, or α to δ Scorpii

2½° = α to β Aquilae

4° = α to β Canis Minoris, or α to β Crucis

5° = α to β Ursae Majoris, or α to β Centauri

The total area of the celestial sphere is 41 253 square degrees.

Astronomical Instruments

BINOCULARS

Binoculars are almost indispensable, and make an ideal first instrument for the newcomer to astronomy, being far more effective than a small, cheap telescope. They need not (in fact should not) be large or high-powered, otherwise the benefits of extreme portability and general ease of use will be lost.

Binoculars have particular combinations of magnification and aperture, and these are given in the form, for example, 8×30, which indicates a magnification of $\times 8$ and an aperture of 30 mm. For astronomical work the larger the aperture the better, although beyond a certain size – about 70 mm – binoculars become rather cumbersome for a hand-held instrument. The 8×40 and 10×50 specifications are popular, but the purchaser is strongly advised to beware of cheap binoculars, allegedly for astronomical use, with excessive magnifications of $\times 20$ or more.

Any good-quality binoculars, even the smallest, will reveal far more stars than are visible with the unaided eye, as well as the brighter star clusters, nebulae and galaxies plotted on the star charts. Their advantages of wide field of view and erect image, which allow easy reference to the naked-eye view and to the charts, simplify the initial location of objects even when a telescope is being used. Some large objects, such as the Pleiades star cluster or rich gatherings of stars in the Milky Way, may be more impressive in binoculars than through any telescope. For some branches of amateur astronomy, variable-star observation and comet-hunting in particular, binoculars of the appropriate size may be the preferred instrument.

A pair of binoculars cannot be used effectively if unsupported: resting the instrument itself or the observer's elbows on a rigid base is necessary to reduce the 'dancing' effect caused by the observer's heartbeat and inevitable muscle tremor. *Image-stabilized* binoculars include components in the optical train that actively damp down small movements. They can show stars up to a magnitude fainter than hand-held binoculars of the same aperture, partly because they often provide a higher magnification than conventional binoculars. However, they are heavier and more expensive than their non-stabilized counterparts.

Binoculars may be used for:

(a) general viewing of the constellations, star fields and the Milky Way

(b) acting as an adjunct to the main telescope, to identify the general location of an object before bringing the finder to bear on it

(c) locating planets or bright stars in twilight, when their precise location in the sky is uncertain

(d) observing the brighter variable stars

(e) observing any large-scale sky feature, such as deep-sky objects, a bright comet or even meteor trains

(f) following artificial satellites.

Field glasses or opera glasses are often found in second-hand or antique shops. Unfortunately their very simple optical design offers a prohibitively small field of view, and they cannot be recommended for astronomical use.

Astronomical Telescopes

These are of three main types: refracting, reflecting and catadioptric, each with its own advantages, but for all three types the effectiveness is determined largely by the *aperture* – the clear diameter of the lens or mirror which receives and focuses light from the object being observed. Bigger apertures offer greater light-gathering power (i.e. they show fainter objects) and higher resolution (i.e. finer detail). For general astronomical purposes, any good-quality telescope of at least 75 mm aperture will permit interesting and useful observational work in a number of fields, while pleasing views of many objects may be obtained with smaller telescopes or even binoculars.

LIGHT-GATHERING POWER OR 'LIGHT-GRASP' The larger the aperture, the greater the amount of light received and focused by the telescope. As light-grasp theoretically increases in proportion to the square of the aperture, a telescope of 75 mm aperture (henceforth referred to as a 75-mm telescope) has twice the light-grasp of a 50-mm telescope, while a 100-mm has nearly twice the light-grasp of a 75-mm.

A telescope's aperture helps to determine the faintest detectable stars (the *limiting magnitude*). The values in Table 10 were calculated from the formula

$$m = 2.7 + 5 \log D$$

where D is the telescope aperture in millimetres. This formula was derived from observational tests on faint stars in the Pleiades with a wide range of apertures. Factors such as atmospheric conditions and the observer's skill influence the limiting magnitude to such an extent that the calculated values must be regarded as very approximate.

RESOLUTION Although all night-time stars appear as immeasurably small points of light, their image at the focus of a lens or mirror has a core of finite radius (the *Airy disk*), which corresponds to an angular distance on the celestial sphere of $116/D$ arcsec (D

TABLE 10. *Limiting magnitude and the Dawes limit. Approximate values for the faintest star visible and the separation of the closest pair of objects distinguishable for telescopes of various apertures.*

CLEAR APERTURE	(mm)	50	60	75	100	112.5	125	150	200	250	300
	(in.)	2.0	2.4	3.0	4.0	4.5	5.0	6.0	8.0	10.0	12.0
LIMITING MAGNITUDE *Faintest star visible*	(mag.)	11.2	11.6	12.1	12.7	13.0	13.2	13.6	14.2	14.7	15.1
DAWES LIMIT *Closest stars resolvable*	(arcsec)	2.3	1.9	1.5	1.2	1.0	0.9	0.8	0.6	0.5	0.4

being the aperture in millimetres). This is the so-called *Dawes limit*. Stars separated in the sky by less than this angular distance cannot be clearly resolved, or shown separated, no matter how high a magnification is used. Hence every telescope has a *resolving limit* determined by its aperture alone. Table 10 gives the Dawes limit for various apertures.

The amount of detail visible on the Moon and planets is largely determined by aperture, and to a lesser extent by the size of the central obstruction (zero for a refractor).

THE REFRACTOR

The refractor essentially consists of two lenses: a large one of long focal length called the *object glass* or *objective*, which forms at its focus an image of the star or other object (Figure 5(*a*)), and a small lens of much shorter focal length, known as the *eyepiece* or *ocular*. The eyepiece, in effect, enables the observer to obtain a close-up view of the image formed by the object glass.

The objective is composed of two or more lenses or *elements*, which combine to minimize the amount of false colour in the image that would be produced by a single lens. The quality of an objective is largely determined by how well it overcomes this *chromatic aberration*. The usual two-element *achromat* inevitably shows a faint bluish halo around bright stars and at the edge of the Moon. This *secondary spectrum* can be reduced by making the lenses from glass with unusual characteristics, such as fluorite, or by adding more lens elements. Such an instrument is known as an *apochromat*. Although more expensive, both these alternatives are becoming increasingly available to amateurs.

The *focal ratio* of an object glass (the focal length divided by the aperture) is usually about 10 (written *f*/10) or slightly longer in order to help minimize secondary spectrum; therefore even a refractor of modest aperture is bulky, and instruments of more than 100 mm aperture can hardly be considered portable.

THE REFLECTOR

In this form of telescope a concave mirror (the *main* or *primary mirror*) does the same job as the refractor's objective lens. Mirrors have the advantage over lenses that they reflect all colours of light

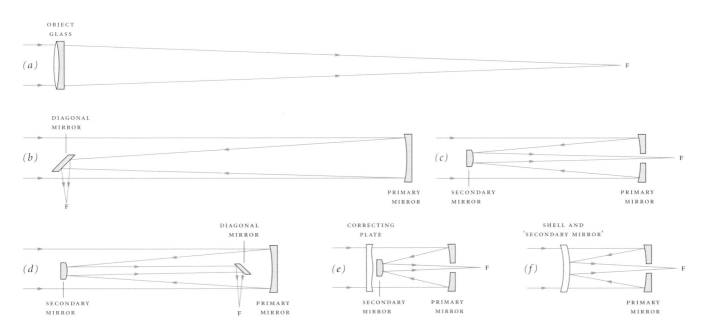

FIGURE 5. *Different types of telescope. For telescopes of similar aperture, the path of light rays from a celestial object to the focus F is shown through (a) a refractor, (b) a Newtonian, (c) a Cassegrain, (d) a Cassegrain–coudé, (e) a Schmidt–Cassegrain and (f) a Maksutov–Cassegrain.*

equally, and therefore do not suffer from chromatic aberration, although coma gets progressively worse with distance from the optical axis. The mirror is held in a cell at the lower end of the telescope tube, and the light rays from the object pass down the tube to be reflected back to a focus at the upper end. What happens then depends on the particular design of the telescope.

THE NEWTONIAN The converging rays are intercepted by a small flat mirror (known as the *flat* or *diagonal*) inclined at 45° to the main mirror, which reflects the light to the side of the tube (Figure 5(*b*)) where the eyepiece is situated. This form is most convenient for small and medium-sized reflectors (up to an aperture of perhaps 400 mm), since their tubes are much shorter than the tubes of refractors of the same aperture (focal ratios usually between $f/5$ and $f/8$). Thus the eyepiece is usually at a convenient height for making observations.

THE CASSEGRAIN After being reflected by the main mirror, the rays are sent back down the tube by a small convex mirror (the *secondary*), either passing through a hole in the primary mirror (Figure 5(*c*)) or being reflected out through the side of the tube by a Newtonian-type diagonal (Figure 5(*d*)). The classical Cassegrain effectively compresses a long focal length, typically $f/20$, into a relatively short tube and can be a very compact instrument. Today most telescopes of Cassegrain design are of the catadioptric type (see below).

TELESCOPE MIRRORS are optically worked and coated on their front surfaces, so that unlike everyday mirrors (which are coated on the back) the light does not pass through the glass at all. This permits the use of opaque materials such as low-expansion glass of the type used for ovenware, or even ceramics, to minimize the effect of temperature change on the extremely accurate optical surface.

The reflective coating is usually of aluminium, deposited by vaporization under vacuum. A clear overcoating, usually of silicon monoxide, is often applied on top of the aluminium to protect the surface and increase its lifetime. Overcoatings have also been developed to protect very delicate silver coatings, which were once the norm but were abandoned in favour of the more durable aluminium. Silver coatings reflect about 93% of visible light, aluminium about 89%. Without a protective layer, however, silver quickly tarnishes and requires replacing, often in six months or less. Enhanced aluminium coatings that reflect up to 96% of the incident light are becoming more commonly available.

CATADIOPTRIC TELESCOPES

These are instruments which use both refraction and reflection to form the image. The two forms of most interest to the amateur are Cassegrain-type designs which use a primary mirror with a spherical curvature, rather than the parabolic curvature that is normal in reflectors. A spherical mirror is much easier to manufacture than a parabolic one, but its images suffer from the defect of *spherical aberration*.

In the *Schmidt* system (Figure 5(*e*)) a thin, specially figured, almost flat lens is placed at the top of the tube. This lens corrects the aberration of the primary mirror and carries the convex secondary mirror at its centre.

In the *Maksutov* system (Figure 5(*f*)) the corrective lens is a sharply curved 'shell' of glass. In many designs the centre of the inner (convex) surface is aluminized to give the effect of a secondary mirror.

Both these telescopes have the advantage of combining a short tube, typically less than twice the aperture, with an effective focal ratio of between $f/10$ and $f/15$. They therefore offer the chance of owning a truly portable instrument with an aperture of up to 300 mm or so.

EYEPIECES AND MAGNIFICATION

An eyepiece acts as a magnifying glass, but whereas the latter is usually a single lens that is convex on both sides, a telescope eyepiece contains two or more lenses of various types. The lens nearest the eye is the *eye lens*; the one farthest from it is the *field lens*.

A good eyepiece should:

(a) correct for chromatic aberration

(b) correct for spherical aberration

(c) have a 'flat' field – in other words, images should be in focus both at the centre and at the edge of the field of view

(d) have a large field of view, with at least reasonable definition near the margins

(e) be free from 'haunting', or ghost images of bright objects.

The performance of an eyepiece is as important as that of the other optical elements in the telescope. A given eyepiece may not produce equally good results on all telescopes. Factors (a) and (b) above depend crucially on the focal ratio of the object glass or mirror; the simpler and cheaper types of eyepiece may work perfectly when used with an $f/15$ refractor, but their inherent aberrations will be all too evident at the focus of an $f/5$ Newtonian.

EYE RELIEF An eyepiece forms an image of the objective or primary mirror known as the *exit pupil*, and the distance between the outer surface of the eyepiece and this image is called the eye relief. Since the pupil of the observer's eye needs to be brought to the plane of the exit pupil if the full field of view of the eyepiece is to be seen, large eye relief is essential if spectacles need to be worn.

FIELD OF VIEW If the telescope is aimed at the bright daylight sky, with the eye at the exit pupil, a circular disk of light is seen. The apparent angular diameter of this disk is the *apparent field of view* of the eyepiece. Its value, for different designs, can range from about 25° to 80°. Dividing the apparent field of view by the magnification of the eyepiece gives the *real field of view*.

INVERTED IMAGE An astronomical telescope, whether a reflector or refractor, gives an inverted image. Binoculars and terrestrial telescopes contain extra lenses or prisms to restore the orientation, but these are omitted from astronomical telescopes in the interests of maximum light transmission and image quality. Some catadioptric telescopes incorporate an 'erect-image' system to allow them to be used on terrestrial objects.

TYPES OF EYEPIECE AND THEIR USES Eyepiece types once thought to be complex and costly have now become common and comparatively cheap. Furthermore, new varieties have become available in response to the popularity of certain telescope designs such as short-focus reflectors and long-focus but compact catadioptrics.

In general the simpler and cheaper designs are perfectly acceptable for instruments of large *f*-number, or as long as quite a small field of view is acceptable to the observer. But with low *f*-numbers, or where a wide field of view free from distortion is needed, the requirements become more demanding.

Eyepiece barrels are usually 1¼-in. (31.7-mm) push-fit, but 0.965-in. (24.5 mm) push-fit versions do exist. A 2-in. (50-mm) diameter push-fit barrel is used for super-wide-angle and for long-focal-length eyepieces. Focal lengths are normally in the range 5–50 mm, while apparent fields of view range from 30° to 50° for standard eyepieces, and from 60° to over 80° for wide-field eyepieces. Various types of eyepiece are shown in Figure 6.

Huygenian. The commonest type for small refractors. Has a somewhat curved field of view, i.e. the field edge may require slight refocusing. Unsuitable for *f*-numbers below 10, which rules out most reflectors. The *Huygenian–Mittenzwey*, of slightly different design, has a wider field of view.

Ramsden. A flatter but narrower field of view than the Huygenian, with the drawback that the field lens can be in focus, revealing any dust on it. Suffers from ghost images, but is usable on instruments of up to *f*/6.

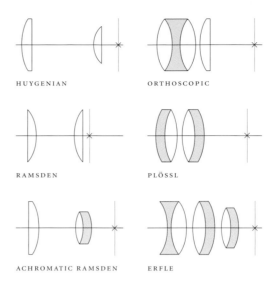

FIGURE 6. *Common types of eyepiece. The Huygenian and Ramsden both consist of two plano-convex lenses. In the achromatic Ramsden one of these is replaced by an achromatic doublet. The orthoscopic eyepiece contains a cemented triple achromat. The Plössl is made up of two achromats. The form of Erfle shown here contains three doublets, but in other forms one or two of these can be replaced by a single-element lens. In each case the position of the exit pupil is marked by a cross.*

Kellner. Very widely used eyepiece for binoculars, as it has a fairly wide and flat field and large eye relief. Virtually all eyepieces described as Kellners are in fact *achromatic Ramsdens*. This is a versatile and popular design which gives excellent results at *f*/8 and is usable down to *f*/4, though with some aberrations at the field edges.

Orthoscopic. Noted for its fairly wide, distortion-free field and large eye relief. Although suitable for most types of telescope, it has now largely been superseded by the Plössl as the standard eyepiece design.

Plössl. Can give wider and flatter fields than orthoscopics, and preferred by many observers. The name is used for several types of eyepiece of slightly differing design. Because of its good reputation some manufacturers have also applied the name to quite different designs.

Erfle. The standard low-power, wide-field eyepiece, with a field of view of around 60° and a focal length of 20 mm or longer. The *König* is a variant of the Erfle, usually with shorter focal length.

Other wide-field eyepieces. There are now a number of special ultra-wide-field eyepieces, some with apparent fields of 80° and more, such as the *Nagler*, suitable only for focal lengths less than 20 mm. In addition to offering 'picture-window' views through a telescope, such eyepieces are especially useful with telescopes that are not equipped with a drive system: the wider field requires the

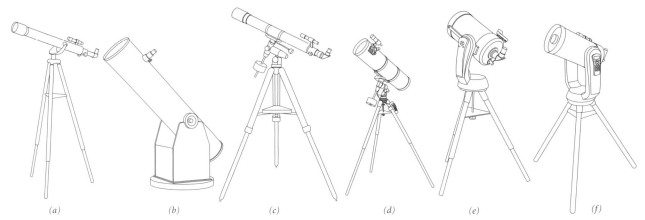

FIGURE 7. *Different types of telescope mounting. (a) A simple altazimuth mounting for a small refractor, (b) a Dobsonian altazimuth mounting for a Newtonian, (c) a German equatorial mounting for a refractor, (d) a German equatorial mounting for a Newtonian, (e) a single-arm altazimuth fork mounting for a computer-controlled Schmidt–Cassegrain, (f) an altazimuth fork mounting for a computer-controlled Schmidt–Cassegrain.*

telescope to be moved less often to keep an object in view. While these eyepieces are generally very well made, with multilayer coatings to maximize the light transmitted through their many elements, they also tend to be quite expensive and heavy.

MAGNIFICATION The magnifying power of a telescope is determined by the ratio of the focal length of its objective lens or primary mirror to that of the eyepiece. For example, a refractor or Newtonian having a focal length of 1500 mm used with an eyepiece of 20 mm focal length will give a magnification or power of 1500/20 = 75 diameters, usually written ×75. Therefore the magnification provided by any eyepiece is not unique, but depends on the focal length of the telescope with which it is used.

High magnification, although it can show more detail than low magnification, has its drawbacks:

(a) the field of view is smaller

(b) the effects of bad seeing are more pronounced

(c) any tremors in the tube, or faults in the drive, become more noticeable.

In addition, low magnification reveals more detail in extensive objects such as comets or large nebulae by concentrating their light. A range of magnifications is therefore necessary for general work, and a minimum of three different powers should be available:

(a) *low power*, about 0.2 to 0.3 times the telescope aperture in millimetres (×20 to ×30 for a 100-mm telescope) for showing the greatest possible area of sky

(b) *medium power*, about 0.5 to 0.8*D* (×50 to ×80 for a 100-mm telescope) for general views

(c) *high power*, about 1.0 to 2.5*D* (×100 to ×250 for a 100-mm telescope) for studying objects such as close double stars and planets.

For larger instruments the medium and high powers may be scaled down somewhat. Powers greater than about ×400 are rarely beneficial because the slight atmospheric tremors that prevail even on the best of nights then begin to affect the image. Very high magnifications also tend to reduce the contrast of lunar and planetary markings, although they can improve the visibility of close double stars.

THE BARLOW LENS is a small concave lens that diverges rather than converges light rays passing through it; it should be achromatic. When placed before the eyepiece it has the effect of increasing the effective focal length of the telescope by a selected factor, usually ×2. Since the magnification of an eyepiece is directly proportional to the effective focal length of the telescope, a Barlow lens doubles the range of magnifications obtainable with a set of eyepieces. It also enables high powers to be obtained without the need to use eyepieces of very short focal length and correspondingly small eye relief.

TELESCOPE MOUNTINGS

However high a telescope's optical quality may be, the instrument will not perform as it should unless it is properly mounted. A good mounting must:

(a) prevent the tube from shaking – the image should not vibrate when the eye is brought to the eyepiece or when the focusing knob is turned

(b) allow smooth and responsive control, whether in moving the telescope to seek an object and bring it to the centre of the field, or in keeping it there as the Earth turns

(c) allow the telescope to be pointed anywhere in the sky, or at least to within a few degrees of the horizon.

TYPES OF MOUNTING Mountings can be divided into two main types: *altazimuth* and *equatorial* (Figure 7).

In the altazimuth type the telescope is linked to two axes at right angles to each other, one permitting motion in altitude (vertically) and the other in azimuth (horizontally). Traditionally the altazimuth has been associated with small, cheap 'beginner's telescopes', but its mechanical simplicity has led to its recent adoption not only as a popular mounting for more advanced amateur and computer-controlled telescopes, but also for the present generation of large professional instruments. Continual adjustment is necessary in both altitude and azimuth in order to follow a celestial object as the Earth spins, although this is an easy matter for telescopes equipped with a computer-controlled drive. The *Dobsonian* mounting is a variation on the altazimuth design. In recent years it

has become very popular for large-aperture, short-focus reflectors, its chief advantage being its simplicity of construction.

The equatorial mounting simplifies matters by having a *polar axis* aligned with the Earth's axis. Driving the instrument around this single axis once a day, in a direction opposite to the Earth's spin, keeps the telescope tube pointing at a given star, as shown in Figure 8. Initial pointing is done by moving the telescope in right ascension (RA), using the polar axis, and in declination, using the *declination axis*. An electric motor usually supplies the drive for the polar axis.

An altazimuth fork mounting can be converted to an equatorial simply by tilting the base at an angle equivalent to the observer's latitude. This is usually accomplished by means of what is termed a *wedge*, usually adjustable over a range of latitudes.

Refractors, reflectors and catadioptrics have different mounting requirements, so they are considered separately.

MOUNTINGS FOR REFRACTORS Since the eyepiece is at the lower end of the tube, and the tube is long, the mounting needs to be high off the ground. The traditional support for a small refractor is a tall tripod or, for apertures greater than about 100 mm, a permanently fixed column. Of the equatorial mounts, the *German* type has the advantage that it allows the long tube to be pointed towards the region of the celestial pole. Figure 9 shows a typical small refractor on a German equatorial mounting. The labelled parts and accessories are also found on other telescope designs.

MOUNTINGS FOR REFLECTORS Since the tube is relatively short, and the eyepiece (in a Newtonian) is near the upper end, a low mounting is most convenient. The Dobsonian altazimuth mounting has the virtues of cheapness and rigidity: the tube is pivoted

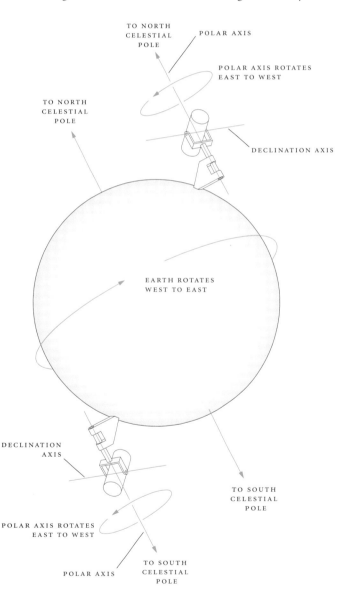

FIGURE 8. *The principle of the equatorial mounting. The polar axis is aligned parallel to the axis of the Earth and is turned east to west to counteract the Earth's rotation.*

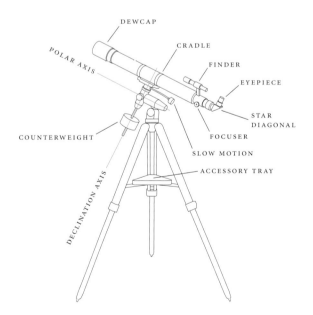

FIGURE 9. *Parts of a typical refractor on a German-style equatorial mounting.*

just above the mirror and swings vertically inside a short, box-like fork which turns horizontally on a base-plate. By using practically frictionless Teflon as a bearing surface for both axes, it is possible to move the tube through small and controlled angles by hand while observing. The success of this simple mounting has led to a new enthusiasm for large-aperture, short-focus Newtonian telescopes, since the low cost of making (or buying) the mounting means that more aperture may be bought for the same total expenditure.

Equatorially mounted Newtonians are usually mounted in the German style on a short pillar; a fork mounting is also suitable. Classical Cassegrains are mounted like refractors.

MOUNTINGS FOR CATADIOPTRICS The tubes of these telescopes are extremely short, so they are usually supplied with a fork mount, which has the advantage of needing no counterweight. The motor drive and short polar axis are concealed in the base, and the instrument may stand on its own legs or be fitted to a tripod or pillar. Their compactness, and the fact that there is little over-hanging weight, makes these instruments extremely steady and a pleasure to use; however, their benefits must be weighed against their drawbacks, which include slightly lower image contrast and a strongly curved focal surface.

DRIVES FOR TELESCOPES

Many telescopes are equipped with manual *slow-motion drives*, which allow objects to be tracked by turning a knob. But for hands-free operation an electric drive is essential. The simplest electric drives for telescopes are synchronous motors, which require a.c. power (as from the domestic supply). Such motors run in synchronism with the a.c. frequency, and the only way to control their rate is to vary the a.c. frequency. This is done by means of a variable-frequency oscillator (VFO), which may draw its power from a battery and convert it to a.c. Hence, when a battery-powered VFO is used, the drive can be operated without the need for the domestic supply. However, this system has been largely superseded by the use of low-voltage d.c. *stepper motors*.

Stepper motors rotate their shaft over a specific and precise angle for each pulse of d.c. power they receive, so they are well suited to telescope drives. A typical motor may require a pulse rate of a few hundred hertz for normal operation. Their control units are designed to deliver a stream of pulses at the appropriate rate for the gear train in use. Such motors also have the advantage that they can be driven from low-voltage portable battery packs or from car batteries, thus allowing the telescopes to be used anywhere. More advanced drive systems use d.c. servo motors rather than stepper motors. They incorporate encoders to measure their rotation and provide feedback to improve their accuracy.

Control units for motor drives typically provide a range of drive rates, such as sidereal, lunar, solar and 'King'. The King rate, named after Edward S. King, who advocated it, is 0′.4 per day slower than the sidereal rate, and helps to compensate for the fact that atmospheric refraction causes objects to move slightly more slowly across the sky than would be the case on an airless world. However, a GO TO telescope controller can allow for refraction in its choice of driving rate. Much faster rates, for moving the telescope from object to object, are known as *slew rates*.

COMPUTER-CONTROLLED (GO TO) TELESCOPES

Telescopes on mountings that include position encoders and electric motors on both axes can be controlled by software to point to any location in the sky. A database of objects is included in the software so that, once correctly oriented, the telescope will go to the chosen object automatically. These are commonly called *GO TO telescopes*, and they are available in a wide variety of types and sizes, from small portable refractors to observatory-sized reflectors. The controllers consist of keypads that plug into the telescope's base unit.

These systems work on the basis that once the direction of any two fixed objects on the celestial sphere is known, the orientation of the telescope with respect to the celestial sphere is fixed, and hence all other objects of known position can be found. The database also contains locations on the Earth's surface and algorithms to calculate the positions of Solar System bodies at any time. So once the time and location have been entered at the start of the observing session, the database can predict the altitude and azimuth of any object in its database. Advanced instruments include a GPS (global positioning system) receiver, precluding the need to enter the time and location before observing.

It then remains to direct the instrument to two specific objects so that it can fix its true orientation. This orientation procedure is the same for both altazimuth and equatorial mounts, and it is not even necessary for the mount to be on a horizontal surface. However, the procedure usually requires the telescope to be levelled and pointed north, so that the database can then direct the telescope to two bright stars in order to establish its orientation. The instrument should then be able to find any other object in its database. A well set-up instrument should place objects within the central third of the field of view of a low-power eyepiece.

It is usually also possible to link the instrument to a computer running appropriate sky-mapping software so that it can be pointed under the control of the computer. This allows for remote observing, in which the observer may be in a separate control room some way from the telescope, and for automated observation of objects such as galaxies for supernova patrols.

On cheaper instruments, in particular, the drive may not allow very precise pointing, and errors can rapidly build up. It is therefore useful to refine the instrument's awareness of its orientation by centring the object that it has found in the field of view and pressing the appropriate keys.

Once an instrument is correctly aligned on the sky, its database can apply the changes in altitude and azimuth to track any object, even if it is on an altazimuth mount. This should enable any object to be observed for a long period without further correction. However, one drawback of an altazimuth mount is that the orientation of the field of view will continually rotate, so long-exposure imaging is not possible. To eliminate this effect, a *field rotator* must be used, which automatically compensates for the change in field orientation.

ACCESSORIES

There are many accessories available for work in different observational fields. Great advances in the sensitivity and precision of professional instruments have been at least matched by the enterprise of advanced amateurs and commercial suppliers. Some of the more common accessories are described below.

DEW-CAP On damp nights, dew forms on exposed surfaces. The outer lens of an object glass, or the correcting lens of a catadioptric telescope, is particularly vulnerable, and must be protected by a dew-cap. This is a cylinder of non-conductive material projecting, ideally, a distance of two to three times the telescope's aperture in front of the lens. A simple tube rolled from thin black card has a minimal effect on the balance of the telescope, and is easily renewed as required. For more extreme conditions, heated dew-caps are available which can be powered by a low-voltage supply.

FILTERS Colour filters are becoming commonly used in lunar and planetary work. Since most astronomical objects are relatively dim and weakly coloured (even the Moon, almost blindingly bright in a telescope, is pale by daytime standards), the eye relies mainly on the non-colour-sensitive rods of the retina for its information. However, if a coloured object is observed through a filter of its own tint, it should appear brighter than if viewed through a filter of the opposite colour (e.g. a green filter for a red object).

Dyed glass filters are available in metal mounts which screw into the end of the eyepiece barrel. A wide range of colours are available which match the spectral characteristics of the original Wratten gelatin filters. Some of these are illustrated in Figure 10. Special interference filters, known alternatively as *nebula filters* or *light-pollution reduction* (LPR) filters, block the worst light-polluting wavelengths and let through wavebands which show nebulae and galaxies well, effectively increasing their contrast against the sky. *Broadband filters* allow colour photography with reasonably well colour-balanced images. *Narrowband filters* are used where light pollution is severe; the most extreme types of

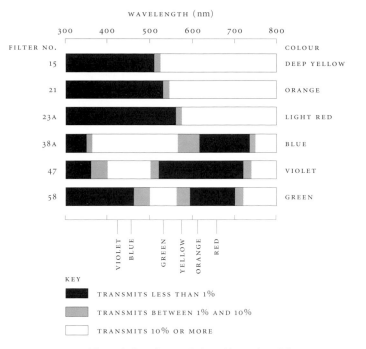

FIGURE 10. *The variation of transmission with wavelength for some filters made of dyed glass which are commonly used to enhance the colour contrast of astronomical objects. All the filters transmit infrared radiation. Filters 47 and 58 are dense, and best suited to apertures over 200 mm.*

narrowband filter are *line filters* (O III and Hβ) which transmit only the emission lines from nebulae and make the sky appear almost black.

The Sun should be observed only with a full-aperture filter; *all other types of filter can be dangerous.* Safe solar filters consist of an alloy coated onto a tough plastic (Mylar) or a glass disk. This diminishes the intensity of sunlight over a wide range of wavelengths – including the infrared – before it is focused, so that intensities are never dangerously high anywhere in the optical system. Air turbulence on a sunny day limits resolution to about 1 arcsec, so many observers with large instruments 'stop' their telescopes down to 125–150 mm and use a solar filter of just this diameter.

FINDERS A good finder may be considered essential for even the smallest astronomical telescope, particularly to an observer locating objects with reference to nearby stars. The finder serves its purpose if it permits the naked-eye view of the region to be rapidly reconciled with what is seen in the main telescope.

Therefore a finder must have a wide field of view (and hence low magnification) which will take in at least one convenient naked-eye star to act as a guide to the field being sought. A magnification of ×5, with a suitable eyepiece, will give an adequate 10° field, and an aperture of 30–40 mm will reveal sufficient stars to allow the object (or the part of sky in which the object lies) to be brought into a low-power view with the main instrument. Whether the finder should give an upright or an inverted image is a matter of individual preference. Large instruments, and those used regularly

Selecting a Telescope

REFRACTORS

Advantages

- Excellent definition and image contrast since there is no central obstruction in the optical path.
- The optics hold their alignment well.
- There are no reflecting surfaces requiring maintenance.
- Simple eyepieces work well.
- Wide, flat focal plane suitable for imaging (apochromatic designs).

Disadvantages

- The bulkiest telescope for its aperture.
- Difficult to mount rigidly.
- Only the smallest apertures are portable.
- All but the costliest objective lenses suffer from false colour.
- Usually not suitable for imaging without filters to reduce secondary spectrum.
- Per unit of aperture, object glasses cost more than reflecting or catadioptric optical systems.
- The eyepiece can reach awkward viewing positions.

If *critical definition* of planetary detail or close double stars is desired, then aperture-for-aperture a refractor is best. However, for the same price can be bought a Newtonian of about twice the aperture of a refractor.

NEWTONIAN REFLECTORS

Advantages

- Per unit of aperture, Newtonians are the cheapest optical system.
- The eyepiece is usually in a convenient viewing position.
- Equally suitable for imaging and visual use.
- Freedom from false colour.
- Fairly compact, and therefore do not require a tall tripod or pillar.
- Can readily be obtained in apertures far larger than those of other systems.

Disadvantages

- Sensitive to optical alignment, which in any case is less permanent than with a refractor.
- The aluminized surfaces gradually deteriorate.
- The diagonal mirror and its support diffract light and reduce image contrast, as well as blocking some of the light reaching the primary mirror.

If *aperture* is the major consideration (i.e. if the aim is the observation of variable stars or deep-sky objects to the faintest possible limit), then a Newtonian will be the choice. A small Newtonian is the ideal instrument for the beginner.

CATADIOPTRICS (CASSEGRAIN-TYPE)

Advantages

- Extremely compact and therefore portable.
- The whole sky can be observed from a comfortable sitting position.
- Keen competition between firms ensures value for money.
- The reflective coatings are well protected in the enclosed tube.
- Equally suitable for photographic and visual use.
- Systems with computer control are widely available to assist the rapid finding of objects.

Disadvantages

- Lower image contrast, as the relatively large secondary obstruction causes more light to be transferred from the image core to the diffraction rings.
- With low powers, the large secondary obstruction produces a 'blind spot' near the centre of the field.

If *portability* and general convenience are required, the compact catadioptric systems have much to recommend them.

---

for variable-star work, may have a second and more powerful finder to bridge the gap between the two apertures.

Non-magnifying finders produce a virtual image of an illuminated graticule at infinity, so that it appears to be projected onto the sky. These are popular for initial pointing of the telescope, and in the case of Dobsonians may be all that is required. Simple rifle-type sighting devices on the telescope have the drawback that the eye cannot focus on both the sight and the sky simultaneously.

GUIDE TELESCOPES Instruments used for photography requiring exposures of more than a few seconds are often equipped with a high-power auxiliary telescope of smaller aperture. By keeping a nearby guide star accurately centred on cross-hairs in the eyepiece, any residual tracking errors may be compensated for. Another method is to guide on a star imaged by the main telescope, using an eyepiece set to one side of the main field.

MICROMETERS In astronomy, a micrometer is used to measure the angular distance between two objects, either at the telescope itself or on an image. The *filar micrometer*, in which two fine webs can be moved a measured distance apart, is the traditional instrument for measuring the separation of double stars, but many other types suitable for amateur construction have been devised. In the *double-image* type a movable divided lens or glass plate permits the images of two stars to be superimposed, the amount of motion required to do so indicating the separation.

The *object-grating* or *diffraction* micrometer has the advantage of being simple to make, if not to use. A grating of several spaced parallel strips in front of the telescope produces a set of spurious images around each star; by manipulating the grating to create a symmetrical star pattern, the separation (and position angles) of the stars may be determined.

In addition to measuring double stars, some amateurs have

been successful in measuring, to professional standards, the positions of minor planets and comets on CCD images. Measurements made using computer software applied to the scanned image have now replaced traditional measuring systems.

PHOTOMETERS   Devices for determining the brightness of a star at the telescope can be divided into *visual* and *photoelectric* kinds. Most visual photometry depends on the equalization principle: the observer has to judge when the star appears equal in brightness to a comparison star. This may be achieved either by using a real comparison star, and dimming one or the other by artificial means such as interposing a wedge of tinted glass; or by using an artificial comparison star projected into the field of view from a suitable instrument, and adjusting its brightness.

In photoelectric photometry a light-sensitive device is used to record the star's brightness. To calibrate the system, measurements of other stars of known magnitude are taken. Photoelectric photometry is more accurate than visual methods, but the associated equipment is more expensive and complex.

However, most amateur photometry is now carried out on CCD images of the variable star and the comparison objects, using appropriate computer software. For precise work, a standard photometric filter should be used and the colour response of the CCD taken into account.

STAR DIAGONALS   Observing objects within about 30° of the zenith with a refractor or Cassegrain-type telescope will cause neck-strain unless a suitable observing chair is available. A star diagonal contains a plane mirror or a totally reflecting prism and allows a more comfortable observing angle, although at the cost of some loss of light and lateral (left-to-right) reversal of the image.

Tests, Adjustments and Maintenance of the Telescope

OPTICAL QUALITY

The accuracy with which a reflecting telescope's mirror focuses light is a measure of its quality. The widely accepted *Rayleigh criterion* states that the wavefront of the light being focused should be within a quarter of the wavelength of yellow light of the ideal. This in turn requires that the mirror's surface should be within one-eighth of a wavelength of its ideal shape. Optics of this quality are known as *quarter-wave* or *diffraction-limited*, in that the limit of detail that the telescope can reveal is set by the diffraction properties of light rather than by the accuracy of the mirror. However, tests show that in practice a mirror of tenth-wave quality – where the surface nowhere deviates by no more than one-twentieth of the wavelength of light – gives superior performance under ideal conditions.

The tolerance of an objective lens is less strict, but harder to define, because with two or more elements there are more surfaces to consider. However, manufacturers' claims about optical quality should be treated with caution. Without facilities for making a proper laboratory test, the only satisfactory way of judging the optical quality of a telescope is by its actual performance on a celestial object.

CHROMATIC ABERRATION   To test for the achromatism of a refractor, observe the bright limb of the Moon or the planet Venus when it is well above the horizon and against a fairly dark sky, using a medium power. A faint blue-violet halo is acceptable, but green or any other colour is not. The test should be repeated with other eyepieces to ensure that what is observed is not an artefact of the particular eyepiece used.

SPHERICAL ABERRATION   means that rays refracted or reflected by different zones of the objective lens or primary mirror do not share a common focus. It is best studied by observing the image of a third-magnitude star under high power, with the eyepiece alternately pushed and pulled a few millimetres inside and outside the position of best focus. The star's image, expanded into a small disk, should appear identical on either side of focus, with no rays or flares extending from its edge. In a Newtonian or Cassegrain-type telescope there will be a round spot at the centre representing the outline of the diagonal or secondary mirror. The secondary spectrum of a refractor will produce a greenish-yellow fringe outside the focus, and a reddish fringe inside the focus.

This test has the advantage over studying the quality of a focused star image or trying to resolve double stars at the limit of resolution (Table 10, p. 29), in that seeing effects are minimal. On many nights a star viewed with high power is a small 'boiling' blur several times larger than it would be in calm air. However, the disadvantage is that slight differences in the appearance of the disks have little or no effect on the performance – in other words, it may be misleadingly sensitive until the observer has gained some experience of testing different telescopes.

MAGNIFICATION AND FIELD SIZE

The magnification of a particular eyepiece may be calculated by dividing its focal length into the focal length of the telescope. However, its focal length might not be marked or might be incorrect, or the focal length of the telescope itself (particularly if it is a Cassegrain) might not be accurately known. It is possible to measure the focal length of a Newtonian or a refractor by focusing an image of the Moon onto a piece of semi-transparent paper stretched over the mouth of the draw-tube, and to measure the distance to the object glass or (via the flat) to the primary mirror.

To find the magnification directly, focus the telescope on a star or a remote terrestrial object, and then point it at an illuminated surface such as a white wall, or even a sheet of card (but *not* the daylight sky, the brilliance of which will produce an erroneous result). Holding a magnifying glass, focus on the exit pupil, which will be seen as a sharp-edged circle of light, either at or a little way behind the eyepiece. The diameter of this circle divided into the aperture of the telescope gives the magnification. A metal scale marked in half-millimetres or a vernier calliper held in the plane of the exit pupil will give the diameter (and hence the magnification) to within a few percent.

TO FIND THE DIAMETER OF THE FIELD OF VIEW set the telescope so that a star near the celestial equator, such as Delta (δ) Orionis or Zeta (ζ) Virginis, may pass through the centre of the field. Measure in minutes and seconds how long the star takes to pass across the field of the stationary telescope, and multiply by 15 to derive the field diameter in minutes and seconds of arc.

OPTICAL ADJUSTMENT

If the optical components are not properly aligned, or *collimated*, the image quality will suffer. Some instruments, particularly small refractors and catadioptric telescopes, are – in theory at least – permanently adjusted by the manufacturer, and must be returned to them should any mishap occur. Larger refractors, and all reflecting telescopes, should have provision for collimating the optics. The object glass of a refracting telescope is much less sensitive to misalignment than is the primary mirror of a Newtonian.

REFRACTORS should have three sets of push–pull screws linking the objective cell to the tube. If the image appears flared to one side, or if the defocused image is elongated rather than circular, adjust the pair of screws nearest the axis of the defect. By trial and error the asymmetry should disappear. If it does not, then the object glass is defective or has not been properly adjusted in its cell, and should be examined by an expert.

NEWTONIAN TELESCOPES have to be adjusted in several stages, and having an assistant can be helpful. It is best done in daylight, with the tube pointing at the sky; an eyepiece is not used until final testing on a star. For a totally misaligned instrument (Figure 11(*a*)), the procedure is as follows:

(a) The diagonal should be central in the tube, except in a very-short-focus reflector where the centre of the diagonal should be offset a small distance away from the eyepiece. The arm or set of vanes ('spider') on which it is mounted should be adjustable for this purpose.

(b) The diagonal should be on the axis of the draw-tube. Again, there should be provision for moving the diagonal along the axis of the main tube. To ensure that the eye is central in the draw-tube, fit a disk with a central viewing hole at the end of the draw-tube, which should be racked out to its fullest extent. With the diagonal holder correctly positioned, the view down the draw-tube should be as in Figure 11(*b*).

(c) The reflection of the primary mirror must be made central in the outline of the diagonal, as in Figure 11(*c*). The adjusting screws at the back of the diagonal holder are used for this purpose.

(d) The optical axis of the primary mirror must pass through the centre of the diagonal. This will have been achieved when the black outline of the diagonal is centred on the bright reflected image of the primary mirror. This dark spot is always most distant from the primary-mirror adjusting screw that should be turned inwards: in Figure 11(*c*) the screw nearest the position marked by the arrow must be turned inwards to bring the outline of the diagonal to the centre, as in Figure 11(*d*).

Final collimation (using the primary mirror's adjusting screws) must await a star test, when very slight adjustment should be all that is necessary to produce a perfectly symmetrical image, both in and out of focus. After adjustment (c), the image of the primary in the diagonal of a short-focus Newtonian (*f*/6 or less) should be displaced slightly down the tube rather than being perfectly concentric, since the converging cone of light from the primary is wider where it strikes the lower part of the diagonal.

CASSEGRAIN telescopes are adjusted in the same manner and sequence as for Newtonians, with the difference that the image of the primary in the secondary mirror must always be perfectly symmetrical.

A PERFECT STAR IMAGE When the air is steady and the instrument carefully adjusted and focused, the image of a fairly bright star will appear under high magnification as a minute disk – almost a point of light – surrounded by two or three tiny concentric rings, called *diffraction rings*. The extreme delicacy of these rings is almost impossible to portray on the printed page. The clear aperture of a refractor gives the most intense central core relative to the brightness of the rings, whereas the obstruction of the secondary mirror in the Newtonian and Cassegrain systems produces an image with a core of slightly smaller diameter, but with much more light present in the rings.

SETTING UP AN EQUATORIAL TELESCOPE

The fundamental requirements of an equatorial mounting are that

(a) the polar and declination axes are at right angles

(b) the axis of the telescope tube is at right angles to the declination axis

(c) the polar axis is parallel to the Earth's axis.

Points (a) and (b) are, or should be, the responsibility of the manufacturer. Here it is assumed that just the polar axis remains to be aligned.

The procedure can be divided into alignment in azimuth and alignment in altitude. In practice, fine adjustments to one are likely to affect the other, so that alternate and increasingly finer tuning is needed.

INITIAL ROUGH ADJUSTMENT in azimuth (i.e. in the north–south direction) may have to be done before the mounting is set up at the observing site to ensure that final precise setting is within the range of the adjusting screws. Sufficiently accurate methods include observation of the Pole Star (for observers in the northern hemisphere), observation of the Sun at local noon (allowing for the equation of time) and using a surveyor's compass (allowing for the current magnetic variation). Having defined the meridian by these or other methods, set the declination axis horizontal, and twist the mounting in azimuth until the telescope tube can be aligned north–south by rotating it in declination alone.

Once this is done, the polar axis can be set approximately in altitude by measuring the angle it makes with the horizontal, which should be equal to the observer's latitude. An adjustable protractor or a home-made wedge can be used in conjunction with a spirit-level to obtain an accuracy of a degree or so, which is sufficient for initial purposes.

FINE ADJUSTMENT for a permanent mounting may have to be spread over more than one night. Some recommended methods assume that the Pole Star is observable, which it is not south of the equator; others require accurate right ascension and declination circles. However, satisfactory adjustment can be obtained by using a high magnification to observe the north–south drift, if any, of stars in different parts of the sky as they are tracked east to west by turning the polar axis.

The principle behind the method is as follows. If the azimuth of the polar axis is too far *west* of the pole, then a star crossing the meridian appears to drift *north* (and vice versa). If the altitude of the polar axis is too *high*, then a star due *east* (ideally with an hour angle of 270° and an altitude of about 40°) will appear to drift *north*,

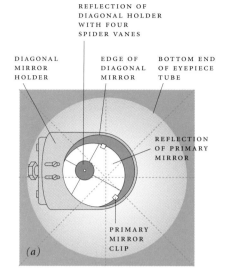

REFLECTION OF DIAGONAL HOLDER WITH FOUR SPIDER VANES

DIAGONAL MIRROR HOLDER

EDGE OF DIAGONAL MIRROR

BOTTOM END OF EYEPIECE TUBE

REFLECTION OF PRIMARY MIRROR

PRIMARY MIRROR CLIP

(a)

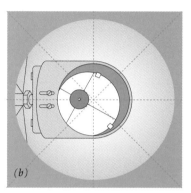

(b)

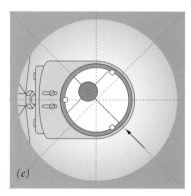

(c)

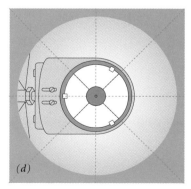

(d)

FIGURE 11. *Collimation of a Newtonian telescope. (a) The view down the empty draw-tube – everything is misaligned. (b) The diagonal holder has been correctly positioned, with the outer edge of the diagonal centred on the edge of the draw-tube. (c) The image of the primary mirror has been centred on the edge of the diagonal. (d) The reflection of the diagonal has been centred on the image of the primary mirror. Source: Alan M. MacRobert,* Sky & Telescope, *Vol. 75, p. 260 (1988).*

while one to the *west* (hour angle 90°) will appear to drift *south* (and vice versa if the altitude of the polar axis is too low).

The observations must be made with the highest possible magnification, using an eyepiece equipped with cross-hairs. Defocusing the star slightly allows the hairs to be projected as black lines on its disk, and any errors of symmetry caused by drifting are readily apparent. By trial and error the drift can be reduced, if not to zero then at least to an acceptable value.

For routine visual work, satisfactory accuracy will have been achieved when the object remains at the centre of the field for the duration of the observation. If the mounting includes circles of sufficient precision for finding faint objects, then the pointers must be calibrated using stars of known RA and declination.

The work most demanding of accurate polar alignment is long-exposure photography of deep-sky objects.

PORTABLE EQUATORIALS, if always used at the same observing site, may be positioned each time fairly accurately if location points are set into the ground to take the tripod legs. Alternatively, a sighting telescope may be designed to fit onto the polar axis so that the position of the celestial pole is centred on the cross-hairs. The telescope itself could also be used for this purpose, provided (a) it can be accurately set at a declination of 90°, and (b) the optical axis of the telescope is then truly parallel to the polar axis. Some telescope mountings come equipped with special built-in sighting telescopes for making quick and accurate polar alignment. These include rotatable graticules that can be oriented according to the time and date, so as to compensate for the offset of the Pole Star (or in the southern hemisphere, nearby stars) from the true pole.

CARE AND MAINTENANCE

Before bringing a portable telescope indoors after a night's observing, cover the object glass or mirror with the cover provided for the purpose, or the optical surface will become dewed in the warmer air. In any case, an instrument should not be stored in a warm room since thermal currents will be set up in the tube when it is taken outside into the cold air. Ideally, the telescope should already be at the outside air temperature before observing is begun, so a garage or secure shaded shed is a suitable place for storage. If a telescope is housed in an observatory, the structure should be opened up as soon as the temperature begins to fall in the evening to allow both the inside of the building and the telescope to achieve thermal equilibrium.

CLEANING LENSES   It is inadvisable to take an eyepiece or objective lens to pieces without expert advice. Only the exposed surfaces should be cleaned; these are of course the most likely to become dirty.

Although scattered specks of dust have little noticeable effect on the image, they can cause permanent marks if rubbed; therefore they must be carefully removed with a soft lens brush. The lens may then be gently wiped with a piece of damp cotton-wool (absorbent cotton). Care must be taken not to make the surface wet, as moisture may penetrate the lens mount. If available, pure (industrial) alcohol is excellent; common methylated spirit (denatured alcohol) contains oils which can leave smears. A really dirty lens may be cleaned with a strong domestic detergent, although industrial detergent, which is free from oily additives, is preferable.

Never press hard on the glass, and use a clean area of the swab for each stroke until the surface appears generally clean. Wipe with gentle lifting strokes in one direction so as to bring the dirt clear of the glass.

Some eyepieces such as the Plössl and Erfle have an outer component of flint glass. Being softer than crown glass, such surfaces need to be treated with particular care to minimize abrasion.

Old objective lenses may be stained between the elements, and the outer surfaces, particularly that of the inner (flint) lens, may have developed permanent dull patches. Such a lens should be sent away for professional treatment.

CLEANING MIRRORS   All the mirrors in a reflecting telescope should be closely covered when not in use, and never allowed to become dewed. Small reflectors often have no provision for protection, apart from perhaps a cap over the mouth of the tube; it is then well worth making lightweight covers (plain cardboard is better than nothing) to protect the mirror surfaces from the air.

Eventually a layer of dust will adhere to the mirror's surface. To remove it, the mirror should be taken from its cell and washed in cool water with detergent. Cotton wool or a similar soft material can be used to swab off any remaining particles. Good mirror coatings will show clear, high-contrast reflections, whereas a milkiness in the coating will scatter light, reducing image quality and masking faint detail.

Astronomical Imaging

With the equipment now available, the modern amateur astronomer can produce pictures that rival those taken by professionals. Many amateurs prefer to begin astronomical imaging using standard photographic film, but recent advances in digital technology offer many new options. The introduction of light-sensitive microchips called *charge-coupled devices* (CCDs) in 1990 brought about significant advances in astronomical imaging. As CCDs and their associated hardware and image-processing software have become cheaper and more powerful, they have been enthusiastically adopted by amateur astronomers.

CCDs consist of an array of picture elements termed *pixels*.

Early CCDs were only a fraction the size of a 35-mm film frame and had much coarser resolution, so film remained preferable for wide-field photography such as portraits of constellations. However, the newest generation of CCDs rivals 35-mm film in size and resolution. Electronic imaging offers several advantages over film, not least the ability to take very deep images with relatively short exposures. Image-processing can extract further detail from the raw CCD frames and can minimize the effects of light-pollution. Most CCDs are monochrome only, though, so to produce colour images three separate exposures through different filters are required.

To get the best images, whether with film or CCD, clean, high-quality optics and a sturdy driven mount are necessary. Longer exposures require a well-aligned drive, good focus and accurate guiding (preferably with a CCD autoguider), although short exposures are possible without drives.

This section first explains the techniques of astrophotography using conventional film and cameras, before going on to discuss electronic imaging.

CHOICE OF CAMERA

A conventional single-lens reflex (SLR) camera is very convenient for astrophotography. However, many modern cameras which have only automatic exposure, no B setting and an electronically controlled shutter are unsuitable.

The ideal SLR camera for astrophotography would have:

(a) manual shutter-speed control

(b) a B shutter speed, which together with a locking cable release allows the shutter to be kept open for any length of time

(c) a wide range of other shutter speeds, especially ones of several seconds' duration

(d) a mirror which can be locked in the 'up' position so as to prevent undue vibration when beginning an exposure

(e) easily replaceable viewing screens, with the availability of a clear screen with a single reticle for focusing by parallax (see below)

(f) a viewing screen magnifier, ideally one which replaces the normal pentaprism viewing system, for more accurate focusing

(g) a mechanical shutter, rather than an electronic one which drains the battery during a long exposure (some electronic shutters will operate on B setting without the battery, however)

(h) light weight, to reduce balance problems when the camera is on the telescope.

CHOICE OF FILM

Films are usually categorized by their *speed* (sensitivity to light); whether they are black-and-white or colour; and whether they are *negative*, for making prints, or *reversal films*, which produce transparencies.

The speeds defined by the International Standards Organization (ISO) are identical with the older ones of the American Standards Association (ASA) and Deutsche Industrie Norm (DIN). A standard film for everyday photography may have a speed of ISO 100/21°, usually simply called ISO 100; one with twice the sensitivity has an ISO speed of twice this figure. In general, faster films have lower contrast and are more grainy than slower ones. A slow film may be ISO 25; a very fast one, at ISO 3200, is 128 times faster and will therefore record stars over five magnitudes fainter in a short exposure.

Graininess is undesirable as it masks fine detail, so the choice of film is a compromise between speed and image quality.

(a) Black-and-white film has the advantage of being easy to process and print at home. There is a wide range of films, including many with special characteristics such as narrow spectral sensitivity. Virtually all black-and-white films are negative films.

(b) Colour print film is very widely available, but with comparatively few different emulsions. Home processing is more complex than with black-and-white film, and commercially made machine prints are often unsatisfactory for astronomical subjects. Colour balance is not maintained, and overall printing density is often inappropriate.

(c) Colour transparency film offers a wide variety of emulsions, and can usually be processed by the user, though commercial laboratories will undertake non-standard 'push' processing for a small extra charge. The reversal process has the advantages that the colour rendering is not subject to variations in print quality and, since no prints are needed, costs are lower. A greater brightness range can be shown on a transparency than on a print.

In general, use slow, contrasty, fine-grain films where detail is important, such as in planetary photography. Fast films are more suited to constellation photography and nova patrols, and where maximum sensitivity is needed in short exposures, such as for aurorae and meteors. For deep-sky work it is generally better to use a slow or medium-speed film, and to expose for longer, than to use fast films (unless the object is very faint).

FOCAL LENGTH AND IMAGE SCALE

Astrophotography gets more difficult as the focal length increases. Every type of lens has its uses: short focal lengths, such as the standard 50-mm camera lens, give small image scales and a wide field of view (28° × 41°) on 35-mm film, ideal for meteor, aurora or constellation photography. Planetary detail, however, requires the large image scale provided by a telescope.

The image scale depends solely on the focal length, the formula being

$$\theta = 57.3/F$$

where θ is the image scale in degrees per millimetre and F is the focal length in millimetres. So for a 50-mm focal-length lens the Moon's image will be on average less than 0.5 mm across, while with a 500-mm telephoto lens it will be 4.4 mm across and the field of view is then 2°.8 × 4°.1, which is appropriate for extended star clusters and associations.

The brightness of the image, however, depends on both focal length and lens diameter – that is, on the focal ratio. This applies to extended objects only; for stars, which are points of light, the image brightness depends solely on the lens diameter.

A photograph of a nebula taken with a focal ratio of $f/2.5$ will show the same amount of nebulosity whether the optical system is a standard camera lens or a large telescope. What changes, however, is the image scale and the faintest star that can be recorded with a given exposure time and film.

USING AN ORDINARY CAMERA

In the simplest form of astrophotography, a firmly mounted camera is focused on infinity. An exposure time of a few seconds with a standard lens at full aperture (say $f/2$) and everyday film (say ISO 100) will show a number of stars in the night sky. The faintest stars in the image will be tiny points of light which are hard to see. Photographs taken at twilight or in moonlight can show foreground details, adding pictorial value.

Increasing the exposure time beyond a few seconds results in trailed star images because of the rotation of the Earth. Slight trailing makes the faint images easier to see, but it soon becomes obvious. Long exposures, in which the images are considerably trailed, can produce interesting and attractive pictures, especially in colour, as the star colours are clearly visible.

The exposure time that gives noticeable trailing varies with the focal length of the lens and the declination of the object. Table 11 provides a useful guide.

With fast films, camera apertures of $f/2.8$ or faster and the exposure times given in Table 11, it is possible to photograph stars fainter than naked-eye visibility. (It is, incidentally, usually worth

TABLE 11. *Star trailing tolerances. Exposure times, in seconds, at which star trailing becomes noticeable for various focal lengths and declinations.*

FOCAL LENGTH (mm)	DECLINATION OF OBJECT			
	0°	40°	60°	80°
20	25	33	50	144
24	21	27	41	120
28	19	23	36	103
35	14	19	29	82
50	10	13	20	58
85	6	8	12	34
100	5	7	10	30
135	4	5	7	21
200	2.5	3	5	14
300	1.7	2.1	3	10
400	1.2	1.6	2.5	7
500	1.0	1.3	2.0	6
750	0.7	0.9	1.3	4
1000	0.5	0.7	1.0	3
1500	0.3	0.4	0.7	1.9
2000	0.2	0.3	0.5	1.4

stopping down the camera lens by at least one stop from full aperture, e.g. $f/2.8$ rather than $f/2.0$, to improve the image quality.) To record fainter objects requires longer exposure times, and for this some sort of driven mounting is needed.

DRIVEN CAMERA MOUNTS

One popular method is to mount the camera piggyback on a driven, equatorially mounted telescope. The mounting's motor or its slow motions can track the stars sufficiently accurately to give crisp pictures with exposures of several minutes using short-focal-length lenses.

An alternative is to use a system which is calculated to give the correct drive rate. Suitable devices include portable battery- or clockwork-driven equatorial camera platforms, and the hand-operated *Scotch mount*, so named by its inventor George Haig, and also called a *screw drive* or *barndoor mount*. In its simplest form the Scotch mount consists of two hinged boards, with the hinge aligned on the celestial pole, as shown in Figure 12. Turning a screw threaded through one of the boards at a controlled rate separates them so that a camera mounted on the top board tracks the stars. Haig's original design uses a screw of pitch 1 mm at a distance of 229 mm from the hinge. Turning this screw once a minute in synchronism with the second hand of a watch gives the correct rate.

The system works well for exposures up to 15 minutes with 50-mm lenses, but the drive errors increase with time because the screw is straight and not the arc of a circle, and for longer exposures more elaborate systems are needed. But the basic Scotch mount has the advantage of being simple, cheap and easily made.

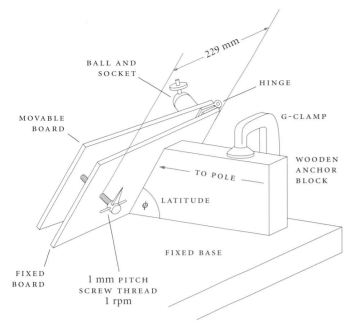

FIGURE 12. *The Scotch mount. The dimension shown is for a screw of pitch 1 mm and a rotation of 1 rpm, but can be adapted for other values. The design can be modified for any latitude, and the massive anchor block is not needed if a tripod bush can be fitted.*

POLAR ALIGNMENT

All equatorial camera platforms must be aligned on the celestial pole. This is made easier by adding a sighting device or small telescope, parallel to the polar axis, which can then be pointed at the true position of the pole with the help of one of the polar region maps (Figure 3 on pp. 24–5). Since there is no bright star exactly at either pole, there is likely to be an alignment error. The maximum polar alignment error allowable with a particular focal length of lens, for similar tolerances to trailing as given in Table 11, is given by the formula

$$394°/Ft$$

where F is the focal length in millimetres and t is the exposure time in minutes.

If the mount is aligned on Polaris or Sigma (σ) Octantis (both about 1° from their respective poles) rather than the true pole, then the maximum exposure time t_{max} (in minutes) without noticeable image distortion can be approximated as

$$t_{max} = 400/F$$

For a 50-mm lens, therefore, exposures of up to 8 minutes can be made before alignment errors become noticeable.

No matter how accurate the polar alignment, however, neither the Scotch mount nor electrically driven camera platforms will give very good results with long focal lengths or long exposure times. For these, more elaborate systems and some form of guiding are necessary.

GUIDED EXPOSURES

For focal lengths longer than about 200 mm, the driving rate should be monitored through an eyepiece, and fine guiding corrections made as necessary. There are several ways of correcting the drive rate of a camera platform or telescope, including:

(a) Varying the rate of a synchronous electric motor by altering the a.c. frequency supplied to it, by means of a variable-frequency oscillator (VFO). This method is suitable for units driven from a domestic electricity supply, although it is possible to get *inverters* which provide a.c. voltage from batteries.

(b) Varying the stepping rate of a stepper motor. These run on low d.c. voltages, and associated electronics will be necessary to provide a string of pulses whose rate can be varied. Because this system can be run from batteries, it is ideal for portable telescopes.

(c) With a worm-and-wheel drive, turning the whole worm and motor slightly, independently of the drive rate.

(d) Using a tangent arm on the declination axis whose position can be adjusted by turning a screw, or alternatively a worm and wheel. Declination corrections should be minor; if they keep having to be made during an exposure it means that the polar axis is misaligned.

(e) Moving the camera or film holder alone, while allowing the telescope to drive uncorrected.

However the drive rate is corrected, it must be possible to monitor the drive by viewing a star through the main telescope, or through its finder or guide telescope using an eyepiece with cross-wires (a reticle), preferably illuminated. Such eyepieces are available commercially, and generally have focal lengths of about 12 or 18 mm.

GUIDING ARRANGEMENTS

If the main instrument is used for guiding, the guide star must be outside the camera's field of view, off the optical axis. At the same time it must be close enough to the axis for it to be focused to an acceptable point. The usual method is to insert a small prism into the optical path just to one side of the frame area. This is a very convenient method, and add-on guiders complete with illuminated reticle eyepieces are available commercially. Another advantage is that, as the guide eyepiece is close to the camera, there is little risk of movement between the two. A drawback, however, is that it can be hard to find a sufficiently bright guide star at the right distance from the object being photographed.

The main alternative method is to use a separate guide

telescope. Ideally this should be on an adjustable mounting so that it is possible to guide either on the object being photographed, if it is suitable, or on a guide star anywhere in the vicinity. The drawbacks of this method are that it is more cumbersome since it requires a separate telescope, with the risk of some flexure in the fixings between the guide eyepiece and the camera's focal plane. The advantages are that the guide star can be chosen, and that the system can be adapted to photograph comets.

Many comets are so diffuse that it is difficult to guide on them. They move relative to the stars, and for exposures of more than a few minutes this can cause trailing of the comet's image. The solution is to use a guide eyepiece which can be shifted slightly by a micrometer screw thread. The comet's motion is calculated beforehand from its ephemeris, and the eyepiece is moved in the opposite direction and at the same rate by shifting it a predetermined distance every minute or so; the guide star is then recentred. This results in trailed stars but a sharp comet image. The steps should be fine enough to produce straight rather than jagged trails.

PHOTOGRAPHY THROUGH A TELESCOPE

Although a camera lens can be pointed through the eyepiece of a telescope, with both focused on infinity, it is more usual to remove the lens of the camera and attach the camera body to the eyepiece mount of the telescope. Adapters to suit most lens mounts and eyepiece fittings are available commercially. To use the full field of the instrument, no eyepiece is used. There must be enough focusing range within the limits of the normal eyepiece position to allow for the thickness of the camera body – about 50 mm. Many Newtonians intended for visual work do not have this range, so it will be necessary either to move the entire secondary and focusing assembly nearer the mirror, or to move the mirror up the tube.

To increase the effective focal length, and hence the image scale, such as for planetary photography, use an eyepiece or Barlow lens to project the image; extension tubes may be needed. Pictures can be taken in this way even if there is not enough focusing range for direct photography.

The effective focal length is obtained by first calculating the magnification M of the image given by the eyepiece. This depends on the distance of the eyepiece in front of the film, d, and on its focal length F:

$$M = (d/F) - 1$$

Measure d from the film plane, shown on most cameras by the symbol $-\ominus-$, to the estimated location of the field stop in the eyepiece. (If a Barlow lens is used rather than an eyepiece, note that its quoted magnification applies only when it is used visually with an eyepiece, and its power when used to project an image into a camera will be different.) The effective focal length is now the telescope's original focal length multiplied by M.

To increase the field of view, giving a smaller image scale, use a *telecompressor* (rich-field adapter) – an achromatic positive lens inside the focus position. These are available commercially, mainly for use with Schmidt–Cassegrain telescopes which otherwise are rather slow, with a comparatively small field of view. The image quality at the field edges may, however, be poor when using a telecompressor and there is bound to be *vignetting*, with the image dimmer at the edges.

FOCUSING METHODS

It is essential to have some means of checking the focus from time to time. With a fast lens or mirror (small f-number) the zone of sharp focus is only a fraction of a millimetre deep, and temperature changes during an observing session can easily defocus an instrument, particularly if it has a long focal length. Suitable focusing methods include a ground glass screen, a parallax focusing screen, a knife-edge and a Ronchi grating.

To focus by parallax requires a clear screen with a reticle or mark on it. With a magnifier, the eye is moved from side to side until the reticle and the image move together. A knife-edge focuser works in the same way on a star image as the same test for telescope mirrors does with an artificial star. It uses a knife-edge at the focal plane of the camera, exactly where the film emulsion will be. It must be possible to look through the back of the camera, so for an SLR the back must be opened before loading the film. Point the telescope at a moderately bright star and move the eye from side to side. The objective should be filled with the star's light, which is cut off instantly right across the objective by the knife-edge only when it is exactly in the focal plane. If it is slightly in front of or behind the focus position, its shadow will be visible moving one way or the other. Those who are more familiar with the Ronchi mirror test can use a Ronchi grating instead of the knife-edge.

EMULSION CHARACTERISTICS

The technical specifications of an emulsion for astrophotography are its characteristic curve, spectral sensitivity, resolving power, graininess and reciprocity failure characteristics.

An emulsion's *characteristic curve* reveals how the film responds to varying exposures to white light. It is a graph of photographic density, the degree of darkening of the developed emulsion, against exposure time. A typical characteristic curve (Figure 13) has a *toe*, where light produces little response. The curve then rises more steeply, following a more-or-less straight line. At the long-exposure end the emulsion reaches maximum density just beyond the *shoulder*.

There is an unavoidable *chemical fog* level, providing a minimum background density to the emulsion, even in the absence of any exposure to light. A faint object produces a small increase in

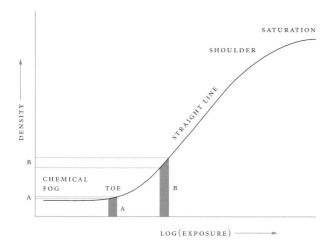

FIGURE 13. *A typical characteristic curve for an emulsion. A small exposure increase at A, on the toe of the curve, produces less of an increase in density on the film than would a similar exposure increase at B, on the straight-line part.*

density over this fog level. On the toe, a small exposure increase produces a rather small density increase, but on the straight-line part, where the contrast is higher, the response to the same small exposure increase is greater. This is why it is better to go for exposures with a certain sky background level – to get onto the bottom of the straight-line, higher-contrast part of the curve and so make faint objects more detectable.

Different emulsions have different characteristic curves. An overall shift to the left means that the emulsion is basically faster. In general, the faster emulsions have lower contrast.

Each emulsion has a recommended development time, chosen by the manufacturer to give pleasing results for everyday photography. But astronomers are often interested in getting the best performance out of the film, so it is common practice to give more development (*push-processing*) or to use a more active developer than is recommended. Increasing the development gives greater contrast, but also raises the chemical fog level. In some cases it can mean that a given exposure produces more response than on an inherently faster film developed normally, i.e. the film's speed seems to have increased.

Colour emulsions have three sensitive layers, responding to red, green and blue light, so they have three characteristic curves rather than one. Each layer may respond differently to push-processing, resulting in a *colour cast*. Reversal film curves are a mirror image of negative ones, so a small exposure increase gives a decrease in the density.

The *spectral sensitivity* curve of a film shows its response to light of different colours. It is usually different from the colour response of the eye, in that the dark-adapted eye is most sensitive to yellow-green light, while film may be most sensitive to blue or red light, with maybe several peaks in its sensitivity distribution. This is why the colours of emission nebulae on film are almost invariably different to the colours seen by a visual observer.

A film's *resolving power* is its ability to reveal fine detail. It is measured by photographing a test object, then studying the images to determine the finest detail that is resolved. The contrast of the test object has a bearing on the resolving power: if a stark black-and-white chart with a contrast ratio of 1000:1 corresponds to resolving a double star, then one of contrast 1.6:1 is representative of fine planetary detail. Resolving power is quoted in lines per millimetre (more accurately, the separation of line pairs). High resolving power for a low-contrast test object is around 100 lines/mm, typical of a slow film, while a fast film might have a value of 40 lines/mm.

Any photographic image, when examined closely, will show a grainy structure, caused by clumping of the developed silver filaments that form the image. The subjective impression of this structure is called the *graininess*, or often just the *grain*, of the film. Manufacturers often quote only a subjective term such as 'fine grain'. It is rare for them to admit to any emulsion having 'coarse grain'.

RECIPROCITY FAILURE In everyday photography, doubling the exposure time (say from ¹⁄₂₅₀ to ¹⁄₁₂₅ second) gives twice the effect on the film. This is the reciprocity law. But at low light levels this no longer happens: the reciprocity law breaks down, and the emulsion is said to suffer from reciprocity failure. In practice this means that after a certain time a doubling of the exposure time produces only a slight improvement in the faintest magnitude recorded.

There are several ways of overcoming reciprocity failure. One is to cool the emulsion to as low as −80°C. Cold cameras are available commercially, but they are inevitably more difficult to use than conventional cameras. Baking the emulsion, to drive off the water and oxygen which desensitize the emulsion, was once another popular technique.

Today, however, the most popular process is *gas hypersensitization*, often called simply *hypering*, which originally meant baking the film in a hydrogen atmosphere. This has a similar effect to straight baking. Hydrogen is a dangerous gas, however, so many people prefer to use *forming gas*, which is nitrogen with a small percentage of hydrogen. It works almost as well, and is much safer. Forming gas is not always readily available in small quantities, but some emulsions are available commercially hypered, and retain their properties for several months if stored in a domestic freezer.

Different commercial emulsions suffer from reciprocity failure to different extents, which can give rise to apparent anomalies. This is why a rather slow film such as Kodak Technical Pan Film (also known as 2415) can give better results than the super-fast films of ISO 1000 and more, although only with a much longer exposure time.

OTHER WAYS OF IMPROVING PERFORMANCE
Pre-flashing the film is one way to get somewhat better results from it, particularly if the skies are so dark that the sky background

remains unexposed. The technique is to give an overall pre-exposure so that any additional photons from true astronomical objects can readily have an effect. The pre-exposure time should be between $\frac{1}{100}$ and 1 second.

Another, quite different way to improve the chances of recording faint objects is to superimpose several weak exposures. This normally works only with negative films. For example, to get an image of a planet dense enough for printing might require a 2-second exposure. Four exposures, each of 0.5 seconds on pre-flashed film, would yield one rather dense image, but with less graininess because there are more, overlapping, grains forming the image. In addition, the shorter exposure times should give better resolution of detail in poor seeing.

There are other ways of getting the most out of a negative image. Some images have an enormous brightness range which cannot be shown in a single print. To reduce this range, a slightly defocused positive copy of low contrast and density is made, and sandwiched together with the original to make a print. Broad areas will have their brightness range compressed, but contrast will be maintained in fine detail.

Alternatively, copying an original onto high-contrast film using diffuse illumination makes faint deep-sky objects more visible, though brighter objects will saturate the film. Sandwiching several such copies from separate originals reinforces any faint objects just at the limit of detectability on the individual exposures.

Electronic digital methods of enhancement such as image addition and tone manipulation can now be applied to images on black-and-white and colour films, producing impressive results.

PHOTOGRAPHY THROUGH FILTERS

Differential reciprocity failure between the layers of colour film can give rise to colour shifts at long exposure times. One way around this is to make separate exposures on black-and-white film through red, green and blue filters which, together with the film's colour sensitivity, cover the visible spectrum. The final image is produced by combining the three separate exposures, through the appropriate colour printing filters, in the darkroom.

Light-pollution-reduction (LPR) filters, designed to block strong emission lines from mercury or sodium streetlights while transmitting strong emission lines of astrophysical interest, are very effective for photographing emission nebulae. But they are expensive, and a cheaper alternative is a red filter such as a Wratten 25 or 29, used in conjunction with a red-sensitive film (Kodak 2415 or colour film). This combination records emission nebulae (H II regions). In badly light-polluted areas, a filter such as the deeper red Wratten 29 or 92, or Lumicon Hα, a longer exposure time and sensitive medium- to high-contrast film are needed for good results.

ELECTRONIC IMAGING SYSTEMS

Imaging systems based on CCDs are now readily available to amateur astronomers. Although the amount of light needed to expose a CCD chip fully is similar to that required by medium-speed film, the efficiency of the CCD is much greater. This enables objects of much lower contrast to be detected, which in practice usually means stars two magnitudes fainter. Furthermore, the chips do not suffer from reciprocity failure and have a *linear response* to light (i.e. doubling the exposure time doubles the recorded signal), which is of great advantage for photometric and astrometric measurements.

A single CCD exposure can produce a megabyte or more of data, but computing power and memory size have more than kept pace with the demands of these devices, allowing small telescopes under light-polluted skies to record objects originally accessible only to large apertures. With 200 mm aperture it is possible to record stars down to 16th magnitude with exposures of under 2 minutes. The only real restriction is in chip area, but with careful matching of image detail to pixel size, and electronic image-addition techniques, results of near-photographic quality can be achieved.

Standard light-pollution-reduction (LPR) filters can be used with CCDs. Many amateurs are also turning to Hα filters, which eliminate light pollution and can be used even with a full Moon in the sky, although longer exposures are required.

True-colour pictures can be produced by taking consecutive monochrome exposures through red, green and blue filters. The three files are then combined in the computer to produce the colour image. CCDs from different manufacturers have their own spectral responses, so the exposure through each filter must be adjusted to produce the optimum result. One common characteristic of CCDs is their extended red sensitivity, so infrared-blocking filters are used to improve the colour balance. Such filters are also essential when refracting optics are used, as the infrared light will usually be focused at a different point from the visible light.

Colour CCDs are also available. They use a matrix of tiny coloured filters placed over the pixels to produce colour files in a single exposure. Both methods have their advantages, but it is generally accepted that tricolour imaging will produce the best results, albeit with more effort.

CCD AUTOGUIDERS Traditionally, the process of monitoring the tracking of a telescope during an exposure and making minor corrections to prevent star images from trailing was done by the astronomer and required tedious observation of a guide star. With the introduction of the CCD autoguider, coupled with a drive system capable of accepting its input, the whole process is automated and is far more accurate. By mounting a separate guidescope on the tracking platform, a suitable guide star can be easily selected.

A simple calibration procedure is required to train the autoguider to apply the appropriate corrections. Once calibrated, the drive is controlled by the autoguider, which quickly senses and corrects any errors. Many CCD cameras are fitted with a second autoguiding chip, allowing imaging and guiding to be done simultaneously. This is a powerful technique as guiding is carried out using the main optics of the telescope, offsetting any flexure (a problem that can occur with a separate guidescope). Further options include autoguiding with the main imaging CCD by taking advantage of 'guider' frames and 'imaging' frames delivered by proprietary inter-laced (or twin-field) CCD cameras.

VIDEO ASTRONOMY Low-light-level video cameras are perfect for recording transient events such as occultations, eclipses and even the passage of Earth-orbiting satellites. The video camera is placed in the telescope's eyepiece holder and linked to a television monitor. Once focused, images can be viewed on the monitor and recordings made on a separate video cassette recorder or camcorder. When coupled with an accurate time-signal, recordings of such events can be used for serious scientific research. Video recordings can be transferred to computer in digitized form and analysed to locate individual video frames taken during moments of good seeing. In this way the best images can be *co-added* ('stacked') to create a superior still image. For planetary imaging, where high magnification is necessary, Barlow lenses or eyepiece projection will produce a larger image scale.

WEBCAMS Lightweight webcams have proved to be ideal, low-cost alternatives to standard astronomical video cameras and are routinely producing some of the best planetary images taken by amateurs. In most cases it will be necessary to remove the camera's own lens and add a suitable adapter before the camera can be attached to the telescope. A choice of frame rates can be selected, a good range being between 15 and 25 frames per second. The camera is focused by watching the image on the computer monitor. Image sequences are then saved to the computer's hard disk for processing.

Several very sophisticated image-processing programs available for free download from the Internet can be used to analyse the sequence and to align, co-add and process the best frames to produce high-quality stills. Brighter astronomical targets such as the planets, the Sun (with appropriate filtration) and the Moon are ideal for webcam imaging as the standard webcam is not designed to give exposures longer than a fraction of a second. It is possible to modify them for longer exposures, but the CCD detectors in webcams are not cooled (unlike traditional astronomical CCD cameras) and so the chips will rapidly be flooded by electronic noise. In some cases just attaching a standard PC cooling fan can make a lot of difference.

FIGURE 14. *Comparison of (above) a single 5-minute CCD exposure of the Eagle Nebula and (below) a set of five combined exposures. The processed image has been contrast-enhanced and filtered to bring out detail in the nebula. The image was taken with a 250-mm Schmidt–Cassegrain telescope and an SBIG ST-8 CCD camera.*

FIGURE 15. *The effect of image-processing on a film original is shown in this example. The upper image of the North America Nebula was taken on ISO 200 film though a 75-mm refractor with a light-pollution reduction (LPR) filter and a 20-minute exposure. The slide was scanned and contrast-enhanced to produce the lower image.*

DIGITAL CAMERAS   Modern consumer digital cameras can also be used for astronomy. In most cases the same limitations apply as for webcams, in that the CCD detectors are not cooled and only short exposures are possible. For portraits of the Moon and planets the digital camera can be held up to the eyepiece and the shutter opened. This is known as the *afocal projection* method. The best results are obtained if a bracket is used to keep the camera fixed in position at the eyepiece. Some *vignetting* (darkening around the edge of the field of view) may be apparent but this can be cropped out at the processing stage. Better cameras feature removable lenses and can be coupled to the telescope via a commercial adapter; these will produce the best images.

IMAGE PROCESSING   With digital images there is no need for darkrooms and chemicals; all processing is carried out on the computer. All CCD cameras are supplied with image-acquisition software and many come with a basic processing package, but more powerful software is widely available. Typical processing routines are brightness and contrast scaling (to show faint details and remove light-pollution), sharpening, cosmetic improvements, co-addition of multiple frames to increase signal-to-noise ratio, and colour imaging. Figure 14 shows the difference between a single raw CCD image and a set of combined processed images.

Ironically, the digital revolution has led to a revival of traditional film imaging. Pictures taken on photographic emulsion can be digitized with high-quality scanners and then subjected to the same processing as CCD images, including co-addition and light-pollution suppression. For this reason emulsion photography has become very popular again, and good results can be obtained even from light-polluted locations. Figure 15 shows a scanned film original and the same image after electronic processing. Additionally, scanned images can be stored on media such as CD-ROMs and DVDs without suffering the effects of discoloration and fading that are inevitable with colour film.

SUMMARY OF TECHNIQUES

A suitable instrument, film and trial exposure time are given below for each subject. The exposure time is only a guide, and in poor conditions a much longer exposure may be needed. For instruments with a lower *f*-number give less exposure, and vice versa.

For CCD imagers, treat as medium-speed film for exposures of less than 10 seconds and as faster film for longer exposures, since there is no reciprocity failure. Because of the extended red sensitivity of CCDs, an infrared-blocking filter may be useful for planetary imaging and to reduce the effects of chromatic aberration in refracting optics.

SUN   Prime focus on medium-sized refractors or catadioptrics, $f/10$ to $f/15$; full-aperture Mylar filter; slow film, $1/30$ second. Same conditions for partial eclipses. *Total eclipses*: long telephoto lens, e.g. 500 mm $f/8$. No filter during totality. Slow film; give exposures from $1/250$ to 1 second.

MOON   *Whole disk*: prime focus with focal length longer than 750 mm. Slow or medium film. Exposures depend strongly on phase. For $f/8$ expose as follows: *full*, $1/125$ second; *quarter*, $1/30$ second; *crescent*, $1/8$ second. *Total eclipses*: telephoto lens, fast film, 10 seconds. *Surface details*: use eyepiece projection to give $f/40$ or more. Medium to fast film; 1 second.

PLANETS   As large a telescope as possible; focal length 1500 mm or longer. Use eyepiece projection to give $f/40$ or more. *Venus*: slow film; $1/8$ second. *Mercury, Mars*: medium-speed film, $1/4$ second. *Jupiter*: medium-speed film, $1/2$ second. *Saturn*: medium to fast film, 1 second.

ASTEROIDS AND OUTER PLANETS   Photograph as stars.

ZODIACAL LIGHT   Wide-angle, fast camera lens; very fast film; 1 to 5 minutes driven or undriven.

AURORA   Standard or wide-angle camera lens, $f/2$, fast film, 5 to 15 seconds.

METEORS   Standard camera lens, $f/2$, fast film, 5 minutes driven or undriven.

CONSTELLATIONS   Camera lens, one stop less than full aperture; slow to very fast; 10 seconds to 30 minutes depending on film and sky brightness. Driven exposures give star images; undriven, star trails.

STARS AND STAR CLUSTERS   Guided exposures using large apertures at prime focus or with eyepiece projection. Medium to fast film; 1 to 15 minutes.

COMETS   As for stars, but guided on comet's motion. Comets with long tails require shorter focal lengths, such as 300 mm $f/4$, preferably on large-format film.

DEEP-SKY OBJECTS   Large objects, such as the Orion Nebula, can be photographed with a telephoto lens and fast film, e.g. 200 mm $f/4$, 5 minutes. For better results, and for most galaxies, use large aperture, fast f-number, at prime focus with slow to medium film, preferably hypered, for 15 minutes.

CHAPTER THREE

The Solar System

The Sun

The Sun is a rather ordinary yellow dwarf star. Because it is over a quarter of a million times closer than its nearest stellar neighbour, we can study it in much greater detail than any other star, even with small instruments. It is blindingly brilliant, so special precautions have to be taken while observing it. *Never look at the Sun directly through any optical instrument. Even staring at the Sun with the naked eye for any length of time can cause damage.* The safest technique is to project the Sun's image onto a card, as described in the section Observing the Sun (p. 53). Physical data for the Sun are given in Table 12.

The Sun's Place in the Galaxy

Our Galaxy (the Milky Way) contains about 10^{11} stars and has a flattened, spiral structure roughly 25 kiloparsecs in diameter. The Sun is located about 8 kiloparsecs from the centre, and just a few parsecs

TABLE 12. *The Sun: physical data.*

Mass: 1.99×10^{30} kg ($= 3.33 \times 10^5$ Earth masses)

Diameter: 1.392×10^6 km

Volume: 1.41×10^{18} km^3 ($= 1.30 \times 10^6$ Earth volumes)

Mean density: 1.41×10^3 kg m^{-3}

Surface gravity: 273.87 m s^{-2} ($= 27.94$ Earth gravity)

Escape velocity: 6.18×10^5 m s^{-1}

Total radiation emitted: 3.83×10^{26} W

Mean distance from Earth: 1.496×10^8 km

Inclination of axis to perpendicular to plane of ecliptic: 7°.25

Sidereal period of axial rotation (at latitude 17°): 25.38 d

Spectral type: G2V

Absolute visual magnitude: +4.82

Apparent visual magnitude: −26.78

Effective temperature: 5770 K

north of the galactic plane. The stars in the solar neighbourhood are, on average, revolving around the centre of the Galaxy with a velocity of about 250 km s^{-1} and make a complete revolution in approximately 220 million years. In addition, the Sun moves with respect to the nearby stars with a velocity of around 19.5 km s^{-1} in the direction of a point on the borders of the constellations Hercules and Lyra, near the bright star Vega. This point is known as the *solar apex*, and its approximate coordinates are RA 18h, dec. +30°.

The Structure of the Sun

THE INTERIOR Gravitational forces create enormous temperatures (about 15×10^6 K) and pressures (about 3×10^{11} atmospheres) at the centre of the Sun. Under these conditions, nuclear reactions occur which turn hydrogen nuclei into helium nuclei, releasing a large amount of energy in the process. It is this energy that has kept the Sun shining for around 4.5×10^9 years, and will continue to power it for a similar time to come.

The high-energy radiation produced in the centre gradually diffuses outwards, taking perhaps 10 million years to reach the surface, in a long series of absorption and re-emission processes. Eventually, about three-quarters of the way from the centre to the surface, convection takes over, conveying heat to the *photosphere* (the visible surface) largely through rising columns of hot gas. The gas cools at the surface, and is carried down again to be reheated. Since hot gas emits more light than cool gas, the pattern of convective cells can be seen at the surface, as *granulation*. Each granule is about 1000 km across.

THE SURFACE AND ABOVE The average temperature of the photosphere is about 5800 K, and the gas pressure perhaps a tenth of that at the Earth's surface. Under these conditions many of the common atoms are at least partially *ionized* (stripped of some of their electrons), forming a *plasma*. As light passes upwards through the Sun's outer layers, the atoms and ions in those layers absorb radiation at certain wavelengths; measurement of the strength of these *absorption lines* in the solar spectrum tells us the composition of the Sun, at least at the surface. It appears to be about 90% hydrogen and 10% helium (by numbers of atoms), with other constituents making up less than 1%.

For about 500 km above the photosphere the temperature continues to fall gradually. However, for the next 1500 km a gentle rise sets in, and the temperature reaches 10 000 K at the top of the *chromosphere*, named after the red-coloured rim which can be seen around the Sun at the time of a total solar eclipse. The colour of the chromosphere is caused by the emission of red light from hydrogen. The spectrum of the chromosphere can be obtained during the few seconds at the beginning and end of a total eclipse when the chromosphere flashes out as the brighter photosphere is covered. This so-called *flash spectrum* consists of a number of bright lines, many of which coincide in wavelength with those seen in absorption in the photosphere.

Above the chromosphere the temperature rises very sharply to over 500 000 K through a zone only a few hundred kilometres thick called the *transition region*; it is best observed in the far ultraviolet part of the spectrum using instruments carried by spacecraft. Above the transition region is the *corona*, a vast envelope of tenuous gas reaching far out from the Sun, which can be seen as a pearly halo of light at the time of a total eclipse. Its shape varies with the sunspot cycle (see p. 51), being more uniform and symmetrical at maximum and exhibiting a much more irregular structure at times of low activity, when it shows long equatorial streamers and polar tufts and plumes. The coronal temperature is between one and two million kelvin and, with the extremely low densities prevailing, highly ionized atoms are present, e.g. iron with up to 15 of its 26 electrons stripped away.

The corona is continually being replenished by material from below while losing material into interplanetary space at the rate of about three million tonnes per second in what is called the *solar wind*, which gusts at speeds of a few hundred kilometres per second. This flow appears to be strongest from *coronal holes*, extensive areas of the corona lower in density and temperature than their surroundings, where the magnetic field flows directly out from the Sun. Coronal holes are the source of the high-speed streams in the solar wind.

Surface Features

The complex interplay of convection, radiation and magnetic fields produces a number of different surface features.

SUNSPOTS Throughout history, when the Sun has been observed rising or setting through a mist, dark patches have been reported periodically on its surface. These sunspots are actually regions lower in temperature than the surrounding area, which thus appear dark by contrast. The *umbra* is the darker central part, with a temperature of about 4500 K, while the *penumbra*, which surrounds it, is only a few hundred degrees cooler than the photosphere. The penumbra of a well-developed sunspot is always larger in area than the umbra. Sunspots vary in size from small *pores*, about the size of granules, which can be seen clearly only in high-resolution images obtained under conditions of good seeing, up to large groups covering several billion square kilometres.

The usual unit for estimating areas of sunspots is a millionth of the visible disk, which corresponds to an area of about 3 million square kilometres. For a spot to be visible to the naked eye, its area must be greater than about 500 millionths; a very large group may cover several thousand millionths. The largest recorded sunspots and their areas are listed in Table 13.

The Scottish astronomer Alexander Wilson noted in 1774 that sunspots appear to be depressions in the solar surface; the effect is most pronounced at the limb, where the part of the penumbra closer to the limb becomes broader and the umbra is displaced away from the centre of the spot. The explanation of this so-called

TABLE 13. *Largest-ever sunspots. The date, active region number and area in millionths of the Sun's visible disk are listed for all sunspots that reached an observed area larger than 3000 millionths. For comparison, the entire surface area of the Earth is 169 millionths of the solar disk.*

DATE	ACTIVE REGION NUMBER	AREA (MILLIONTHS)
1892 February 10	2421	3038
1905 February 2	5441	3339
1917 February 14	7977	3590
1917 August 9	8181	3178
1926 January 19	9861	3716
1937 July 28	12455	3303
1937 October 5	12553	3340
1938 January 12	12673	3627
1938 July 20	12902	3379
1938 October 12	13024	3003
1939 September 5	13394	3054
1941 September 21	13937	3088
1946 February 7	14417	5202
1946 July 29	14585	4720
1947 March 5	14851	4554
1947 April 3	14886	6132
1951 May 19	16763	4865
1968 February 1	21482	3202
1982 June 14	3776	3100
1989 March 17	5395	3600
1989 August 29	5669	3080
1990 November 18	6368	3080

Source: David Hathaway, NASA/Marshall Space Flight Center.

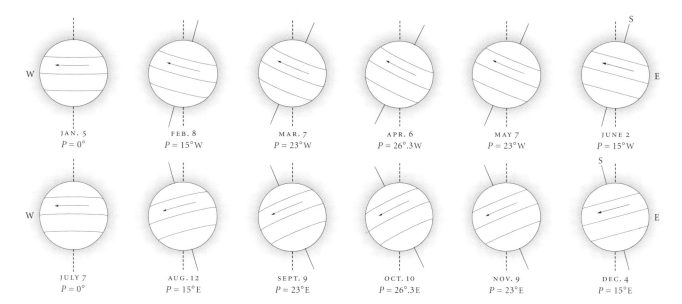

JAN. 5	FEB. 8	MAR. 7	APR. 6	MAY 7	JUNE 2
$P = 0°$	$P = 15°W$	$P = 23°W$	$P = 26°.3W$	$P = 23°W$	$P = 15°W$
JULY 7	AUG. 12	SEPT. 9	OCT. 10	NOV. 9	DEC. 4
$P = 0°$	$P = 15°E$	$P = 23°E$	$P = 26°.3E$	$P = 23°E$	$P = 15°E$

FIGURE 16. *The apparent motion of sunspots across the Sun's disk at various times of the year. The dashed vertical line is the line of the hour circle. The diagram shows apparent motion as viewed in an inverting telescope from the northern hemisphere; for use in the southern hemisphere the diagram should be inverted.*

Wilson effect is that the gas in and above the sunspot is cooler and less dense than at the normal surface, so that we can see deeper into the photosphere in the region of the umbra. This gives the impression that the umbra is depressed.

Sunspots never appear near the poles, and are generally confined between latitudes 35° north and south, although they are seldom found right on the equator. Most individual sunspots last for a few days, although large ones and groups can persist for several weeks. The daily motion of spots across the disk from east to west demonstrates the rotation of the Sun, which is faster at the equator than at the poles. The synodic period (i.e. as viewed from the Earth as it revolves around the Sun) is 27.275 d at a latitude of 17° (sometimes referred to as the mean synodic period). For other latitudes ϕ the synodic period P, in days, is given by

$$P \approx 26.75 + 5.7 \sin^2\phi$$

The corresponding sidereal period for latitude 17° is 25.38 d, which is adopted as the mean rotation period of the Sun. Each rotation of the Sun is identified by a number in a series called *Carrington rotations* started by the English astronomer Richard Carrington. Rotation No. 1 began on 1853 November 9.

The Sun's axis of rotation is inclined at 7°.25, and the Earth's axis of rotation at 23°.5, to the vertical to the plane of the ecliptic; hence the direction of the Sun's axis as viewed from the Earth changes throughout the year. In consequence, the apparent motion of sunspots in transit across the disk changes with the seasons, as illustrated in Figure 16. To determine the true latitude and longitude of a spot at a particular time the observer needs to know the position angle of the Sun's axis and the heliographic latitude of the apparent centre of the Sun's disk; these data are given in Tables 14 and 15 (see also Observing sunspots, p. 54).

THE SOLAR CYCLE Observations carried out over more than three centuries have shown that the number of sunspots rises and falls with an average period of 11.1 years, although the actual figure can vary between about 8 and 16 years. Activity can vary appreciably from one cycle to another. The rise to maximum is normally more

TABLE 14. *Variation of the position angle P of the north point of the Sun's disk during the year.*

DATE	P	DATE	P
Jan. 5, July 7	0°	July 7, Jan. 5	0°
Jan. 16, June 26	5°W	July 18, Dec. 26	5°E
Jan. 27, June 15	10°W	July 30, Dec. 16	10°E
Feb. 8, June 2	15°W	Aug. 12, Dec. 4	15°E
Feb. 23, May 18	20°W	Aug. 28, Nov. 20	20°E
Mar. 7, May 7	23°W	Sept. 9, Nov. 9	23°E
Mar. 18, Apr. 25	25°W	Sept. 21, Oct. 29	25°E
Apr. 6	26°.3W	Oct. 10	26°.3E

TABLE 15. *Variation of the heliographic latitude B_0 of the centre of the Sun's disk during the year.*

DATE	B_0	DATE	B_0
Dec. 7, June 6	0°	June 6, Dec. 7	0°N
Dec. 16, May 29	1°S	June 14, Nov. 30	1°N
Dec. 23, May 20	2°S	June 23, Nov. 21	2°N
Jan. 1, May 11	3°S	July 2, Nov. 13	3°N
Jan. 10, May 2	4°S	July 11, Nov. 4	4°N
Jan. 20, Apr. 21	5°S	July 21, Oct. 25	5°N
Feb. 1, Apr. 10	6°S	Aug. 3, Oct. 12	6°N
Feb. 19, Mar. 21	7°S	Aug. 23, Sept. 22	7°N
Mar. 5	7°.2S	Sept. 8	7°.2N

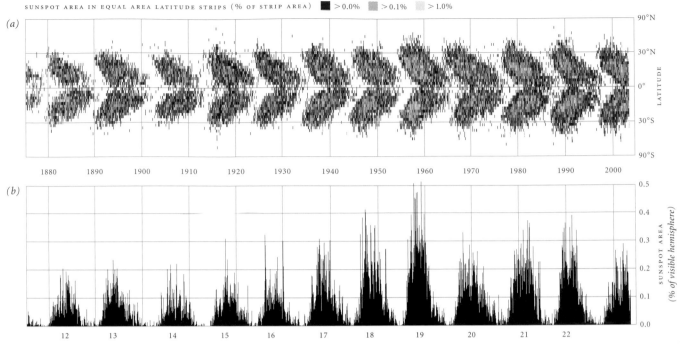

FIGURE 17. *(a) Butterfly diagram of sunspots since 1874, and (b) corresponding plot of area covered by sunspots. Both plots are of daily sunspot data averaged over individual rotations. The sunspot cycles are numbered (below the plot (b)) in a sequence beginning with the cycle that peaked around 1760. Source: David Hathaway, NASA/Marshall Space Flight Center.*

rapid than the decline, taking about 4.8 years. At sunspot minimum no spots may be seen for weeks on end, while large spots occur most frequently around maximum. At the beginning of a cycle spots tend to occur at the higher latitudes, both north and south, but as the cycle develops they are to be found progressively nearer the equator. This effect (known as *Spörer's law*) is shown beautifully by the *butterfly diagram*, first plotted in 1904 by Walter Maunder of the Royal Observatory, Greenwich, in which the number and latitude of spots are shown as functions of time. Figure 17(*a*) is plotted from sunspot data collected at the Royal Greenwich Observatory from 1874 to 1976, and subsequently by the U.S. Air Force and the U.S. National Oceanic and Atmospheric Administration (NOAA). The corresponding sunspot areas are plotted in Figure 17(*b*), which shows that different cycles are far from identical.

Although an 11-year periodicity is evident in sunspot activity, studies of magnetic fields on the Sun have shown that the true cycle of activity is actually twice as long, i.e. 22 years. Sunspots, particularly those in groups, are regions of intense magnetic activity, and the magnetic polarities of the spots reverse in each successive 11-year cycle. For example, if in the northern hemisphere of the Sun the preceding spot of a pair has a north magnetic polarity, then the following spot has a south magnetic polarity; in the southern hemisphere the polarities of the preceding and following spots are the other way round. This pattern of polarities persists in each hemisphere throughout a given cycle, but when the spots of the next cycle start to appear at high latitudes they have the opposite magnetic polarities to those of the previous cycle.

The strong, localized magnetic fields on the Sun hold the key to the existence of sunspots, for they seem to constrain the movement of hot gas and prevent the convection of heat locally, thereby keeping the area affected by the magnetic fields cooler than the rest of the photosphere.

SUNSPOT NUMBERS For objectively recording sunspot activity, the International Sunspot Number is used; it was formerly known as the Wolf, or Zurich, Relative Sunspot Number, as it was originated by Rudolf Wolf of Zurich Observatory in the nineteenth century. It is now published by the Sunspot Index Data Center (SIDC) in Brussels, Belgium. It is computed daily and takes into account sunspot groups as well as individual spots. If g is the number of sunspot groups, f the total number of individual spots (including those in the groups) and k a factor depending on the instrument used and the observer, then the Wolf number R is given by

$$R = k(10g + f)$$

For telescopes of about 100 mm in aperture, k may be taken to be equal to 1. Hence, if an observer notes 3 sunspot groups and 11 individual spots on a given day, then the sunspot number for that day is $30 + 11 = 41$. The International Sunspot Numbers are usually averaged for a year, and have ranged from 0 at minimum to 190.2 at the highest recorded maximum, in 1957. The highest monthly mean was 253.8 in October that year.

Another sunspot index, the daily Boulder Sunspot Number, is computed by the NOAA Space Environment Center in Boulder,

Colorado. The Boulder number incorporates data from different observatories and is usually about 25% higher than the SIDC number. Dividing either of the official sunspot numbers by 15 gives the approximate number of individual sunspots that can be seen on the projected image of the solar disk with a small telescope.

Other Surface Features

FACULAE are hotter and hence brighter patches in the photosphere which usually appear near sunspots, and are best seen against the darkened limb of the Sun's disk. They are often the precursors of sunspots and may also be found after the spots have disappeared. Faculae not associated with spots can be found near the poles; such polar faculae appear as bright points of light, a few arc seconds in diameter, above 60° latitude. Unlike the faculae at lower latitudes, they do not group together to form luminous patches but are scattered at random. Faculae lie in the upper reaches of the photosphere and correspond roughly to *plages* – clouds of hot gas in the overlying chromosphere which can be observed in the light of Hα and the calcium H and K lines.

PROMINENCES are jets of gas that rise above the chromosphere and are shaped by magnetic fields, often forming arcs. They are clearly seen above the limb during solar eclipses. Prominences appear to be relatively cool, dense clouds of gas which can exist in the corona for many hours, steadily pouring material back into the photosphere. Prominences, along with *sprays*, *surges* and *loops*, are more common on an active Sun at times of sunspot maxima, although long-lasting *quiescent prominences* display their curtain-like form at other times. When seen in silhouette against the Sun's bright disk they are called *filaments*.

SPICULES are jets of gas with a spiky, fiery appearance that last for a few minutes. They are found in the lower chromosphere, particularly at the edge of granulation patterns, where they line up. Spicules are well seen in Hα light.

FLARES are another short-lived phenomenon of the Sun's active regions. A sudden release of energy accelerates charged particles in the plasma, which proceeds to emit radiation across the whole spectrum – from X-rays to radio waves. Particles are ejected from the Sun and travel out through the Solar System, sometimes causing disturbances in the Earth's ionosphere, including aurorae, a day or so later. Flares are rarely seen in white light but are easily visible in Hα light.

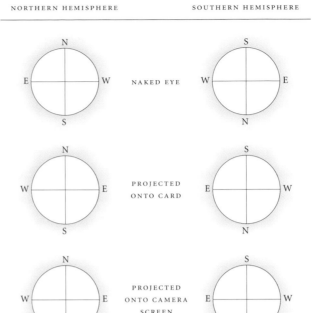

FIGURE 18. *Orientation of the Sun's image when projected.*

Observing the Sun

It is extremely dangerous to attempt to observe the Sun unless proper precautions are taken. If they are not, blindness may well be the penalty. Although special filters are available that cover the full aperture of the telescope, by far the safest method is to support a smooth, white card 30 cm behind the eyepiece of the telescope and to focus the image of the Sun projected onto it. To avoid looking at the Sun while pointing the telescope, use the shadow of the telescope cast on the card as a guide. The cardinal points are shown in Figure 18 for two different methods of projection.

The image of the Sun appears noticeably darker towards the edges. This effect of *limb darkening* occurs because the Sun is gaseous, so that at the centre of the disk we are looking farther into the hotter (and hence brighter) interior than at the rim. Under high power, granulation over the photosphere may be seen.

ACTIVE AREAS The main observational task is to count the number of sunspots visible. There are two principal methods of making sunspot counts: one is based on the number of *active areas* (AAs) seen on the disk, while the other is the International Sunspot Number, mentioned earlier. Every spot, however small, counts as a separate active area if it is at least 10° in latitude or longitude from its nearest neighbour. The same rule applies to sunspot groups. A large group, however extensive, is still one active area unless it has distinct centres of activity at least 10° apart. However, from time to time distinct groups do break out within 10° of each other. When such clearly separate groups occur they should be counted as separate AAs. By counting the number of AAs observed daily, and at the end of each month dividing the number seen by the number of observing days, the Mean Daily Frequency (MDF) is derived.

OBSERVING SUNSPOTS In studying the motion of sunspots across the disk, the inclination of the Sun's axis must be accounted for because the apparent path of sunspots varies with the time of year (Figure 16). They move in straight lines only around June 6 and December 7, when the Earth lies in the plane of the Sun's equator. From January to May, sunspots follow a curved path to the north because the south pole of the Sun is then tilted towards the Earth. From July to November, sunspot paths curve towards the south because the Sun's north pole is leaning towards us. The paths of sunspots across the disk show the greatest curvature near March 5 and September 8, when the south and north poles respectively are tilted towards us at their maximum of $7°.25$.

Publications such as *The Astronomical Almanac* and the *Handbook* of the British Astronomical Association give the following information necessary for determining the true heliographic coordinates of a point on the disk:

P, the position angle of the northern end of the Sun's axis

B_0, the heliographic latitude of the centre of the disk

L_0, the heliographic longitude of the centre of the disk.

The Moon

The Moon is the Earth's natural satellite, orbiting at a mean distance of 384 400 km; its diameter is 3475 km, only 29% smaller than that of the planet Mercury (for other data see Table 21, p. 69). Although the Moon appears bright at night its surface rocks are in fact very dark, reflecting less than 10% on average of the light falling on them. The Moon has essentially no atmosphere or magnetic field, although the magnetization of its surface rocks indicates that a strong magnetic field existed early in its geological history. Our knowledge of the Moon has come from Earth-based observations, from spacecraft pictures and data, and from experiments set up on the lunar surface and samples brought back for analysis, mainly by the Apollo astronauts but also by unmanned Soviet spacecraft.

The Moon's orbit around the Earth is noticeably elliptical, with a mean eccentricity of 0.055, although the actual value at any time ranges between 0.026 and 0.077 because of perturbations by the Sun. As a result of this changing orbital eccentricity, the Moon's closest distance to Earth (*perigee*) and its farthest (*apogee*)

FIGURE 19. *Difference in apparent size of the Moon at a typical apogee (left) and a typical perigee (right).*

range between extremes of about 356 400 km and 406 700 km, which affects the Moon's apparent size as seen from the Earth. At a typical perigee, the Moon appears about 12% larger than at a typical apogee (see Figure 19).

The Motions of the Moon

The orbital motions of the Moon and Earth, together with their axial rotations, have several effects on the observation of lunar surface features.

ROTATIONAL PHENOMENA The Moon and Earth orbit around their common centre of mass, or *barycentre*, with a period of 27.321 66 d (a sidereal month). Because the Earth is 81.3 times more massive than the Moon, this centre of mass lies within the body of the Earth itself, about 1700 km below its surface. The Moon rotates on its axis in the same length of time as it takes to make one orbit of the centre of mass; this is described as a *captured* or *synchronous* rotation, and is a consequence of tidal friction. The result is that the Moon presents the same face to the Earth, although with slight irregularities caused by libration (see p. 55).

PHASES As the Moon orbits the Earth we see varying amounts of its illuminated hemisphere, producing the familiar cycle of phases: new moon (when all of the illuminated side faces away from the Earth), crescent, half moon, gibbous and full moon, followed by the same sequence in reverse order. One cycle of phases (e.g. from one new moon to the next) is called a *lunation* or a synodic month.

It has a mean length of 29.530 59 d, longer than a sidereal month because the Earth moves in its own orbit around the Sun during that time, so the Moon has to travel farther to reach the same relative position as at the start of the month. The actual value of the synodic month can vary from about 29.27 to 29.83 days, largely because of the varying speed of the Earth in its elliptical orbit.

A considerable amount of light is reflected from the Earth onto the Moon's surface, and at the crescent phase near new moon this makes the Moon's night side faintly visible. This effect is called *earthshine*, and the appearance it produces is known popularly as 'the old moon in the new moon's arms'.

LIBRATION Although the Moon keeps essentially the same hemisphere turned towards the Earth, the alignment is not perfect. Two major 'rocking' motions are observed, called *libration in longitude* and *libration in latitude*. Libration in longitude is due to the slight ellipticity of the Moon's orbit, which causes a variation in the speed at which the Moon moves around the Earth. However, the speed of rotation of the Moon on its axis is constant, and the net result is that the Moon appears to rotate back and forth around its axis. Perturbations of the Moon's orbit caused by the gravitational attraction of the Sun cause the maximum values of libration in longitude to vary between about 4°.5 and 8°.1.

Libration in latitude occurs because the Moon's equator is inclined to its orbital plane by about 6°.7, the maximum values ranging between about 6°.5 and 6°.9. Tables showing the amount of both forms of libration at a given time are published in yearly almanacs. An additional effect called *diurnal libration* is caused by the changing viewing position of the observer as the Earth rotates. At moonrise, up to an additional 1° of the Moon's eastern limb is visible; at moonset, up to 1° more of the western limb is presented. Together, these three forms of libration allow us to see a total of about 59% of the Moon's surface (although only 50% at any one time, of course).

Surface Features

The lunar maps (Figure 20 on pp. 56–63) show the part of the Moon visible from the Earth. The maps are presented with south at the top; this is the orientation as seen by a northern-hemisphere observer in a standard, inverting astronomical telescope. West is then shown to the right, following the modern International Astronomical Union definition of west on the Moon (by the earlier convention, west on the Moon corresponded to west in the sky as seen from the Earth, i.e. towards the western horizon).

Most major surface features are named after famous people, particularly astronomers and other scientists, a practice begun by the Italian astronomer Giovanni Riccioli in 1651. Small features are labelled by the name of a nearby large feature, followed by one or more letters, although individual names have been given to certain small features of particular interest.

MARIA These are the darker and smoother parts of the surface. They were so named by early observers who thought that they were expanses of water (*maria* is Latin for 'seas'; singular *mare*). Some of the maria fill lowland basins that are nearly circular (e.g. Mare Crisium or Mare Imbrium), whereas others are quite irregular in outline (e.g. Mare Frigoris). The dark, relatively smooth material appears in most cases to be basaltic lava that erupted from the Moon's deep interior mainly between 4000 million and 3000 million years ago. The maria form the 'Man in the Moon' pattern visible to the naked eye on the Moon's near side. The lunar far side has been surveyed by spacecraft; it contains no large mare areas, but craters and bright highlands abound.

CRATERS These are the most abundant features on the Moon. Craters are roughly circular or polygonal in outline, with a central depression surrounded by raised walls. Numerous craters can be seen through even a low-power pair of binoculars. The largest craters have diameters of over 200 km, while the smallest are too small to be visible in even the largest telescopes. Where craters overlap, it is almost always the smaller crater that is superimposed on the larger one.

Small craters are commonly bowl-shaped, while large craters have wide, flat floors, often with a mountain or group of mountains at their centre. The largest craters have an even more complex interior structure, consisting of concentric rings of mountains. There appears to be no basic difference – other than size – between the largest craters and the circular mare basins.

There has been much controversy over the origin of lunar craters. Currently, most researchers accept that most of the craters were formed by the impact of meteoroids and small asteroids striking the lunar surface at speeds in excess of $10\,km\,s^{-1}$. However, many craters show evidence of later modification by volcanic activity or by forces of tension or compression at the surface. A few craters are almost certainly of volcanic origin; some of these are found at the heads of sinuous rilles or on domes (see p. 64). Generally they are less than 10 km in diameter, have subdued rims, are often elongated in shape and tend to appear in chains.

HIGHLANDS The brighter parts of the lunar surface are systematically higher and rougher than the mare surfaces, and are known to be generally older. Many of the Apollo rock samples from the bright areas are more than 4000 million years old. The highland topography is dominated by large and small craters in various stages of preservation, and the most prominent mountain ranges often form the rims of the circular mare basins, notably around Mare Imbrium and Mare Serenitatis.

(text continued on p. 64)

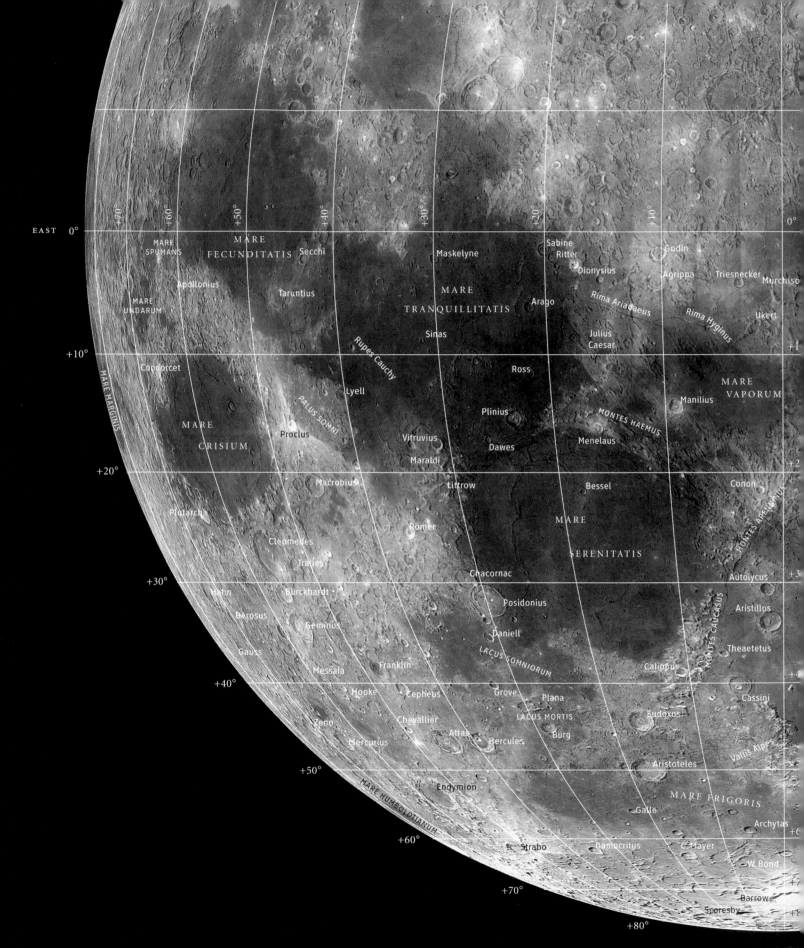

EAST 0°

+70° +60° +50° +40° +30° +20° +10° 0°

MARE SPUMANS

MARE FECUNDITATIS   Secchi   Maskelyne   Sabine   Godin
                                            Ritter
Apollonius   Dionysius   Agrippa   Triesnecker   Murchiso
MARE UNDARUM   Taruntius   MARE   Arago   Rima Ariadaeus   Rima Hyginus   Ukert
                                   TRANQUILLITATIS

Condorcet   Sinas   Julius   +1
                                                                Caesar

+10°   Rupes Cauchy   Ross   MARE
                                                                                  VAPORUM
                        Lyell   Manilius
                                            Plinius   MONTES HAEMUS
MARE   PALUS SOMNI   Vitruvius   Menelaus
CRISIUM   Proclus   Dawes   +2

                                      Maraldi

+20°   Macrobius   Littrow   Bessel   Conon

Plutarch   Römer   MARE
                                                             SERENITATIS   MONTES APENNINUS
              Cleomedes
              Tralles   Chacornac   Autolycus   +3
+30°   Aristillus
Hahn   Burckhardt   Posidonius
Berosus   Theaetetus
                          Geminus   Daniell   LACUS SOMNIORUM MONTES CAUCASUS
Gauss   Calippus
              Messala   Franklin   +4
                                                 Grove   Plana   Cassini
+40°   Hooke   Cepheus   Eudoxus
              Zeno   Chevallier   LACUS MORTIS
         Mercurius   Atlas   Hercules   Bürg   Aristoteles   Vallis Alpes
                                                                                  MARE FRIGORIS
+50°   MARE HUMBOLDTIANUM   Endymion   Galle   Archytas
                                                                                              +6
                                            Strabo   Democritus   C. Mayer
                                                                                  W. Bond
+60°
+70°   Barrow   +8
                                                                Scoresby
+80°

NORTH

FIGURE 20 *(pages 56 to 63). Map of the Moon divided into four quadrants: (a) Quadrant I (NE), (b) Quadrant II (NW), (c) Quadrant III (SW) and (d) Quadrant IV (SE), with a grid of selenographic coordinates and names of major features. South is at the top, as it appears in an inverting telescope. The accompanying lists give the longitude and latitude of each named crater to the nearest degree, and its diameter in kilometres. Since many craters are highly irregular in shape, these diameters are only approximate. Other named features are also listed. For linear features such as valleys and mountain ranges, the figure listed under 'Diameter' is the maximum length. For these features and for the maria, the position given is that of the approximate centre.*

(a) *Quadrant I: North-east*

CRATER	LONGITUDE	LATITUDE	DIAMETER (km)	CRATER	LONGITUDE	LATITUDE	DIAMETER (km)
Agrippa	10 E	4 N	46	Murchison	0	5 N	58
Apollonius	61 E	4 N	52	Plana	28 E	42 N	44
Arago	21 E	6 N	26	Plinius	24 E	15 N	42
Archytas	5 E	59 N	32	Plutarch	79 E	24 N	68
Aristillus	1 E	34 N	55	Posidonius	30 E	32 N	100
Aristoteles	17 E	50 N	87	Proclus	47 E	16 N	28
Atlas	44 E	47 N	87	Ritter	19 E	2 N	31
Autolycus	1 E	31 N	39	Römer	36 E	25 N	40
Barrow	8 E	71 N	93	Ross	22 E	12 N	26
Berosus	70 E	33 N	74	Sabine	20 E	1 N	30
Bessel	18 E	22 N	16	Scoresby	14 E	78 N	56
W. Bond	4 E	65 N	158	Secchi	43 E	2 N	25
Burckhardt	56 E	31 N	57	Sinas	32 E	9 N	12
Bürg	28 E	45 N	40	Strabo	54 E	62 N	55
Calippus	11 E	39 N	31	Taruntius	46 E	6 N	56
Cassini	5 E	40 N	56	Theaetetus	6 E	37 N	25
Cepheus	46 E	41 N	40	Tralles	53 E	28 N	43
Chacornac	32 E	30 N	51	Triesnecker	4 E	4 N	26
Chevallier	51 E	45 N	52	Ukert	1 E	8 N	24
Cleomedes	56 E	28 N	126	Vitruvius	31 E	18 N	28
Condorcet	70 E	12 N	74	Zeno	73 E	45 N	65
Conon	2 E	22 N	22				
Daniell	31 E	35 N	27	*Other Features*			
Dawes	26 E	17 N	18				
Democritus	35 E	62 N	39	Lacus Mortis	27 E	45 N	150
Dionysius	17 E	3 N	18	Lacus Somniorum	31 E	37 N	500 × 120
Endymion	57 E	54 N	125	Mare Crisium	59 E	17 N	420 × 550
Eudoxus	16 E	44 N	67	Mare Fecunditatis	51 E	9 S	900 × 600
Franklin	48 E	39 N	56	Mare Frigoris	7 E	56 N	1100 × 150
Galle	22 E	56 N	21	Mare Humboldtianum	82 E	58 N	280
Gauss	79 E	36 N	177	Mare Marginis	86 E	13 N	400 × 110
Geminus	57 E	34 N	86	Mare Serenitatis	18 E	27 N	660 × 600
Godin	10 E	2 N	35	Mare Spumans	65 E	1 N	120
Grove	33 E	40 N	27	Mare Tranquillitatis	30 E	8 N	540 × 780
Hahn	74 E	31 N	84	Mare Undarum	69 E	7 N	300
Hercules	39 E	47 N	67	Mare Vaporum	4 E	13 N	180 × 300
Hooke	55 E	41 N	37	Montes Apenninus	4 W	19 N	700
Julius Caesar	15 E	9 N	90	Montes Caucasus	10 E	40 N	540
Littrow	31 E	21 N	31	Montes Haemus	13 E	17 N	330
Lyell	41 E	14 N	32	Palus Putredinis	0	27 N	190 × 70
Macrobius	46 E	21 N	64	Palus Somni	44 E	15 N	240 × 360
Manilius	9 E	14 N	39	Rima Ariadaeus	13 E	7 N	240
Maraldi	35 E	19 N	40	Rima Hyginus	7 E	8 N	220
Maskelyne	30 E	2 N	24	Rupes Cauchy	37 E	9 N	160
C. Mayer	17 E	63 N	38	Sinus Medii	1 E	2 N	200 × 120
Menelaus	16 E	16 N	27	Vallis Alpes	3 E	49 N	180
Mercurius	66 E	47 N	68				
Messala	60 E	39 N	124				

0°

-10°

-20°

-30°

-40°

-50°

-60°

-70°

0° WEST

SINUS MEDII

Gambart

Reinhold

Kunowsky

Hevelius

Murchison

Encke

Hedin

Pallas

Hortensius

Cavalerius

Bode

Kepler

Reiner

Olbers

+10°

Copernicus

+10°

SINUS AESTUUM

OCEANUS PROCELLARUM

Gay-Lussac

Marius

Cardanus

Eratosthenes

MONTES CARPATUS

MONTES APENNINUS

T. Mayer

Krafft

+20°

Pytheas

+20°

Euler

Seleucus

Struve

Herodotus

Aristarchus

Schiaparelli

Prinz

Timocharis

Lambert

Vallis Schröteri

Briggs

PALUS PUTREDINIS

Diophantus

Krieger

Delisle

+30°

Archimedes

Wollaston

+30°

MARE IMBRIUM

Heis

MONTES SPITZBERGENSIS

Carlini

C. Herschel

Naumann

+40°

+40°

Piton

Le Verrier

Helicon

Mairan

Rümker

SINUS IRIDUM

MONTES JURA

Harding

Pico

Louville

Sharp

MONTES ALPES

Bianchini

+50°

RORIS

Repsold

+50°

Plato

Maupertuis

Bouguer

La Condamine

Harpalus

Markov

SINUS

Oenopides

+60°

MARE FRIGORIS

Babbage

+60°

Timaeus

J. Herschel

Birmingham

Fontenelle

Pythagoras

Epigenes

Carpenter

+70°

+70°

Anaxagoras

Philolaus

+80°

+80°

NORTH

CRATER	LONGITUDE	LATITUDE	DIAMETER (km)	CRATER	LONGITUDE	LATITUDE	DIAMETER (km)
Anaxagoras	10 W	73 N	51	Maupertuis	27 W	50 N	46
Archimedes	4 W	30 N	83	Murchison	0	5 N	58
Aristarchus	47 W	24 N	40	Naumann	62 W	35 N	9
Babbage	57 W	60 N	144	Oenopides	64 W	57 N	69
Bianchini	34 W	49 N	38	Olbers	76 W	7 N	71
Birmingham	11 W	65 N	98	Pallas	2 W	5 N	49
Bode	2 W	7 N	19	Philolaus	32 W	72 N	71
Bouguer	36 W	52 N	23	Plato	9 W	51 N	101
Briggs	69 W	26 N	37	Prinz	44 W	25 N	52
Cardanus	72 W	13 N	50	Pythagoras	62 W	63 N	128
Carlini	24 W	34 N	11	Pytheas	21 W	20 N	20
Carpenter	51 W	69 N	60	Reiner	55 W	7 N	30
Cavalerius	67 W	5 N	60	Reinhold	23 W	3 N	42
C. Herschel	31 W	34 N	13	Repsold	78 W	51 N	107
Copernicus	20 W	10 N	93	Schiaparelli	59 W	23 N	24
Delisle	35 W	30 N	25	Seleucus	67 W	21 N	43
Diophantus	34 W	28 N	18	Sharp	40 W	46 N	40
Encke	37 W	5 N	28	Struve	77 W	23 N	183
Epigenes	5 W	67 N	55	Timocharis	13 W	27 N	34
Eratosthenes	11 W	14 N	58	Timaeus	1 W	63 N	33
Euler	29 W	23 N	28	T. Mayer	29 W	16 N	33
Fontenelle	19 W	63 N	38	Wollaston	47 W	31 N	10
Gambart	15 W	1 N	25				
Gay-Lussac	21 W	14 N	26	*Other Features*			
Harding	72 W	43 N	22	Mare Frigoris	7 E	56 N	1100 × 150
Harpalus	43 W	53 N	39	Mare Imbrium	16 W	33 N	1150
Hedin	76 W	3 N	143	Mons Pico	9 W	46 N	25
Heis	32 W	32 N	14	Mons Piton	1 W	41 N	25
Helicon	23 W	40 N	25	Mons Rümker	58 W	41 N	70
Herodotus	50 W	23 N	35	Montes Alpes	0	46 N	240
Hevelius	68 W	2 N	115	Montes Apenninus	4 W	19 N	700
Hortensius	28 W	6 N	14	Montes Carpatus	27 W	15 N	450
J. Herschel	41 W	62 N	156	Montes Jura	37 W	47 N	450
Kepler	38 W	8 N	31	Montes Spitzbergensis	5 W	35 N	60
Krafft	73 W	17 N	51	Oceanus Procellarum	58 W	20 N	2000 × 1300
Krieger	46 W	29 N	22	Palus Putredinis	0	27 N	190 × 70
Kunowsky	32 W	3 N	18	Sinus Aestuum	8 W	12 N	210
La Condamine	28 W	53 N	37	Sinus Iridum	31 W	44 N	250
Lambert	21 W	26 N	30	Sinus Medii	1 E	2 N	200 × 120
Le Verrier	21 W	40 N	21	Sinus Roris	45 W	53 N	400 × 200
Louville	46 W	44 N	36	Vallis Schröteri	51 W	26 N	150
Mairan	43 W	42 N	40				
Marius	51 W	12 N	41				
Markov	63 W	53 N	41				

SOUTH

-80°

-70°

-60°

-50°

-40°

-30°

-20°

-10°

0° WEST

80°

70°

-60°

-50°

-40°

-30°

-20°

-10°

0°

-10°

-20°

-30°

-40°

-50°

-60°

-70°

Short
Casatus
Moretus
Gruemberger
Kircher
Cysatus
Blancanus
Bettinus
Rutherfurd
Scheiner
Zucchius
Clavius
Rost
Deluc
Phocylides
Porter
Bayer
Schiller
Nasmyth
Wargentin
Maginus
Longomontanus
Inghirami
Saussure
Mee
Schickard
Tycho
Wilhelm
Huggins
Hainzel
Drebbel
Orontius
Lexell
Ball
Walter
Deslandres
Wurzelbauer
Capuanus
Gauricus
Cichus
Ramsden
Hell
Vitello
Fourier
Pitatus
Mercator
Vieta
Regiomontanus
Campanus
Doppelmayer
Palmieri
Kies
Purbach
Hippalus
Liebig
de Gasparis
König
Cavendish
MARE
NUBIUM
MARE
HUMORUM
Thebit
Birt
Bullialdus
Mersenius
Nicollet
Arzachel
Agatharchides
Alpetragius
Lubiniezky
Gassendi
Lassell
Crüger
Alphonsus
Darney
Billy
Sirsalis
Davy
Guericke
Hansteen
Letronne
OCEANUS PROCELLARUM
Ptolemaeus
Bonpland
Parry
Euclides
Grimaldi
Herschel
Fra Mauro
Damoiseau
Lalande
Flamsteed
Riccioli
SINUS
MEDII
Mösting
Lansberg

Rupes Recta
Rimae Hippalus
Rimae Mersenius
MONTES RIPHAEUS
Rima Sirsalis

(c) *Quadrant III: South-west*

CRATER	LONGITUDE	LATITUDE	DIAMETER (km)	CRATER	LONGITUDE	LATITUDE	DIAMETER (km)
Agatharchides	31W	20S	49	Longomontanus	22W	50S	145
Alpetragius	4W	16S	40	Lubiniezky	24W	18S	44
Alphonsus	3W	13S	118	Maginus	6W	50S	185
Arzachel	2W	18S	96	Mee	35W	44S	132
Ball	8W	36S	40	Mercator	26W	29S	47
Bayer	35W	52S	47	Mersenius	49W	21S	82
Bettinus	45W	63S	71	Moretus	6W	71S	114
Billy	50W	14S	46	Mösting	6W	1S	26
Birt	8W	22S	17	Nasireddin	0	41S	52
Blancanus	22W	64S	110	Nasmyth	56W	50S	77
Bonpland	17W	8S	60	Nicollet	12W	22S	15
Bullialdus	22W	21S	59	Orontius	4W	40S	105
Campanus	28W	28S	48	Palmieri	48W	29S	40
Capuanus	27W	34S	60	Parry	16W	8S	47
Casatus	30W	73S	110	Phocylides	57W	53S	114
Cavendish	54W	25S	50	Pitatus	13W	30S	105
Cichus	21W	33S	40	Porter	10W	56S	52
Clavius	14W	59S	225	Ptolemaeus	2W	9S	153
Crüger	67W	17S	46	Purbach	2W	25S	115
Cysatus	6W	66S	49	Ramsden	32W	33S	24
Damoiseau	61W	5S	36	Regiomontanus	1W	28S	124
Darney	23W	15S	15	Riccioli	74W	3S	140
Davy	8W	12S	35	Rost	34W	56S	48
de Gasparis	51W	26S	30	Rutherfurd	12W	61S	50
Deluc	3W	55S	47	Saussure	4W	43S	56
Deslandres	5W	32S	235	Scheiner	27W	60S	110
Doppelmayer	41W	28S	64	Schickard	55W	44S	227
Drebbel	49W	41S	30	Schiller	40W	52S	165 × 65
Euclides	29W	7S	13	Short	7W	75S	71
Flamsteed	44W	4S	21	Sirsalis	60W	12S	42
Fourier	53W	30S	51	Thebit	4W	22S	55
Fra Mauro	17W	6S	94	Tycho	11W	43S	85
Gassendi	40W	17S	110	Vieta	56W	29S	87
Gauricus	13W	34S	80	Vitello	37W	30S	42
Grimaldi	68W	5S	220	Wargentin	60W	50S	84
Gruemberger	10W	67S	94	Wilhelm	21W	43S	107
Guericke	14W	11S	60	Wurzelbauer	16W	34S	85
Hainzel	33W	41S	70	Zucchius	50W	61S	64
Hansteen	52W	11S	45				
Hell	8W	32S	33	*Other Features*			
Herschel	2W	6S	41				
Hippalus	30W	25S	58	Mare Cognitum	23W	10S	360 × 240
Huggins	1W	41S	65	Mare Humorum	39W	24S	370
Inghirami	69W	47S	90	Mare Nubium	16W	20S	650
Kies	22W	26S	44	Montes Riphaeus	28W	8S	180
Kircher	45W	67S	72				
König	25W	24S	22	Oceanus Procellarum	58W	20N	2000 × 1300
Lalande	9W	4S	24	Rimae Hippalus	29W	25S	270
Lansberg	27W	0	40	Rimae Mersenius	45W	19S	260
Lassell	8W	15S	23	Rimae Sirsalis	61W	15S	280
Letronne	42W	11S	120	Rupes Recta (Straight Wall)	8W	22S	120
Lexell	4W	36S	63				
Liebig	48W	24S	38	Sinus Medii	1E	2N	200 × 120

SOUTH

−80°

−70°

−60°

−50°

−40°

−30°

−20°

−10°

EAST 0°

Boussingault   Boguslawsky   Simpelius

Curtius

Manzinus   Pentland

Mutus   Zach

Pontécoulant   Hagecius   Nearch   Jacobi

Rosenberger   Hommel   Lilius

Biela   Vlacq   Baco   Cuvier

Watt   Pitiscus   Heraclitus

Steinheil   Ideler   Licetus

Lockyer   Clairaut

Janssen   Barocius

Fabricius   Nicolai   Maurolycus   Faraday

Metius   Stöfler   Nasireddi

Büsching   Buch   Miller

Rheita   Fernelius

Riccius   Kaiser

Furnerius   Stiborius   Rabbi Levi   Gemma Frisius   Walter

Adams   Stevinus   Lindenau   Zagut   Goodacre

Neander   Rothmann   Allacensis

Reichenbach

Snellius   Piccolomini   Pontanus   Werner

Humboldt   Apianus

Petavius   Sacrobosco   Playfair

Wrottesley   Borda   Polybius   Azophi   Abenezra

Hecataeus   Santbech   Fracastorius

Holden   Monge   Geber   Airy

Vendelinus   Cook   Beaumont   Catharina   Almanon

Lamé   Colombo   Tacitus

Lohse   MARE   Abulfeda

Ansgarius   Magelhaens   NECTARIS   Cyrillus   Albategnius

La Pérouse   Goclenius   Gaudibert   Mädler   Descartes

Kästner   Langrenus   Theophilus   Kant

Gutenberg   Capella   Isidorus   Andel

Alfraganus   Hind   Halley

Torricelli   Hipparchus   Horrocks

Messier   Delambre

Rhaeticus

+70°   +60°   +50°   +40°   +30°   +20°   +10°   0°

MARE AUSTRALE

VALLIS RHEITA

Oken

RUPES ALTAI

MONTES PYRENAEUS

MARE FECUNDITATIS

MARE SMYTHII

(d) *Quadrant IV: South-east*

CRATER	LONGITUDE	LATITUDE	DIAMETER (km)	CRATER	LONGITUDE	LATITUDE	DIAMETER (km)
Abenezra	12 E	21 S	42	Lockyer	37 E	46 S	34
Abulfeda	14 E	14 S	65	Lohse	60 E	14 S	42
Adams	68 E	32 S	66	Mädler	30 E	11 S	28
Airy	6 E	18 S	37	Magelhaens	44 E	12 S	39
Albategnius	4 E	11 S	136	Manzinus	27 E	68 S	98
Alfraganus	19 E	5 S	21	Maurolycus	14 E	42 S	114
Aliacensis	5 E	31 S	80	Messier	48 E	2 S	10 × 16
Almanon	15 E	17 S	49	Metius	43 E	40 S	88
Andĕl	12 E	10 S	35	Miller	1 E	39 S	61
Ansgarius	80 E	13 S	94	Monge	48 E	19 S	37
Apianus	8 E	27 S	63	Mutus	30 E	64 S	78
Azophi	13 E	22 S	48	Nasireddin	0	41 S	52
Baco	19 E	51 S	69	Neander	40 E	31 S	50
Barocius	17 E	45 S	82	Nearch	39 E	58 S	75
Beaumont	29 E	18 S	53	Nicolai	26 E	42 S	42
Biela	51 E	55 S	76	Oken	76 E	44 S	72
Boguslawsky	43 E	73 S	97	Pentland	12 E	65 S	56
Borda	47 E	25 S	44	Petavius	61 E	25 S	177
Boussingault	55 E	70 S	131	Piccolomini	32 E	30 S	89
Buch	18 E	39 S	54	Pitiscus	31 E	50 S	82
Büsching	20 E	38 S	52	Playfair	8 E	23 S	48
Capella	35 E	8 S	45	Polybius	26 E	22 S	41
Catharina	23 E	18 S	104	Pontanus	14 E	28 S	58
Clairaut	14 E	48 S	75	Pontécoulant	66 E	59 S	91
Colombo	46 E	15 S	78	Rabbi Levi	24 E	35 S	81
Cook	49 E	17 S	47	Reichenbach	48 E	30 S	71
Curtius	5 E	67 S	95	Rhaeticus	5 E	0	46
Cuvier	10 E	50 S	75	Rheita	47 E	37 S	70
Cyrillus	24 E	13 S	95	Riccius	26 E	37 S	71
Delambre	17 E	2 S	53	Rosenberger	43 E	55 S	96
Descartes	16 E	12 S	48	Rothmann	28 E	31 S	42
Fabricius	42 E	43 S	78	Sacrobosco	17 E	24 S	98
Faraday	9 E	42 S	69	Santbech	44 E	21 S	64
Fernelius	5 E	38 S	65	Simpelius	15 E	73 S	70
Fracastorius	33 E	21 S	124	Snellius	56 E	29 S	83
Furnerius	61 E	36 S	150	Steinheil	46 E	49 S	67
Gaudibert	38 E	11 S	30	Stevinus	54 E	32 S	74
Geber	14 E	19 S	45	Stiborius	32 E	34 S	43
Gemma Frisius	13 E	34 S	88	Stöfler	6 E	41 S	126
Goclenius	45 E	10 S	55 × 75	Tacitus	19 E	16 S	40
Goodacre	14 E	33 S	46	Taylor	17 E	5 S	42
Gutenberg	41 E	9 S	71	Theophilus	26 E	11 S	110
Hagecius	47 E	60 S	76	Torricelli	28 E	5 S	20 × 30
Halley	6 E	8 S	36	Vendelinus	62 E	16 S	155
Hecataeus	79 E	22 S	127	Vlacq	39 E	53 S	89
Heraclitus	6 E	49 S	90	Walter	1 E	33 S	128
Hind	7 E	8 S	29	Watt	49 E	49 S	66
Hipparchus	5 E	5 S	150	Werner	3 E	28 S	70
Holden	62 E	19 S	47	Wrottesley	57 E	24 S	57
Hommel	33 E	55 S	125	Zach	5 E	61 S	71
Horrocks	6 E	4 S	31	Zagut	22 E	32 S	84
Humboldt	81 E	27 S	210				
Ideler	22 E	49 S	39				

Other Features

Isidorus	33 E	8 S	42

Mare Australe	92 E	47 S	980
Mare Fecunditatis	51 E	8 S	820 × 660
Mare Nectaris	35 E	15 S	350
Mare Smythii	87 E	1 N	370

Jacobi	11 E	57 S	68
Janssen	41 E	45 S	180 × 240
Kaiser	7 E	36 S	52

Montes Pyrenaeus	41 E	15 S	280

Kant	20 E	11 S	33
Kästner	79 E	7 S	105

Rupes Altai	23 E	24 S	530

Lamé	64 E	15 S	84
Langrenus	61 E	9 S	133
La Pérouse	77 E	11 S	78

Vallis Rheita	51 E	42 S	500

Licetus	7 E	47 S	75
Lilius	6 E	55 S	61
Lindenau	25 E	32 S	53

Sketching the Moon

When sketching the Moon, choose a small area of the lunar surface and make the drawing a reasonable size, say 100 mm across. First lightly sketch the basic outline of the crater using a soft pencil, as in the top illustration above. This outline can be drawn at the eyepiece, or copied from a suitable photograph or lunar atlas indoors. Next, fill in the darker areas by applying layers of pencil, exerting as little pressure on the paper as possible (centre). Finally, add the more subtle shading and finer detail (bottom). A detailed lunar drawing can take more than an hour to complete. This is the crater Bullialdus (22°W, 21°S – see Figure 20(c)).

RAYS These systems of bright streaks radiate across the lunar surface for up to 1000 km from young, prominent craters such as Tycho and Copernicus. They dominate the appearance of the full moon but are scarcely visible under low illumination. Rays are apparently splash patterns from impact craters, and consist either of deposits of material ejected from the main crater or of small secondary impact craters that have churned up the pre-existing surface, exposing lighter-coloured material.

RILLES AND FAULTS There is a variety of well-defined elongated valleys on the lunar surface to which the word 'rille' is applied. Some are sites where part of the lunar surface has subsided between two nearly parallel faults or fractures – these are called *linear rilles*. In places the fractures are curved, being controlled by the stresses around some large central structure such as a circular mare basin, producing what are known as *arcuate rilles*. Faults sometimes occur as individual features, the most famous being Rupes Recta (also known as the Straight Wall) in Mare Nubium, traceable for at least 120 km. It is actually a scarp about 250 m high. The shadow cast by the fault causes the Straight Wall to show as a dark line before full moon; after full moon the face of the fault scarp shows as a bright line.

Perhaps the most striking lunar valleys are the *sinuous rilles*, meandering valleys up to several hundred kilometres long which usually originate in a rimless crater-like structure at their high end, and become shallower and narrower as they wind their way downhill. These are volcanic features, caused either by fast-moving lava flows or by the collapse of roofs over underground lava channels, as occurs on the Earth on a smaller scale. In 1971 the Apollo 15 astronauts David Scott and James Irwin drove to the edge of Hadley Rille at the foot of the lunar Apennines and took photographs of the stratification revealed in the valley walls.

DOMES These are low, rounded hills with slopes of only a few degrees, often with a small crater at their summit. They tend to occur in groups, generally in areas that have been volcanically active. Their distinctive shape suggests that they formed from lava that was more viscous (stickier) than the surrounding surface, perhaps with a higher silica content.

CHANGES ON THE SURFACE The ages of the rocks returned by the Apollo missions imply that almost all of the Moon's volcanic activity took place more than 3000 million years ago. Also, the present rate of formation of craters by meteoroid impact is so low that changes are unnoticeable, even through the largest telescopes. Old reports of small permanent changes are now discredited, but more intriguing are the reports of temporary appearances of coloured patches on the surface or occasional obscurations of normally distinct features. These events, known as *transient lunar phenomena* or *lunar transient phenomena* (TLPs or LTPs), appear in particular areas, notably the craters Aristarchus, Gassendi and Alphonsus. Whether these events are real, perhaps some kind of gas release from the lunar interior, or whether they are an artefact of unusual observing conditions is still debatable.

Observing the Moon

The appearance of a lunar surface feature changes greatly according to the direction from which sunlight falls upon it. The illumination varies over the course of each lunation as the Sun rises, culminates and sets over a given feature. Changes in appearance are the most dramatic when the feature is near the terminator, since shadows are then at their longest and slight changes in relief are easily picked out; it is like observing a road surface at night by the light from a car's headlights.

Libration produces small variations in illumination and viewing geometry from one lunation to the next; these changes are most noticeable for features near the limb. The net effect is that a great deal of information can be deduced about the topography of an area by observing it systematically over many lunations around local sunrise and sunset.

By contrast, virtually no topographic detail is discernible near full moon because there are no shadows. However, this is the time to observe albedo differences on the surface, which generally indicate variations in surface structure and composition. Lunar rays are the most obvious high-albedo features, but there are more subtle albedo differences across the dark surfaces of maria. A crater which is prominent near the terminator may appear to vanish close to full moon if its walls and floor are of similar surface structure and composition to their immediate environment. Only craters which are young and bright (e.g. Aristarchus or Kepler), or older craters substantially filled with mare lavas to give dark floors (e.g. Grimaldi or Plato), are readily identifiable under all illuminations.

POSITION OF THE TERMINATOR It is useful to quote the position of the terminator at the time of an observation. Yearly almanacs list the longitude of the morning terminator on the Moon, known as the Sun's selenographic colongitude, symbol S. If libration is ignored, the selenographic colongitude is 270° at new moon, 0° at first quarter, 90° at full moon and 180° at last quarter. The selenographic colongitude increases by about 0°.5 per hour, and by 12°.2 per day.

REPETITION OF THE SAME PHASE   The mean lunation lasts just over 29.5 days, but the length of a given lunation can range between 29.27 and 29.83 days. Hence the same phase of illumination near the same time of night recurs after two lunations (59 d), 1½ hours later in the evening on average; then again after four lunations and 3 hours later in the evening, and so on up to 15 lunations (442 d 23 h), when the phase recurs 1 hour earlier in the evening.

The mean interval from one lunar perigee to the next, known as the mean anomalistic month, is 27.554 55 d, and recurs at the same phase after 14 lunations, or about 1½ months later in the following year, so that optimum conditions gradually disappear for a time.

OBJECTS NEAR THE LIMB   Features near the limb are best placed for observation when a suitable viewing phase coincides with a favourable libration that brings them as far as possible onto the visible disk. Limb features are nearest the centre of the disk when their position angle (measured from the Moon's north pole) added to the position angle of the Moon's axis is closest to the position angle of maximum libration, as tabulated, for example, in the *Handbook* of the British Astronomical Association.

POSITION ANGLE OF THE MOON'S AXIS   This oscillates some 25° each side of the hour circle over the course of a month, the extremes occurring as the Moon crosses the celestial equator, i.e. when the Moon's RA is 0h or 12h. The position angle is zero when the Moon is around RA 6h or 18h. The position angle of the Moon's axis of rotation is tabulated for each day in *The Astronomical Almanac*.

BEST ALTITUDE CONDITIONS   For any given phase, the Moon is highest in the sky (and therefore visible for longer under better seeing conditions) at one time of the year only. Table 16 indicates the most favourable times for observing the principal phases from the northern and southern hemispheres.

The Planets and Their Satellites

Nine major planets orbit the Sun, and all but Mercury and Venus have at least one natural satellite. In addition there are a great many minor planets, commonly known as asteroids, plus countless comets and smaller pieces of orbiting debris.

The four inner planets, Mercury, Venus, Earth and Mars, are relatively small, rocky bodies that are collectively termed the *terrestrial planets*. Jupiter, Saturn, Uranus and Neptune are often known as the *gas giants* because of their composition and size. Pluto, wandering at the edge of the Solar System, is in many ways unlike any of the other planets.

Mercury and Venus are also known as the *inferior planets*, because their orbits are closer to the Sun than that of the Earth. Likewise the planets from Mars outwards are known as the *superior planets*, since their orbits are outside the orbit of the Earth.

The orbits of all the planets are elliptical, some more so than others. The ellipticity of Pluto's orbit is so great that it actually crosses that of Neptune. The distances of the planets from the Sun are often expressed in terms of the Earth's distance, the astronomical unit.

ASTRONOMICAL UNIT (AU), the average distance of the Earth from the Sun, is the basic unit of length in the Solar System, and is also the base-line for measurements of the parallax of stars. As defined by the International Astronomical Union, the astronomical unit is 149 597 870 km. The term *unit distance* is sometimes used to refer to a distance of 1 AU. The *light time for unit distance* is the time taken for a beam of light to cover 1 AU; it is almost exactly 499 seconds (8.3 minutes). Another important quantity in measuring the scale of the Solar System is the *solar parallax*, which is the angle subtended by the Earth's equatorial radius as seen from a distance of 1 AU. As defined by the International Astronomical Union, the solar parallax is 8.794 148 seconds of arc.

TABLE 16. *The most favourable times for observing the Moon at its principal phases.*

	THIN CRESCENT (3–4 DAYS)	FIRST QUARTER	FULL MOON	LAST QUARTER	THIN CRESCENT (25–26 DAYS)
Northern hemisphere	End of April	March	December	September	End of July
Southern hemisphere	End of October	September	June	March	End of January

Planetary Orbits

The planets, and indeed all orbiting bodies, obey the three *laws of planetary motion* established by the German mathematician Johannes Kepler between 1609 and 1618:

1. The orbit of each planet around the Sun is an ellipse, with the Sun at one focus.

2. Each planet moves along its orbit so that the radius vector (the line joining the planet and the Sun) sweeps out equal areas in equal times; this means that the closer the planet is to the Sun, the faster it moves.

3. The square of the orbital period of each planet in years equals the cube of its mean distance from the Sun in astronomical units.

In the elliptical planetary orbit shown in Figure 21, the line AB across its greatest diameter is the major axis, and the line DE across its smallest diameter is the minor axis; they are at right angles to each other, and meet at the centre of the ellipse, C. AC or BC is the semi-major axis; this is the mean distance of the planet from the Sun. S is the focus of the ellipse at which the Sun lies; F is the empty focus, where nothing lies. P is a planet moving along its orbit. The line PS is the radius vector.

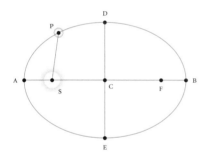

FIGURE 21. *An elliptical planetary orbit*

The eccentricity of an ellipse is given by dividing the distance SF by AB. For most planets the eccentricity is very small, being 0.017 for the Earth, and 0.007 for Venus, the lowest of all. Only Mercury (0.206) and Pluto (0.25) have orbits that depart appreciably from circles.

PERIHELION AND APHELION The planet reaches its closest point to the Sun at A in Figure 21. This is termed *perihelion*. The planet is farthest from the Sun at B, known as *aphelion*. The *perihelion distance* (symbol q) and *aphelion distance* (symbol Q) are the distances between the Sun and the planet on these occasions. For an orbit around the Earth, the closest and farthest points are called *perigee* and *apogee*, respectively. For an orbit around Jupiter the corresponding points are termed *perijove* and *apojove*; similar terms are used for other planets.

ORBITAL ELEMENTS An orbit is described by six quantities, known as *elements*. They are the semi-major axis a, the eccentricity e, the inclination i, the longitude of the ascending node Ω, the longitude of perihelion ϖ, and the time of perihelion pas-

sage T or the longitude L of the orbiting body at some other time. For double stars the orbital period P is also given. The semi-major axis and the eccentricity define the size and shape of the orbit, the inclination and the longitude of the ascending node together define the plane of the orbit, and the longitude of perihelion defines the orientation of the orbit. (Sometimes the argument of perihelion, ω, is given; this is added to the longitude of the ascending node to find the longitude of perihelion.)

For orbits that are subject to perturbations, elements are given for a specific time; these are called *osculating elements*, and they allow the position of the orbiting object to be calculated for times close to the epoch of osculation.

ASPECTS OF THE PLANETS As seen from the Earth, the planets reach certain positions relative to the Sun that are known as *aspects*. For the superior planets, the two most significant aspects are opposition and conjunction (Figure 22).

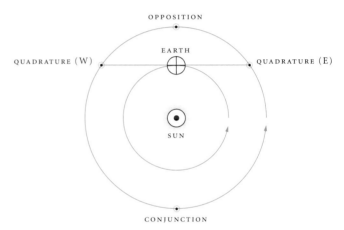

FIGURE 22. *Aspects of the orbit of a superior planet.*

At *opposition* a planet is opposite the Sun in the sky, i.e. its celestial longitude and the Sun's differ by 180°. At opposition a planet is visible all night, and lies on the meridian at midnight. Opposition is the best time to observe the superior planets, since they are then at their closest to the Earth. As seen through a telescope, the apparent size of a planet which has a markedly elliptical orbit, such as Mars, varies considerably depending on whether opposition occurs near the time of the planet's perihelion or aphelion. Perihelic oppositions of Mars are the best times for observation, although the planet may then be low in the sky for northern observers.

At *conjunction*, a planet has the same celestial longitude as the Sun and so lies on the far side of the Sun as seen from the Earth. The planet is then obscured by the Sun's glare. When a planet or the Moon is in line with the Sun, at either opposition or conjunction, it is said to be at *syzygy*; the Moon at syzygy is either new or full.

An additional, less important aspect of the superior planets is

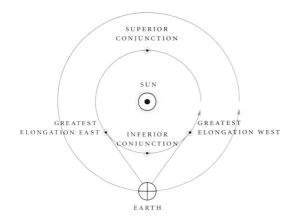

FIGURE 23. *Aspects of the orbit of an inferior planet.*

quadrature, when the angle between the planet and the Sun is 90°. At quadrature the superior planets can show a slight phase effect; the phase is most noticeable for Mars, which appears distinctly gibbous around this time.

The inferior planets, Mercury and Venus, cannot come to opposition or quadrature, but they have two types of conjunction: *inferior conjunction*, when they lie between the Earth and the Sun, and *superior conjunction*, when they lie on the far side of the Sun (Figure 23). The widest angular separation of Mercury and Venus from the Sun is known as *greatest elongation*: either greatest elongation west (in the morning sky) or greatest elongation east (in the evening sky).

Planetary Motions

The planets orbit the Sun from west to east, known as *direct* motion. This is the usual direction of motion in the Solar System, although some comets orbit the Sun in the opposite direction, known as *retrograde* motion. Some of the moons of the outer planets also have retrograde orbits.

Around the time of opposition, a superior planet can appear to move temporarily retrograde as the Earth, with its faster orbital motion, catches the planet up and overtakes it. The planet seems to perform a backwards loop in the sky. The points where the planet is changing from direct motion to retrograde motion and back again are called the *stationary points*.

ORBITAL PERIODS A planet's orbital period is usually measured relative to the background stars on the celestial sphere; this is known as its *sidereal period*, and can be thought of as the planet's 'year'. The planet's orbital period as seen from the Earth is called its *synodic period*; this is the time taken for the planet to return to a specific aspect, such as conjunction or opposition. The synodic period differs from the sidereal period because the Earth is itself moving in orbit around the Sun. For a moon, the synodic period is the time between successive conjunctions or oppositions as would be seen from its parent planet.

Data for the periods of the planets are given in Table 17.

Visibility of the Planets

Of the planets in the Solar System, only Mercury, Venus, Mars, Jupiter and Saturn are conspicuous naked-eye objects – indeed, they were the only planets known to ancient astronomers. Uranus, which can reach mag. 5.5, is just visible to the naked eye, provided one knows where to look; it is easily found with binoculars. The minor planet Vesta can also be seen with the naked eye when at its best (mag. 5.5), and is well within the reach of binoculars, as are some other minor planets. Neptune, at mag. 8, is a binocular object, but Pluto, at mag. 14, needs a large telescope to be seen at all.

BRIGHTNESS The magnitudes of the planets are measured on the same magnitude scale as the stars (see p. 107). The magnitude

(text continued on p.70)

TABLE 17. *Planetary periods and motions.*

PLANET	MEAN SIDEREAL PERIOD	MEAN SYNODIC PERIOD (d)	SIDEREAL ORBITAL VELOCITY (km s⁻¹)	SIDEREAL MEAN DAILY MOTION (DEGREES)	PERIOD OF AXIAL ROTATION	INCLINATION OF EQUATOR TO ORBIT (DEGREES)
Mercury	87.969 d	115.88	47.87	4.092	58.646 d	0.01
Venus	224.701 d	583.92	35.02	1.602	243.019 d (R)	177.36
Earth	365.256 d	—	29.79	0.986	23.934 h	23.44
Mars	686.980 d	779.94	24.13	0.524	24.623 h	25.19
Jupiter	11.863 y	398.88	13.07	0.083	9.842 h	3.13
Saturn	29.447 y	378.09	9.67	0.033	10.233 h	26.73
Uranus	84.017 y	369.66	6.84	0.012	17.240 h (R)	97.77
Neptune	164.79 y	367.49	5.48	0.006	16.110 h	28.32
Pluto	247.92 y	366.72	4.75	0.004	6.387 d (R)	122.53

R = retrograde. The rotation period given for Jupiter is at its equator (System I). The rotation periods for Uranus and Neptune are those of their magnetic fields.

TABLE 18. *Observational data for the planets.*

| PLANET | AT UNIT DISTANCE (1 AU) " | ANGULAR EQUATORIAL DIAMETER | | AT MEAN GREATEST ELONGATION " | AT MEAN OPPOSITION DISTANCE " | MEAN VISUAL MAGNITUDE[a] " |
| | | AT MINIMUM DISTANCE | AT MAXIMUM DISTANCE | | | |
		(FOR THE PERIOD 2000–2100) "	"			
Mercury	6.7	12.3	4.6	7.3	—	0.0
Venus	16.7	63.1	9.6	24.2	—	−4.4
Mars	9.4	25.1	3.5	—	17.9	−2.0
Jupiter	197.1	49.9	30.5	—	46.9	−2.7
Saturn	166.2	20.7	15.0	—	19.5	+0.7[b]
Uranus	70.5	4.1	3.3	—	3.9	+5.5
Neptune	68.3	2.4	2.2	—	2.3	+7.8
Pluto	3.3	0.1	0.1	—	0.1	+15

[a] Greatest elongation for Mercury and Venus, mean opposition distance for Mars to Pluto.

[b] With rings closed; −0.3 with rings open.

TABLE 19. *Planetary orbital data (Keplerian elements for epoch 2000.0)*

PLANET	MEAN DISTANCE FROM SUN (AU)	MEAN DISTANCE FROM SUN (10⁶ km)	MAXIMUM DIST. FROM SUN (AU)	MINIMUM DIST. FROM SUN (AU)	ECCENTRICITY	INCLINATION TO ECLIPTIC (DEGREES)
Mercury	0.387	57.9	0.307	0.467	0.206	7.00
Venus	0.723	108.2	0.718	0.728	0.007	3.39
Earth	1.000	149.6	0.983	1.017	0.017	0.00
Mars	1.524	227.9	1.381	1.666	0.093	1.85
Jupiter	5.203	778.4	4.952	5.455	0.048	1.31
Saturn	9.537	1426.7	9.021	10.054	0.054	2.48
Uranus	19.191	2871.0	18.286	20.096	0.047	0.77
Neptune	30.069	4498.3	29.811	30.327	0.009	1.77
Pluto	39.482	5906.4	29.658	49.305	0.249	17.14

TABLE 20. *Physical data for the planets.*

PLANET	EQUATORIAL DIAMETER[a,b] (km)	MASS (EARTH = 1)	VOLUME (EARTH = 1)	MEAN DENSITY (10³ kg m⁻³)	OBLATENESS	SURFACE GRAVITY[b] (EARTH = 1)	ESCAPE VELOCITY (km s⁻¹)	GEOMETRICAL ALBEDO	COLOUR INDEX B−V
Mercury	4879	0.06	0.06	5.43	0	0.378	4.44	0.11	0.93
Venus	12104	0.82	0.86	5.24	0	0.907	10.36	0.65	0.82
Earth	12756	1.00	1.00	5.52	0.0034	1.000	11.19	0.37	—
Mars	6792	0.11	0.15	3.94	0.0059	0.377	5.03	0.15	1.36
Jupiter	142984	317.83	1321	1.33	0.065	2.364	59.5	0.52	0.83
Saturn	120536	95.16	764	0.69	0.098	0.916	35.5	0.47	1.04
Uranus	51118	14.54	63	1.27	0.023	0.889	21.3	0.51	0.56
Neptune	49528	17.15	58	1.64	0.017	1.125	23.5	0.41	0.41
Pluto	2390	0.0022	0.0066	1.8	0	0.067	1.3	0.3	0.80

[a] Polar diameters: Earth 12714 km, Mars 6752 km, Jupiter 133708 km, Saturn 108728 km, Uranus 49946 km, Neptune 48682 km.

[b] At the 1-bar level in the atmosphere for Jupiter, Saturn, Uranus and Neptune.

TABLE 21. *Planetary satellites.**

PLANET AND SATELLITE	MEAN DISTANCE FROM CENTRE OF PRIMARY		ORBITAL PERIOD (d)	INCLINATION[a] (DEGREES)	ECCENTRICITY[b]	DIAMETER (km)	MASS (PLANET = 1)	DENSITY ($10^3\,\mathrm{kg\,m^{-3}}$)	GEOMETRIC ALBEDO	MEAN OPPOSITION MAGNITUDE
	(10^3 km)	(PLANETARY RADII)								
EARTH										
Moon	384.4	60.27	27.3217	5.16	0.055	3475	0.0123	3.34	0.12	−12.7
MARS										
I Phobos	9.38	2.76	0.319	1.0	0.015	27 × 22 × 18	1.65×10^{-8}	1.87	0.07	11.3
II Deimos	23.46	6.91	1.262	0.9–2.7	0.000	15 × 12 × 10	3.71×10^{-9}	2.25	0.08	12.4
JUPITER										
XVI Metis	128	1.79	0.295	0.02	0.001	43	0.5×10^{-10}	3.0?	0.05	17.5
XV Adrastea	129	1.80	0.298	0.05	0.002	20 × 16 × 14	0.1×10^{-10}	3.0?	0.05	19.1
V Amalthea	181	2.53	0.498	0.4	0.003	250 × 146 × 128	38×10^{-10}	0.86	0.07	14.1
XIV Thebe	222	3.11	0.675	0.8	0.015	116 × 98 × 84	4×10^{-10}	3.0?	0.04	15.7
I Io	422	5.90	1.769	0.04	0.004	3643	4.70×10^{-5}	3.53	0.63	5.0
II Europa	671	9.39	3.551	0.47	0.009	3124	2.53×10^{-5}	3.01	0.67	5.3
III Ganymede	1070	15.0	7.155	0.21	0.002	5265	7.80×10^{-5}	1.94	0.44	4.6
IV Callisto	1883	26.3	16.689	0.51	0.007	4819	5.67×10^{-5}	1.83	0.20	5.7
XVIII Themisto	7387	103.3	129.71	45.67	0.204	8	—	2.6?	0.03?	21.7
XIII Leda	11127	155.6	239.79	27.47	0.179	10	0.03×10^{-10}	2.6?	0.07	20.2
VI Himalia	11480	160.6	250.57	27.63	0.158	170	50×10^{-10}	2.6?	0.03	14.8
X Lysithea	11686	163.5	258.07	27.35	0.141	24	0.4×10^{-10}	2.6?	0.06	18.4
VII Elara	11737	164.2	259.65	24.77	0.207	80	4×10^{-10}	2.6?	0.03	16.8
XII Ananke	21269	297.5	633.68 (R)	150.53	0.358	20	0.2×10^{-10}	2.6?	0.06	18.9
XI Carme	23350	326.6	728.93 (R)	164.95	0.223	30	0.5×10^{-10}	2.6?	0.06	18.0
VIII Pasiphae	23500	328.7	735 (R)	145	0.378	36	1×10^{-10}	2.6?	0.10	17.0
IX Sinope	23700	331.5	758 (R)	153	0.275	28	0.4×10^{-10}	2.6?	0.05	18.3
XVII Callirrhoe	24314	340.1	774.55 (R)	143.15	0.125	8	—	2.6?	0.06?	20.8
SATURN										
XVIII Pan	133.58	2.22	0.575	0.0	0.000	20	—	0.63?	0.5?	19.4
XV Atlas	137.67	2.28	0.602	0.3	0.000	37 × 34 × 27	—	0.63?	0.8?	18?
XVI Prometheus	139.35	2.31	0.613	0.0	0.003	148 × 100 × 68	—	0.63?	0.5?	16?
XVII Pandora	141.70	2.35	0.629	0.0	0.004	110 × 88 × 62	—	0.63?	0.7?	16?
XI Epimetheus	151.42	2.51	0.694	0.34	0.009	138 × 110 × 110	9.5×10^{-10}	0.61	0.8?	15?
X Janus	151.47	2.51	0.695	0.14	0.007	194 × 190 × 154	3.38×10^{-9}	0.66	0.9?	14?
I Mimas	185.52	3.08	0.942	1.53	0.020	397	6.6×10^{-8}	1.17	0.5	12.9
II Enceladus	238.02	3.95	1.370	0.00	0.005	499	1×10^{-7}	1.33	1.0	11.7
III Tethys	294.66	4.89	1.888	1.86	0.000	1060	1.1×10^{-6}	0.99	0.9	10.2
XIII Telesto	294.66	4.89	1.888	1.16	0.001	30 × 25 × 15	—	1.0?	1.0?	18.5?
XIV Calypso	294.66	4.89	1.888	1.47	0.001	30 × 16 × 16	—	1.0?	1.0?	18.7?
IV Dione	377.40	6.26	2.737	0.02	0.002	1120	1.93×10^{-6}	1.50	0.7	10.4
XII Helene	377.40	6.26	2.737	0.0	0.005	36 × 32 × 30	—	1.5?	0.7?	18?
V Rhea	527.04	8.74	4.518	0.35	0.001	1528	4.06×10^{-6}	1.24	0.7	9.7
VI Titan	1221.8	20.27	15.945	0.33	0.029	5150[c]	2.37×10^{-4}	1.88	0.22	8.3
VII Hyperion	1481.1	24.58	21.277	0.43	0.104	328 × 260 × 214	4×10^{-8}	1.1	0.3	14.2
VIII Iapetus	3561.3	59.09	79.330	14.72	0.028	1436	2.8×10^{-6}	1.27	0.05–0.5	10.2–11.9
XXIV Kiviuq	11205	185.9	442.80	48.74	0.154	14?	—	2.3?	0.06?	22.7
XXII Ijiraq	11430	189.7	456.22	49.10	0.364	10?	—	2.3?	0.06?	23.1
IX Phoebe	12952	214.9	550.48 (R)	177	0.163	220	7×10^{-10}	1.3	0.06	16.5
XX Paaliaq	14943	247.9	681.98	47.24	0.464	20?	—	2.3?	0.06?	21.7
XXVII Skadi	15755	261.4	738.32 (R)	148.51	0.206	6?	—	2.3?	0.06?	24.1
XXI Tarvos	17207	285.5	842.71	34.86	0.619	12?	—	2.3?	0.06?	22.7
XXV Mundilfari	18131	300.8	911.50 (R)	169.41	.284	6?	—	2.3?	0.06?	24.4
XXVIII Erriapo	18160	301.3	913.64	33.50	0.625	8?	—	2.3?	0.06?	23.5
XXX Thrym	20295	336.7	1079.46 (R)	174.98	0.513	6?	—	2.3?	0.06?	24.4

[a] Orbital inclinations are relative to the planet's equator, except for the Moon and Phoebe, which are relative to the ecliptic. The Moon's inclination relative to the Earth's equator ranges from 18°.28 to 28°.58. The inclination of the Moon's equator to the ecliptic is constant at 1°.54.

[b] The orbital elements of the outer satellites, particularly their eccentricities, are subject to considerable perturbations.

  R = retrograde.

* There are in addition a number of small satellites of Jupiter, Saturn, Uranus and Neptune whose orbital elements are not well determined.

TABLE 21 (continued). *Planetary satellites.* *

PLANET AND SATELLITE		MEAN DISTANCE FROM CENTRE OF PRIMARY $(10^3\,km)$	(PLANETARY RADII)	ORBITAL PERIOD (d)	INCLINATION (DEGREES)	ECCENTRICITY[a]	DIAMETER (km)	MASS (PLANET = 1)	DENSITY $(10^3\,kg\,m^{-3})$	GEOMETRIC ALBEDO	MEAN OPPOSITION MAGNITUDE
URANUS											
VI	Cordelia	49.77	1.95	0.335	0.08	0.000	26	—	1.3?	0.07?	24.1
VII	Ophelia	53.79	2.10	0.376	0.10	0.010	30	—	1.3?	0.07?	23.8
VIII	Bianca	59.17	2.32	0.435	0.19	0.001	42	—	1.3?	0.07?	23.0
IX	Cressida	61.78	2.42	0.464	0.01	0.000	62	—	1.3?	0.07?	22.2
X	Desdemona	62.68	2.45	0.474	0.11	0.000	54	—	1.3?	0.07?	22.5
XI	Juliet	64.35	2.52	0.493	0.07	0.001	84	—	1.3?	0.07?	21.5
XII	Portia	66.09	2.59	0.513	0.06	0.000	108	—	1.3?	0.07?	21.0
XIII	Rosalind	69.94	2.74	0.558	0.28	0.000	54	—	1.3?	0.07?	22.5
XIV	Belinda	75.26	2.94	0.624	0.03	0.000	66	—	1.3?	0.07?	22.1
XV	Puck	86.01	3.37	0.762	0.32	0.000	154	—	1.3?	0.075	20.2
V	Miranda	129.39	5.06	1.413	4.2	0.003	472	0.08×10^{-5}	1.20	0.27	16.3
I	Ariel	191.02	7.47	2.520	0.3	0.003	1158	1.55×10^{-5}	1.67	0.35	14.2
II	Umbriel	266.30	10.42	4.144	0.36	0.005	1169	1.35×10^{-5}	1.40	0.19	14.8
III	Titania	435.91	17.06	8.706	0.14	0.002	1578	4.06×10^{-5}	1.71	0.28	13.7
IV	Oberon	583.52	22.83	13.463	0.10	0.001	1523	3.47×10^{-5}	1.63	0.25	13.9
XVI	Caliban	7170.4	280.5	579.6 (R)	139.8	0.081	60?	—	1.5?	0.07?	22.4
XX	Stephano	7942.4	310.7	675.71 (R)	141.5	0.146	22?	—	1.5?	0.07?	24.6
XVII	Sycorax	12216	478.0	1289 (R)	152.7	0.512	120?	—	1.5?	0.07?	20.9
XVIII	Prospero	16089	629.5	1948.13 (R)	146.3	0.328	32?	—	1.5?	0.07?	23.7
XIX	Setebos	17988	703.8	2303 (R)	148.3	0.512	30?	—	1.5?	0.07?	23.8
NEPTUNE											
III	Naiad	48.23	1.95	0.294	4.74	0.000	58	—	1.3?	0.06?	24.7
IV	Thalassa	50.07	2.02	0.311	0.21	0.000	80	—	1.3?	0.06?	23.8
V	Despina	52.53	2.12	0.335	0.07	0.000	148	—	1.3?	0.06	22.6
VI	Galatea	61.95	2.50	0.429	0.05	0.000	158	—	1.3?	0.06	22.3
VII	Larissa	73.55	2.97	0.555	0.20	0.001	208×178	—	1.3?	0.06	22.0
VIII	Proteus	117.65	4.75	1.122	0.04	0.000	$436 \times 416 \times 402$	—	1.3?	0.06	20.3
I	Triton	354.76	14.33	5.877 (R)	157.35	0.000	2705[b]	2.09×10^{-4}	2.07	0.77	13.5
II	Nereid	5513.4	222.6	360.136	27.6	0.751	340	2×10^{-7}	1.5?	0.4	18.7
PLUTO											
I	Charon	19.6	16.4	6.387 (R)	99	0.00	1186	0.125	1.85	0.5	16.8

Sources: The Astronomical Almanac *and* Jet Propulsion Laboratory Solar System Dynamics Group.

[a] *The orbital elements of the outer satellites, particularly their eccentricities, are subject to considerable perturbations.*

[b] *Diameter of solid body; diameter at cloud top is 5550 km.*

R = *retrograde.*

* *There are in addition a number of small satellites of Jupiter, Saturn, Uranus and Neptune whose orbital elements are not well determined.*

of a planet can vary widely, depending on its distance from both the Earth and the Sun, and also – for an inferior planet – on its phase. Table 18 gives the magnitudes of the superior planets when at mean opposition distance; the actual values can vary somewhat depending on whether the planet and the Earth are near perihelion or aphelion at the time of opposition (for example, Mars at its best can outshine Jupiter). The brightness of Saturn is strongly affected by the orientation of its rings; when they are tilted towards us at their maximum angle the planet appears over twice as bright as it does when the rings are presented edge-on.

The values given for the inferior planets, Mercury and Venus, are the magnitudes at mean greatest elongation; they depend on

the actual distances of the planets from the Sun at the time. At mean greatest elongation Mercury and Venus are seen exactly half-illuminated, but this is not the time when they are at their brightest. Mercury attains its maximum magnitude around the time of superior conjunction, when its fully illuminated side is turned towards us; however, this is of only theoretical interest since at these times Mercury is almost impossible to see, being obscured by the Sun's glare.

Venus, on the other hand, is impossible to miss at greatest brilliancy, when it outshines all objects except the Moon and the Sun. This happens between the times of inferior conjunction and greatest elongation, when the planet shows a crescent phase. On these

TABLE 22. *Planetary ring systems.*

| PLANET AND RING | MEAN DISTANCE FROM CENTRE OF PLANET | | WIDTH (km) |
	(10^3 km)	(PLANETARY RADII)	
Jupiter			
Halo	100.0–122.8	1.40–1.72	22800
Main ring	122.8–129.2	1.72–1.81	6400
Gossamer ring	129.2–214.2	1.81–3.0	85000
Saturn			
D ring	67.0–74.5	1.11–1.24	7500
C ring	74.5–92.0	1.24–1.53	17500
B ring	92.0–117.5	1.53–1.95	25500
Cassini Division	117.5–122.2	1.95–2.03	4700
A ring	122.2–136.8	2.03–2.27	14600
F ring	140.4	2.33	30–500
G ring	170.0	2.82	8000
E ring	180–480	3–8	300000
Uranus			
Rings	38.0–51.1	1.49–2.00	
Neptune			
Galle	41.9	1.69	15
Le Verrier	53.2	2.15	15
Lassell	55.4	2.24	—
Arago	57.6	2.33	—
Adams	62.9	2.54	<50

Source: USGS Gazetteer of Planetary Nomenclature.

occasions Venus is nearer the Earth than it is at greatest elongation, which more than compensates for its reduced phase. The contrasting behaviour of Mercury and Venus in this respect is accounted for by their very different visible surfaces: bare rock for Mercury, highly reflective cloud for Venus.

Both Mercury and Venus go through a complete cycle of phases from new (at inferior conjunction) to full (at superior conjunction) and back again as they orbit the Sun. The superior planets do not show such a cycle of phases; at most they can appear slightly gibbous around the time of quadrature. The effect is most noticeable with Mars, which can show a phase as pronounced as 84% at quadrature.

The planets are always to be found close to the ecliptic. Hence any bright 'star' near the ecliptic that does not appear on the maps in this Atlas will be a planet (although it could, just possibly, be a nova or a supernova). The coordinates of the planets for any particular night can be found from an almanac, and can then be plotted on the appropriate charts in this Atlas.

Data for the orbits of the major planets are given in Table 19, and physical data for them in Table 20. Table 21 lists the planetary satellites for which orbits are well established, while Table 22 lists planetary ring systems.

Mercury

Mercury is the innermost planet, and the least conspicuous of those visible to the naked eye. It is only 40% greater in diameter than our Moon. Mercury orbits the Sun in 88 d and spins on its axis in 58.6 d, exactly two-thirds of its orbital period.

As seen from the Earth, Mercury's apparent movement consists of a periodic oscillation of small amplitude from one side of the Sun to the other, so that it is always close to the Sun, preceding or following it by no more than 2¼ hours. It can be glimpsed without a telescope, near the horizon in the twilight at dusk or dawn. Owing to the high eccentricity of its orbit, the elongation east or west of the Sun varies considerably, ranging from 17° 50′ at perihelion to a maximum of 27° 50′ at aphelion. Mercury's actual distance from the Sun ranges from 46 million km at perihelion to 70 million km at aphelion.

From north and south temperature latitudes, Mercury is best observed in the evening sky in the spring (eastern elongation) or in the morning sky in the autumn (western elongation). Unfortunately Mercury is near perihelion during its most favourable elongations for northern observers; elongations near aphelion occur when the planet is south of the celestial equator, so observers in the southern hemisphere have the best opportunity to study the planet.

Normally, Mercury is a naked-eye object only near greatest elongation. The planet usually appears about ten days before greatest elongation, disappearing six to seven days later as it recedes into the glare of the Sun. The same intervals, in reverse order, apply to western elongation. In the tropics, where the ecliptic is almost overhead and twilight is of brief duration, Mercury is regularly seen without optical aid. But in higher latitudes it is more fugitive, because the inclination of the ecliptic to the horizon is so acute that even at greatest elongation it is difficult for anyone in a built-up area to spot.

To find Mercury requires a clear sky when the planet is near its greatest elongation, and a flat, unobstructed horizon. The best method is to sweep the area with binoculars. Once located, Mercury will then become a relatively easy naked-eye object, scintillating like a bright star – for which it might be mistaken – at a low altitude about half an hour immediately after sunset or before sunrise.

Movements of Mercury

The Earth is moving in the same orbital direction as Mercury, though less rapidly, and the two planets return to the same relative position after 116 days. This is the mean synodic period of Mercury (the interval between successive inferior conjunctions). In one terrestrial year there are up to seven elongations of Mercury – three or four in the evening, and three or four in the morning.

Starting at greatest elongation east, Mercury is well seen after sunset. It drops back towards the Sun, and reaches inferior conjunction when its phase is new and it cannot be seen. Mercury then moves away from the Sun into the morning sky until it reaches greatest elongation west. The interval from greatest elongation east via inferior conjunction to greatest elongation west is 44 days. After the morning apparition Mercury turns back towards the Sun, swings behind it and sweeps through superior conjunction, when it is directly opposite the Earth, at full phase, but lost in the glare of the Sun. It then reappears in the evening sky and returns to greatest elongation east. The interval from greatest elongation west to greatest elongation east, through superior conjunction, is about 72 days.

At full phase Mercury subtends an apparent angular diameter of around 4.5 arcsec. This increases to as much as 13.0 arcsec at inferior conjunction but, as already mentioned, Mercury cannot be seen at these positions. At greatest elongation, when about half the illuminated hemisphere is turned towards the Earth, it subtends about 7.0 arcsec. Mercury's magnitude at greatest elongation varies from −0.7 to +0.7, a result of its changing distance from Earth and phase, which can range from 37% to 64% on such occasions.

Transits of Mercury

If its orbit were exactly in the plane of the ecliptic, Mercury would *transit* (pass in front of) the Sun once in each synodic revolution. But since its orbit is inclined 7° to the ecliptic, the planet usually passes above or below the Sun as seen from the Earth. A transit occurs only when Mercury is near one of its nodes at inferior conjunction. The ascending node corresponds to the position of the Earth on November 10, the descending node to the position on May 8. If inferior conjunction falls near one of these dates, Mercury will be seen as a tiny black spot, 10–12 arcsec wide, moving slowly across the face of the Sun (see Figure 24).

November transits are more common, repeating at intervals of 7, 13 and 46 years, according to circumstances. May transits recur at intervals of 13 and 46 years. The last transit was on 2003 May 7.

FIGURE 24. *A transit of Mercury. The silhouetted disk of Mercury near the limb of the Sun at the transit of 2003 May 7. Source: The Royal Swedish Academy of Sciences.*

Those for the remainder of the 21st century are listed in Table 23. The length of the chord traversed by the planet in its passage across the Sun's disk determines the duration of a transit, which can last for up to 9 hours.

Timings of the instants of ingress and egress of the planet can be made. The times required are those of first contact (ingress of preceding limb of planet), second contact (ingress of following limb), third contact (egress of preceding limb) and fourth contact (egress of following limb). Timing may be difficult because of the so-called *black drop* that links the silhouetted following limb of the planet with the Sun's limb for a few seconds after apparent second contact; a similar effect may be seen a few seconds before third contact. The black drop is caused by the combined effects of atmospheric seeing, diffraction in the telescope and the limb-darkening of the Sun; it is also seen at transits of Venus.

Observing Mercury

Mercury is best observed by telescope in daylight, since when it is visible to the naked eye it will be close to the horizon, and viewed under unsatisfactory conditions. But locating the planet in full daylight is fraught with difficulty, and the would-be observer must adopt the most stringent safety precautions so as not to meet the blinding glare of the Sun. Daylight observation should be undertaken only if the observer has access to a well-adjusted equatorial fitted with accurate setting circles. Even then success is not guaranteed, chiefly because of the ever-present glare of the Sun and the bad seeing in daylight hours.

While experienced observers using apertures as small as 100 mm have been able to identify the principal surface features, including bright spots which are coincident with the bright ray

TABLE 23. *Forthcoming transits of Mercury in the 21st century.*

DATE	DURATION	DATE	DURATION
2006 November 8	4h 58m	2062 May 10	6h 41m
2016 May 9	7h 30m	2065 November 11	5h 24m
2019 November 11	5h 29m	2078 November 14	3h 57m
2032 November 13	4h 26m	2085 November 7	3h 44m
2039 November 7	2h 57m	2095 May 8	7h 30m
2049 May 7	6h 41m	2098 November 10	5h 22m
2052 November 9	5h 13m		

Source: Fred Espenak, NASA/Goddard Space Flight Center.

systems photographed by the U.S. probe Mariner 10, the general experience is one of frustration and disappointment. The poor conditions under which the planet is observed and its small size are usually blamed, but the real reason is that small apertures are quite incapable of resolving the soft smudges that mottle the surface of the planet. Observers with exceptionally good eyesight have achieved some remarkable results with small telescopes, but for most observers an aperture of at least 200 mm is necessary for serious study. When used with digital imaging equipment, such an instrument is capable of capturing albedo features on the tiny disk. Given a good instrument and a reasonable observing site, Mercury can be a challenging and rewarding subject.

The Surface of Mercury

Mariner 10 photographed nearly half of Mercury's surface in 1974–75, revealing a heavily cratered, lunar-like terrain, with one huge impact basin, the Caloris Basin, 1300 km across. There are extensive scarps which suggest that the planet contracted during its formation. The surface temperature can vary from −180°C at night to over 400°C in the daytime.

Venus

Venus, the brightest of the planets, is the second in order from the Sun, and easier to observe than Mercury. It is often visible in full daylight, and when at its greatest brilliancy it will cast a distinct shadow under clear, dark skies away from artificial lights.

Of all the planets Venus is the most like the Earth in size, with a diameter of 12 100 km, 95% of the Earth's diameter. It can come closer to the Earth than any other planet, within 40 million km, yet we can see nothing of its surface because the planet is permanently covered with cloud. This cloud gives Venus an albedo of 65%, greater than for any other planet in the Solar System, which in conjunction with its proximity accounts for the planet's brilliance in our skies.

Venus spins on its axis in a retrograde direction in 243 d, longer than its sidereal period of 224.7 d.

Movements of Venus

Inferior conjunctions of Venus occur at intervals of 584 days; greatest elongations occur about 72 days before and after inferior conjunction, when the planet is 45° to 47° from the Sun. The interval between eastern and western elongations is thus a little over 20 weeks, and that between western and eastern elongations is about 63 weeks.

At superior conjunction, when Venus shows full phase, the apparent diameter is only about 10 arcsec; at inferior conjunction it is over 65 arcsec. At greatest elongation, when the phase is 50% (±1%), the apparent diameter is about 24 arcsec. Greatest brilliancy

occurs about 36 days from inferior conjunction – that is, about five weeks after greatest elongation east and five weeks before greatest elongation west. Venus is then the brightest object in the sky, after the Sun and Moon, reaching a magnitude of up to −4.7. At greatest brilliancy, the planet has a phase of around 27%. The relative sizes of Venus at different phases are shown in Figure 25.

Maximum brightness occurs about every eight years, when the planet reaches perihelion at the end of December. At such times it is south of the equator and is best seen as a morning star in the southern hemisphere. The most favourable time for observers in the northern hemisphere is when Venus reaches perihelion in mid-March; although the planet is then rather fainter than at a December perihelion, it is much higher in the sky.

The phases of Venus are visible in any small telescope or good binoculars. There are even reports of the crescent being seen with the naked eye. But in general the phase is all there is to see.

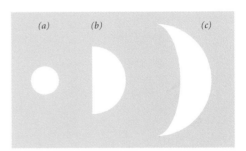

FIGURE 25. *Phases of Venus as it orbits the Sun. (a) Full, at superior conjunction. The planet is on the far side of the Sun from Earth, but cannot be seen because of solar glare. (b) Half phase (dichotomy). The planet reaches half phase at, or close to, the time of greatest elongation. (c) Crescent. The planet appears at its maximum brilliancy in our skies when at a phase of 27%, as shown here. The phases repeat in reverse order after the planet has passed through inferior conjunction (between Sun and Earth). In this diagram, the relative sizes of the planet at each phase are shown to scale.*

Observing Venus

Venus, like Mercury, is probably best observed in daylight, although excellent views can be obtained in bright twilight. Neither planet can be regarded as a good object to observe. Venus shows a much larger disk, but telescopes reveal only a thick cloud blanket upon which the practised eye may detect faint patterns of shading with brighter areas near the cusps.

The markings originate in the upper level of the cloud deck, about 65 km above the surface. They are prominent in the ultraviolet and show variations, often over the course of a few hours; some spots, however, last for many days. From detailed studies of Y-shaped dark markings centred on the equator it has been found that the upper clouds rotate in a retrograde direction around the planet every 4 days or so. A careful drawing should be made of all markings seen. It is usual to draw Venus with a full-disk diameter of 50 mm.

From time to time areas of the planet appear to brighten; any bright patches seen should be accurately recorded. The most contrasty and prominent markings often appear at the horns of the crescent (the *cusps*). These generally take the form of bright patches known as *cusp caps*, occasionally surrounded by dark collars or bands. The cusp caps and dark collars show fluctuations in size and brightness.

The terminator of Venus is often irregular in shape. Sometimes part of the terminator appears to be flat rather than curved at both crescent and gibbous phases, while near *dichotomy* (50% phase) the northern half of the terminator may appear concave, and the southern half convex. Such effects may arise from cloud features near the terminator.

It has long been known that the observed phase of Venus differs from that predicted. This is known as the *Schröter effect*, after the German astronomer Johann Schröter, who described it in the 1790s. In particular it has been found that the calculated date of dichotomy does not agree with the date on which Venus is seen to have a straight terminator. Dichotomy at evening elongation is normally earlier than predicted, while morning dichotomy is later. The cause is unknown. Some authorities attribute it to observational error, others to atmospheric effects.

Another phenomenon that has not been objectively proven is the *ashen light*, in which the unilluminated (night) side of the planet seems to glow with a feeble luminosity. This is usually seen when Venus is a fairly narrow crescent, near inferior conjunction. It is difficult not to be subjective in observing this appearance, for the eye tends to connect the two cusps, giving the impression that the entire disk can be seen when in fact it cannot. The most useful observations of the ashen light are those made with an occulting bar in the telescope field, so that the illuminated crescent is hidden.

Colour filters are often used in the observation of Venus. Such observations are of value only if carried out on a carefully controlled basis, and are the province of the experienced observer. Ultraviolet imaging of the cloud markings can be done with digital equipment and suitable filters, using reflectors or Schmidt–Cassegrains with apertures of 200 mm or over.

Transits of Venus

These are very rare, and occur when inferior conjunction is within a few days of June 7 (descending node) or December 8 (ascending node). Five synodic periods of Venus total almost 8 years, so transits take place in pairs 8 years apart. Such pairs occur at intervals of 105.5 and 121.5 years alternately; the last pair took place in 1874 and 1882. The next pair of June transits occur on 2004 June 8 and 2012 June 5/6 (with a duration of over 6 hours in each case), and the next December pair in 2117 and 2125.

As with transits of Mercury, timings of second and third contact are affected by the black drop effect. In addition, the atmosphere of Venus may be seen as a bright ring of light surrounding the black disk of the planet during the transit, extending beyond the Sun's limb between first and second contacts and between third and fourth contacts. This is a result of the refraction of sunlight in the planet's atmosphere.

The Surface and Atmosphere of Venus

Soviet lander probes have returned images direct from the surface of Venus, and U.S. and Soviet orbiters have mapped the surface by radar to reveal a world of vast plains, slashed by a deep chasm, scarred by extensive cratering, and dominated by two major uplands which tower prodigiously in one place to 11 km above the average level of the surface.

The surface temperature is over 400°C, and the atmospheric pressure is 90 times greater than on Earth at sea level. The atmosphere is chiefly composed of carbon dioxide, which creates a greenhouse effect, while the dazzling clouds contain large quantities of corrosive sulphuric acid. Wind speeds in the upper atmosphere exceed 350 km h^{-1} but at the surface they are only about 1% of that.

Mars

Mars is not an easy object to observe telescopically. Unlike the Moon, Jupiter and Saturn, the planet's small angular diameter makes it disappointing for the beginner. Even at its very best, when at perihelic opposition, the apparent diameter of Mars is only half that of Jupiter. However, observations may be made for many months either side of opposition, the exact duration of observability being governed by the experience of the observer and the size of telescope used. The planet's actual diameter is just over half that of the Earth, and its axial rotation period is about half an hour longer than the Earth's.

The mean interval between oppositions of Mars is longer than for any other planet: 780 days, or a little under 26 months. Thus oppositions occur on average every other year. The distance of the planet at opposition varies between 101 million km for aphelic oppositions, which occur around February, and 56 million km for perihelic oppositions, which occur around August. Because of the inclination of the planet's orbit to the ecliptic, the date of closest approach of Mars to Earth may differ from the opposition date by several days. The apparent diameter of the Martian disk at the most favourable perihelic oppositions is 25 arcsec, whereas at aphelic oppositions it is just under 14 arcsec. At conjunction the apparent diameter is less than 4 arcsec.

When close to opposition the planet is a bright naked-eye object; at perihelic opposition Mars can reach a magnitude of −2.9, similar to Jupiter at its best, and even at aphelic opposition the magnitude is −1.0. Mars appears strongly orange in colour because its surface rocks contain large amounts of iron oxide.

Extremely favourable oppositions of Mars occur at intervals of 15 and 17 years, the last being 2003 August and the next in 2018 July. Oppositions of the planet until 2020 are listed in Table 24. Mars

TABLE 24. *Oppositions of Mars 2005–2020.*

DATE	DISTANCE FROM EARTH (10^6 km)	APPARENT EQUATORIAL DIAMETER (ARCSEC)	MAGNITUDE
2005 November 7	70.35	19.9	−2.3
2007 December 24	88.64	15.8	−1.6
2010 January 29	99.38	14.1	−1.2
2012 March 3	100.87	13.9	−1.2
2014 April 8	92.94	15.1	−1.5
2016 May 22	76.19	18.4	−2.1
2018 July 27	57.72	24.3	−2.8
2020 October 13	62.63	22.4	−2.6

can show a phase which is most pronounced near quadrature, when 84% of the disk is illuminated.

Mars has two tiny satellites, Phobos and Deimos. Both are faint, magnitudes 10.5 and 11.5 at best, and they never stray far from the overwhelming glare of their parent planet. To see them in a moderate aperture, Mars must be placed outside the field of view or be hidden by an occulting bar. Both satellites are irregular in shape, and spacecraft observations have shown that they appear to be captured asteroids.

The thin atmosphere of Mars is composed of about 95% carbon dioxide, 3% nitrogen and 1–2% argon, with a pressure of about 7 millibars at the surface, similar to the pressure of the Earth's atmosphere at an altitude of 30 km. Surface temperatures range from about +23°C down to −133°C.

Observing Mars

At perihelic opposition the south pole of Mars is tilted towards the Sun so that features in the southern hemisphere of the planet are more favourably presented for observation than are those in the north. At aphelic opposition the situation is reversed, the north pole then being tilted towards the Sun.

Bright and dark markings are visible on the orange surface of Mars, as are one or both of the white polar caps. Under favourable circumstances some markings may be visible in telescopes as small as 60 mm, but for serious study of the planet larger instruments are preferable, such as a 200-mm reflector.

The system of nomenclature for the bright and dark areas, using names taken from terrestrial geography and mythology, was devised in the nineteenth century by the Italian astronomer Giovanni Schiaparelli. Since then numerous observers have added to his work, culminating in the official chart of Mars published by the International Astronomical Union in 1957, which shows 128 named features. However, these features are simply differences in albedo and do not always coincide with the real topographic features – the craters, mountains and valleys recorded by spacecraft.

The U.S. Geological Survey has prepared charts showing the topographical surface features of Mars. Albedo charts like the one shown in Figure 26 (overleaf) place south at the top, which is the normal telescopic view, but the topographic maps place north at the top.

The exploration of Mars by spacecraft has meant that Mars is no longer the province of the telescopic observer. However, amateur astronomers continue to contribute to our knowledge of Mars because the surface and atmosphere of the planet undergo continual change.

CHANGES IN SURFACE MARKINGS The broad configuration of the albedo markings on Mars has remained more or less the same during the time that the planet has been under telescopic observation, but numerous small-scale changes occur from one opposition to the next. Certain features such as the dusky Pandorae Fretum and Hellespontus sometimes darken during the Martian spring to become quite prominent during the summer, fading again in winter. Several other features such as Syrtis Major and Solis Lacus have undergone subtle long-term changes over the years. These changes on Mars are due to surface dust being blown from place to place, uncovering some areas of rock and obscuring others.

WHITE CLOUDS AND HAZES Although only traces of water vapour are detectable in the thin Martian atmosphere, temperatures are so low that saturation frequently occurs and clouds of volatiles such as water and carbon dioxide ice are quite common. These are particularly noticeable near the polar caps, which may appear enlarged as a result. The general haziness that is commonly seen near the morning or evening terminator is a result of low-level mists and thin layers of surface frost. White clouds often form over the giant volcano Olympus Mons and the volcanoes on the Tharsis Ridge, particularly on Martian spring and summer afternoons. Cloud also frequently forms in the large impact basins such as Hellas.

DUST STORMS Sometimes called yellow clouds because of their colour, small dust storms of limited extent and duration may become visible as they obscure some familiar dark marking, revealing their nature by their movement or by the reappearance of the marking over the course of a few days. During perihelic oppositions, when Mars is closest to the Sun and therefore temperatures are highest, much larger and more spectacular dust storms may occur. Planet-encircling storms occurred in 1909, 1924, 1956, 1971, 1973, 1975, 1977, 1982 and 2001. These great dust storms originate in the southern hemisphere. It is important to document the times and positions of the storms as they spread, which can sometimes happen quite rapidly over a few days.

POLAR CAPS Often the most prominent changes visible on Mars are the seasonal growth and shrinking of the polar caps throughout a Martian year. As autumn arrives in one hemisphere

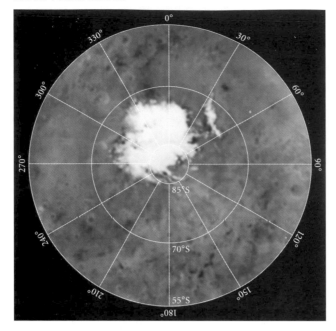

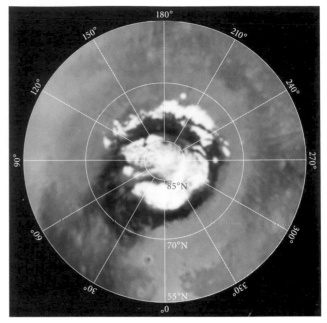

the cap expands under a veil known as the *polar hood*, which consists of clouds of carbon dioxide and water crystals; this hood persists throughout the winter. With the onset of spring the mantle of clouds disperses to reveal a brilliant white cap that shrinks rapidly as summer approaches, but never quite disappears, always leaving a small residual cap. The shrinking cap is bordered by a conspicuous dark collar. Dark rifts may appear in the cap, such as the Rima Borealis in the north polar cap. In addition, detached portions of the caps may appear, such as Novus Mons in the south polar cap. The seasonal caps are composed of carbon dioxide ice that sublimes into the atmosphere rather than melts, whereas the residual cap, in the northern hemisphere at least, is composed of water ice. Mars is at perihelion during summer in the southern hemisphere, and

the residual south polar cap shrinks to less than half the size of the northern one. Variations in weather patterns mean that the seasonal changes are never quite the same from year to year, and differences should be recorded.

METHODS OF OBSERVING   For any serious study of Mars, regular observations must be made throughout an apparition. Only in this way can the observer become familiar with the planet and thus recognize changes such as clouds or dust storms. For many years, drawings were the only practical means by which the amateur could record Martian detail, but the development of CCD imaging has made it possible for amateurs to take high-quality images of Mars through telescopes of 150 mm or over.

FIGURE 26 *(opposite). Albedo map of Mars, simulating the planet's appearance through Earth-based telescopes under the best conditions. The map was constructed from pictures taken by the Mars Global Surveyor spacecraft in 1999 (source: Malin Space Science Systems). The principal markings on Mars are listed below with their areographic longitude in degrees; these coordinates are only approximate since many markings are irregular and ill-defined.*

FEATURE	LONG. (W)	LAT.	FEATURE	LONG. (W)	LAT.	FEATURE	LONG. (W)	LAT.
Achilles Fons	50	30 N	Euxinus Lacus	155	40 N	Noachis	0	45 S
Aeolis	200	5 S	Ganges	65	10 N	Nodus Alcyonius	265	30 N
Aeria	310	10 N	Gehon	0	20 N	Olympia	200	80 N
Aetheria	230	45 N	Hellas	294	44 S	Ophir	65	10 S
Aethiopis	235	10 N	Hellespontica Depressio	340	60 S	Ortygia	350	65 N
Amazonis	140	10 N	Hellespontus	340	50 S	Oxia Palus	17	8 N
Amenthes	255	5 N	Hephaestus	245	20 N	Oxus	15	30 N
Aonius Sinus	105	45 S	Hesperia	225	20 S	Panchaïa	200	65 N
Arabia	330	30 N	Hyperboreus Lacus	50	75 N	Pandorae Fretum	350	25 S
Araxes	130	25 S	Iapygia	290	15 S	Phaethontis	140	45 S
Arcadia	100	50 N	Icaria	115	35 S	Phasis	110	30 S
Argyre	42	51 S	Idaeus Fons	50	35 N	Phison	320	15 N
Arnon	330	48 N	Isidis Regio	265	20 N	Phoenicis Lacus	100	10 S
Aurorae Sinus	50	10 S	Ismenius Lacus	330	40 N	Phrixi Regio	65	30 S
Ausonia	250	45 S	Jamuna	45	5 S	Promethei Sinus	280	65 S
Baltia	80	65 N	Juventae Fons	63	5 S	Propontis	180	40 N
Boreosyrtis	275	50 N	Lemuria	200	75 N	Pyrrhae Regio	30	15 S
Callirrhoe	345	58 N	Libya	265	0	Scandia	150	65 N
Candor	73	3 N	Lunae Lacus	70	15 N	Sinaï	75	15 S
Casius	260	40 N	Maeotis Palus	125	63 N	Sinus Meridiani	0	5 S
Castorius Lacus	157	52 N	Mare Acidalium	30	45 N	Sinus Sabaeus	340	10 S
Cebrenia	210	45 N	Mare Australe	40	65 S	Solis Lacus	90	28 S
Cecropia	315	65 N	Mare Boreum	60	60 N	Styx	195	20 N
Ceraunius	100	35 N	Mare Chronium	225	58 S	Syria	90	12 S
Cerberus	215	13 N	Mare Cimmerium	210	20 S	Syrtis Major	290	10 N
Chalce	15	50 S	Mare Erythraeum	40	25 S	Tanaïs	70	50 N
Chersonesus	280	60 S	Mare Hadriacum	270	45 S	Tempe	70	35 N
Chryse	25	10 N	Mare Serpentis	315	30 S	Tharsis	100	5 N
Chrysokeras	100	55 S	Mare Sirenum	155	30 S	Thaumasia	80	30 S
Claritas	97	20 S	Mare Tyrrhenum	240	30 S	Thyle I	155	70 S
Coprates	65	13 S	Margaritifer Sinus	17	5 S	Thyle II	210	70 S
Cyclopia	225	5 S	Memnonia	150	15 S	Thymiamata	13	5 N
Cydonia	350	50 N	Meroe	285	35 N	Tithonius Lacus	85	5 S
Daedalia	120	30 S	Moab	350	20 N	Tractus Albus	90	15 N
Deltoton Sinus	310	10 S	Moeris Lacus	270	8 N	Trivium Charontis	195	13 N
Deucalionis Regio	345	15 S	Nectar	70	20 S	Uchronia	240	70 N
Deuteronilus	350	40 N	Neith Regio	270	35 N	Umbra	290	50 N
Diacria	180	45 N	Nepenthes	250	15 N	Utopia	255	50 N
Dioscuria	315	50 N	Nereidum Fretum	55	40 S	Vulcani Pelagus	25	32 S
Edom	345	3 S	Niliacus Lacus	30	30 N	Xanthe	45	10 N
Electris	190	45 S	Nilokeras	55	30 N	Yaonis Regio	325	45 S
Elysium	210	20 N	Nilosyrtis	285	40 N	Zephyria	190	5 S
Eridania	210	45 S	Nix Olympica	133	18 N			

For most amateurs, though, the usual method of recording detail is to make a drawing of the disk on a blank 50 mm in diameter. Sketch in the main details and then make a note of the time (UT) before entering the finer features. The longitude of the central meridian should be added, obtainable from tables in publications such as the *Handbook* of the British Astronomical Association or *The Astronomical Almanac*. It is also worth estimating the relative intensity of the visible features on a scale of 0 for the bright polar caps to 10 for the black of the night sky, 2 being the usual brightness of most desert areas.

Colour filters can be useful for emphasizing particular features on Mars. The established range of filters is the Kodak Wratten series (see Figure 10, p. 35). The 15 yellow and 25 red filters will enhance the visibility of the dark markings, and the 25 will assist in the recognition of dust storms, through which they appear brighter. The 44A blue, 47 violet and 58 green filters aid the detection of white clouds and surface frosts by making them appear brighter than in white light. Usually the dark areas are very difficult to see through the 47 filter, but on rare occasions they can be seen quite well. Filter observations should be made only under conditions of good seeing and with telescopes of suitable size. The 47 filter, for example, is too dense for visual use with telescopes under 200 mm aperture. CCD images should be made separately through matched red, green and blue filters so that colour composites can be assembled.

The Surface of Mars

From space missions we know that most of the southern hemisphere of Mars consists of impact-cratered highlands, whereas the northern hemisphere tends to be smoother and 1–2 km lower. The albedo markings seen through telescopes do not always coincide with topographical features – the dark Syrtis Major, for instance, corresponds to an unremarkable slope at the edge of higher ground. Some bright areas, such as Hellas and Argyre, are deep impact basins (the bottom of Hellas is the lowest area on the planet, more than 4 km deep) whereas the bright markings Elysium and Nix Olympica (corresponding to the volcanic structure Olympus Mons) are areas of high volcanoes. Part of the Valles Marineris, a giant rift valley system 4 km deep and 4000 km long, is visible from Earth as the dark streak Coprates.

Jupiter

Jupiter is the most rewarding planet for amateur astronomers to observe. As well as being physically the largest planet, with an equatorial diameter of 142984 km, for most of the time it is also the one with the largest disk size – always more than 30 arcsec in diameter, and ranging from 44 to 50 arcsec at opposition.

Its mean synodic period is 399 d (just over 13 months), so that oppositions occur about one month later each year. Jupiter's distance from Earth varies from about 670 million km at aphelic oppositions to 590 million km at perihelic oppositions, which occur about every 12 years, in September or October. The opposition magnitude ranges between −2.3 and −2.9. Apart from one or two months either side of conjunction, Jupiter can be observed for most of the year. Oppositions of Jupiter until 2020 are listed in Table 25.

The planet's visible surface is the top of a deep and cloudy atmosphere that consists mainly of hydrogen and helium, with substantial amounts of ammonia, methane and simple hydrocarbons. The clouds are believed to consist largely of ammonia crystals, ammonium hydrosulphide and water ice. The cloud layers are marked by dark belts and bright zones that run parallel to lines of latitude.

TABLE 25. *Oppositions of Jupiter 2004–2020.*

DATE	DISTANCE FROM EARTH (10^6 km)	APPARENT EQUATORIAL DIAMETER (ARCSEC)	MAGNITUDE
2004 March 4	662.1	44.5	−2.5
2005 April 3	666.7	44.2	−2.5
2006 May 4	660.1	44.7	−2.5
2007 June 5	644.0	45.8	−2.6
2008 July 9	622.5	47.4	−2.7
2009 August 14	602.6	48.9	−2.9
2010 September 21	591.5	49.9	−2.9
2011 October 29	594.3	49.6	−2.9
2012 December 3	609.0	48.4	−2.8
2014 January 5	630.0	46.8	−2.7
2015 February 6	650.1	45.4	−2.6
2016 March 8	663.4	44.5	−2.5
2017 April 7	666.3	44.3	−2.5
2018 May 9	658.1	44.8	−2.5
2019 June 10	640.9	46.0	−2.6
2020 July 14	619.4	47.6	−2.7

A 75-mm telescope will reveal some irregularities and spots in the clouds, and a 150-mm telescope will show enough detail to permit useful observations. The spots move visibly within 10 minutes, carried around by the planet's rapid rotation; the period is just under 10 hours. Even the smallest telescopes or binoculars will show the planet's disk and its four main moons as they orbit the planet. Since the axis of rotation of Jupiter is inclined at only 3°, the polar regions are never well seen.

Jupiter is visibly flattened under its rapid rotation, the ratio of the polar to the equatorial diameter being 15:16; the rotation is also responsible for drawing out the clouds into the distinctive pattern of zones and belts. Most atmospheric motions are channelled along lines of latitude, so that the atmosphere is dominated by winds blowing east or west at different latitudes.

Infrared observations have revealed variations in temperature across the planet's disk, representing different altitudes. The highest and coldest features are the white and orange ones – i.e. the zones and the Great Red Spot. Lower and warmer are the brown belts, and deepest and warmest of all are the dark bluish-grey patches on the North Equatorial Belt's south edge. The planet must have a hot interior as it emits twice as much heat as it receives from the Sun.

Since Jupiter's outer layers are gaseous, the planet does not rotate as a solid body. A fast-moving jet-stream embraces the equatorial region, while at higher latitudes the visible features move more slowly, so two distinct systems of longitude are used when analysing observations. System I of longitude applies to the region about 9° either side of the equator, from the southern edge of the North Equatorial Belt to the northern edge of the South Equatorial Belt; its adopted period is 9h 50m 30.003s (877°.90 per day). System II applies to the rest of the planet and has an adopted period of 9h 55m 40.632s (870°.27 per day), very close to the average for the Great Red Spot. Radio astronomers have defined System III of longitude, with a period of 9h 55m 29.711s; this is the rotation period of the planet's magnetic field, which is thought to correspond to the solid core of the planet.

Features of Jupiter

Long-term changes in the intensity, colour or structure of particular belts or zones of Jupiter occur over a number of years. There are large, long-lived features such as the Great Red Spot whose appearance and motion evolve over decades. There is also plenty of activity on time-scales from days to months, some of it spectacular. Many of these phenomena seem to fall into regular patterns over several years, to be replaced by new patterns over several decades. Consistent observations of these phenomena provide the raw material from which the physicist can begin to deduce the nature of the planet's subsurface layers and the processes at work in them.

An extraordinary event was the collision of Comet Shoemaker–Levy 9 with Jupiter during the week of 1994 July 16–22. This comet, which had been captured into an orbit around Jupiter, broke up into at least 20 fragments, which burned up in Jupiter's atmosphere at about latitude 48°S. The impacts produced dark clouds of dust that were easily visible from Earth through small telescopes. Over the following weeks the spots spread and merged to form a dark belt which remained visible for over a year.

BELTS AND ZONES Figure 27 shows the names of the standard belts and zones. They are usually abbreviated by their initial letters, except that 'Tropical' is abbreviated as 'Tr' to avoid confusion with 'Temperate'. Although the belts always revert to this pattern, year-to-year variations are common. Often a belt may be double, in which case the components are labelled (N) for north and (S) for south, as in SEB(N) and SEB(S). The space between them, in this example, is referred to as the SEBZ. (Suffixes, as in SEB_n and SEB_s, have also been used to denote belt components, but the British Astronomical Association now reserves these to denote belt edges.) Sometimes there is a narrow belt within one of the standard zones, termed a 'band', such as the STZB in the STZ. Sometimes a belt may be missing, replaced by white material; this is particularly likely to happen to the NTB, the SEB and sections of the STB. Sometimes there is a substantial colour change; belts can range from slate-grey to reddish-brown, and some zones (particularly the Equatorial Zone) can temporarily adopt colours such as yellow or ochre.

The observer will see many types of feature associated with the belts and zones. They are generally referred to as 'spots', but a

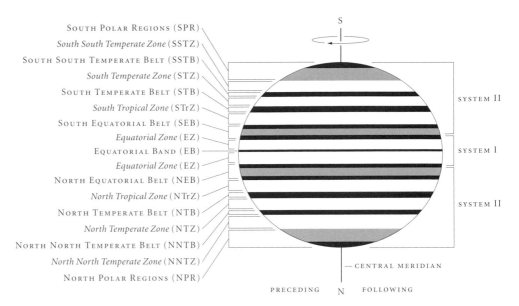

| SOUTH POLAR REGIONS (SPR) |
| South South Temperate Zone (SSTZ) |
| SOUTH SOUTH TEMPERATE BELT (SSTB) |
| South Temperate Zone (STZ) |
| SOUTH TEMPERATE BELT (STB) |
| South Tropical Zone (STrZ) |
| SOUTH EQUATORIAL BELT (SEB) |
| Equatorial Zone (EZ) |
| EQUATORIAL BAND (EB) |
| Equatorial Zone (EZ) |
| NORTH EQUATORIAL BELT (NEB) |
| North Tropical Zone (NTrZ) |
| NORTH TEMPERATE BELT (NTB) |
| North Temperate Zone (NTZ) |
| NORTH NORTH TEMPERATE BELT (NNTB) |
| North North Temperate Zone (NNTZ) |
| NORTH POLAR REGIONS (NPR) |

S
SYSTEM II
SYSTEM I
SYSTEM II
— CENTRAL MERIDIAN
PRECEDING   N   FOLLOWING

FIGURE 27. *The nomenclature for belts and zones on Jupiter and Saturn. The features, shown here schematically, are subject to considerable change. The following abbreviations are used: S = South; N = North; P = Polar; T = Temperate; Tr = Tropical; E = Equatorial; B = Belt or Band; Z = Zone. On Saturn, the SSTZ, SSTB, NNTB and NNTZ are rarely seen. Jupiter's Great Red Spot lies on the southern edge of the South Equatorial Belt.*

variety of more descriptive names are also used: dark 'projections' and bright 'bays' on the edges of belts; white 'ovals' in zones or on the edges of belts; 'barges' (very dark bars, particularly on the NEB_n edge); and 'festoons' (long, curving, dark streaks). Some of the most prominent are described below.

Although most of these features move with currents having rotation periods that differ little from System I or II, outbreaks of spots are occasionally seen which move eastwards or westwards at much greater speeds, of several degrees of longitude per day relative to System II. These speeds define the 'jet-streams'. Amateur records of these outbreaks from before encounters with space probes indicate that jet-streams occur at fixed latitudes on the edges of the standard belts, and the results obtained by space probes have confirmed that there is actually a regular, permanent pattern of jet-streams. They blow in the preceding direction (prograding, decreasing System II longitude) at the equatorward edge of each belt and in the following direction (retrograding, increasing longitude) at the poleward edge of each belt. Only a few of them have ever become observable from Earth, but when they do they produce some of the most impressive events to be seen on the planet, such as the revivals of the SEB (see p. 80).

The following are some of the most conspicuous features and events that may be seen.

THE GREAT RED SPOT This, the most famous feature on the planet, was probably observed by Giovanni Cassini in 1665, although continuous records of it go back only as far as 1831. It is a huge oval marking which spans the STrZ and indents the SEB south edge. It lies within a great bay in the SEB$_s$ edge, called the *Red Spot Hollow*, and the Hollow is always visible even when the Spot itself is not. The Spot first attracted widespread attention in 1878, when it darkened and developed a vivid brick-red colour; an equally impressive coloration lasted from 1969 to 1975. At other times, the Spot often has a pale fawn or grey colour, or it may be replaced by a white oval within a thin dark ring, or it may even disappear entirely. It appears that the oval structure is permanent, but the colour within it is variable.

The Great Red Spot drifts slowly and irregularly in longitude, and over the past century has performed about three circuits of the planet back and forth. Thus it cannot be attached to any fixed point in the interior of the planet. In the 1960s, professional photography revealed that it was circulating anticlockwise with a period of 6 to 12 days, vividly confirmed by the Voyager probes in 1979. The Spot thus appears to roll like a ball-bearing between the jet-streams on the SEB$_s$ and STB$_n$. It is now believed that the Great Red Spot is a giant, self-sustaining anticyclonic disturbance in the atmosphere.

WHITE OVALS ON THE STB   These ovals are similar to the Great Red Spot, both in their position with respect to a belt and in their internal circulation. Three of them, called FA, BC and DE, came into being around 1940 by the subdivision of the STZ into three long sectors. These sectors contracted rapidly to form the three ovals. They continued to contract slowly, and to move slowly into the STB. Following interactions with the Great Red Spot, white ovals BC and DE joined to form BE in 1998. Two years later, ovals BE and FA overtook the Great Red Spot and merged to form oval BA. In 2002 BA overtook the Great Red Spot again and subsequently became less prominent; in time it may disappear.

SOUTH TROPICAL DISTURBANCE   A South Tropical Disturbance is a grey barrier across the STZ, lasting for months or years and having a slow drift in decreasing System II longitude. The greatest such Disturbance on record appeared in 1901 as a small dark spot, and expanded greatly in longitude before it finally faded away in 1939. It repeatedly interacted with the Great Red Spot as it passed it. The Disturbance also gave rise to a remarkable circulating current in which retrograding spots on the SEB$_s$ jet-stream, approaching the concave preceding end of the Disturbance, were diverted around it so as to end up on the STB$_n$ jet-stream, travelling back the way they had come.

Since 1939 at least six shorter-lived phenomena of the same type have been seen, usually starting at the preceding end of the Great Red Spot.

SEB REVIVAL   This is the most spectacular phenomenon to be seen on Jupiter, and recurs at intervals of between 3 and 30 years. The scene is set by the disappearance of the SEB(S) component and the intensification of the Great Red Spot. The Revival begins with a small dark streak across the whitened SEB latitudes. From this focus, many dark and bright spots are poured out, some retrograding on the SEB$_s$ jet-stream and others prograding on the SEB$_n$ (equatorial) jet-stream. A scene of violent activity develops, the Great Red Spot fades, and eventually the SEB is reconstituted.

PROJECTIONS AND PLUMES ON THE NEB SOUTH EDGE
Some of these are usually present, and are among the most conspicuous features on the planet. Often they are associated with white spots; a common structure is the 'plume', consisting of a dark projection streaming into an equatorial festoon with a bright white spot at its base. All these features generally move with System I and last for months, but more rapid motions and changes are sometimes observed.

Observing Jupiter

The most important kinds of observation that amateurs can make are those aimed at determining the longitudes of spots. To learn about the various phenomena associated with spots, it is necessary to identify individual spots and the currents in which they are moving; and because the spots can change and move rapidly, this can be done only by precise measurements of longitude. (Measurements of latitude are less important, and can be left to specialist observers; latitude can be described relative to the belts, and changes are small.)

Longitude is easy to measure, requiring no more than an accurate clock or watch, and of course a telescope of adequate size – at least a 100-mm refractor or 150-mm reflector. A spot's longitude is measured as the planet's rotation carries it across the *central meridian*, an imaginary line of longitude running down the exact centre of the planet's disk, perpendicular to the belts. The observer simply notes the time, to the nearest minute, at which the spot is judged to be on the central meridian. This is the *transit time*, and can be judged to within a minute or two (1 minute corresponds to 0°.6 of longitude.)

The observer should record the transit time, along with a description of the feature observed. It is often helpful to sketch the planet or the feature so that the description can be checked in case of doubt. As for all planetary observations, the telescope and the seeing should also be recorded. After the observing session the observer can use published ephemerides to convert the transit times into longitudes: System I for equatorial features (SEB$_n$, EZ, NEB$_s$) and System II for features in other latitudes.

When a sufficient number of transits have been recorded, the results can be plotted on a graph of longitude against time to reveal the motions of the spots. For example, several transits over several weeks will show clearly that the motion of the STB white ovals differs from that of the Great Red Spot by about 0°.4 a day.

Disk drawings are valuable, though less important than transit measurements. Their main value is to record the detailed appearance and location of spots that were transited, to record the existence of spots whose transits were missed, and to provide a record of the changing aspects of the belts and zones. The observer may make sketches of the whole disk or of a particular region. Observations over several hours may conveniently be recorded on a 'strip-map', which is added to continuously as the planet rotates, so as to produce a cylindrical-projection map of the planet.

Complete disk drawings are best made on printed blanks of the flattened disk of the planet; the standard has an equatorial diameter of 64 mm. The observer should look at the planet carefully before starting a drawing, and begin with a quick outline sketch that locates the main features. The details can then be filled in. It is important to record the latitudes, widths and intensities of the belts and zones as faithfully as possible, although for clarity it is advisable to exaggerate the contrast of the markings.

Estimates of the intensities and colours of the belts and zones can also be useful. The intensity of a feature is estimated by allocating it a number on a scale that runs from 0 (brightest) to 10 (black sky). Because they are subjective estimates, they are worth while only if done systematically and repeatedly, so that any changes on the planet can be demonstrated from the accumulated observations of an individual. Colour estimates are usually simple descriptions, and the observer should be familiar with the spurious causes of colour on the disk. (In particular the 'traffic-lights effect' – fringing with red and blue-green as a result of atmospheric refraction – can apply to each zone as well as to the whole disk.)

CCDs have now made it possible for amateurs to obtain detailed images with modest telescopes. Images obtained through red and blue filters can give useful information on the colours of features.

The Satellites of Jupiter

Jupiter has over 60 known satellites, of which four are of mag. 5 to 6 while the others are no brighter than mag. 14. The four great moons discovered by Galileo – Io, Europa, Ganymede and Callisto – would be visible to the naked eye but for the glare of the planet; they can, however, be seen in binoculars. Because they orbit very close to the equatorial plane of Jupiter, which we see almost edge-on, they usually appear strung out in a line, and they repeatedly pass in front of and behind the planet's disk. A small telescope will show the *transits* of the satellites and their shadows across the face of Jupiter, their *eclipses* by Jupiter's shadow, and their *occultations* by the planet's disk. The four satellites look different when in transit across the planet: Io invisible or a faint grey, and Europa usually invisible, but Ganymede and Callisto very dark. Every six years, during a period of a few months when the Earth passes through the plane of the

TABLE 26. *Disappearances of Jupiter's four brightest satellites.*

DATE	(UT)	
	BEGINS	ENDS
2008 May 22	3h 51m	4h 10m
2009 September 3	4h 44m	6h 30m
2019 November 9	12h 17m	12h 56m
2020 May 28	11h 18m	13h 12m
2021 August 15	15h 40m	15h 48m
2033 July 28	3h 08m	5h 01m
2038 May 22	9h 10m	10h 49m
2038 December 9	8h 20m	10h 36m
2049 October 15	3h 47m	4h 01m
2050 May 28	17h 23m	18h 34m

Source: Jean Meeus, Mathematical Astronomy Morsels *(Willmann-Bell, 1997).*

satellites, they can be seen to eclipse and occult each other in *mutual phenomena*. All the phenomena of the satellites are predicted in ephemerides. Amateurs' timings of these phenomena can be useful in refining the orbits of Jupiter's satellites.

Sometimes two of the inner three Galilean satellites undergo the same phenomenon (e.g. transit) on the same evening, while the third undergoes a different phenomenon (e.g. eclipse or occultation). Such occasions exemplify a mathematical relationship between the orbits, such that the mean planetary longitude of Io, minus three times the longitude of Europa, plus twice the longitude of Ganymede, is always equal to 180°. Thus they can never all line up radially on one side of the planet. On rare occasions, all four of Jupiter's main moons may simultaneously be invisible, being either in front of the planet's disk (in transit), behind it (occulted) or in eclipse. Table 26 gives the occasions until the middle of the 21st century on which Jupiter is predicted to appear without its four brightest satellites.

Surface markings on the Galilean satellites are barely visible, even through large telescopes. However, space probes have revealed extraordinary variety in their surfaces. Io displays continuous volcanic activity, due to the tidal forces induced by the coupled motions of the satellites; the effects of this activity can also be detected by professional Earth-based photometry. Europa has a smooth icy surface covered by a maze of linear markings. Ganymede and Callisto have heavily cratered ice surfaces.

The other satellites fall into three groups, and are too faint for most amateur observers. One group, orbiting closer to the planet than Io, includes the 250 × 146 × 128 km satellite Amalthea, plus three others which are less than 120 km across; these three small moons are visible only to spacecraft, as is the tenuous dusty ring which coincides with the innermost satellite orbit. The other two groups of satellites include only one sizeable one (Himalia, 170 km in diameter, mag. 15); all the others are mag. 17 or fainter. They orbit very far from the planet, in highly inclined and perturbed orbits, and are believed to be captured asteroids.

Saturn

Saturn, the second-largest planet (equatorial diameter 120 536 km), is much farther from us than Jupiter and its disk appears less than half Jupiter's size in the telescope. Saturn's distance at opposition ranges from a minimum of 1197 million km when at perihelion to a maximum of 1357 million km at aphelion. The apparent equatorial diameter of Saturn at opposition ranges from 18.4 arcsec at aphelion (June opposition) to 20.7 arcsec at perihelion (December opposition). Oppositions of Saturn until 2020 are listed in Table 27.

Saturn's mean synodic period is 378 d, so that oppositions occur about a fortnight later each year. The planet is observable at some time of the night for nine or ten months every year. However, oppositions vary tremendously in suitability for observation from a particular location, depending on the planet's declination. Oppositions in December or January are the most favourable for observers in the northern hemisphere, but least favourable for observers in the southern hemisphere, because Saturn is then well north of the celestial equator. For oppositions in June or July the situation is reversed, since the planet is then well south of the celestial equator. During one orbit of Saturn around the Sun, which takes 29.5 years, it is therefore usual for an observer at one location to experience relatively favourable apparitions for about 14 years, followed by an equal period of unfavourable apparitions.

The polar diameter of Saturn is about 10% less than the equatorial diameter, the greatest oblateness of any planet in the Solar System. The marked oblateness is connected with the planet's low density, which is less than that of water.

Features of Saturn

Saturn is similar to Jupiter in that it has dark belts and bright zones, and the same nomenclature is adopted as for Jupiter (see Figure 27). Spots may be seen on the belts and in the zones from time to time, but they are usually short-lived, which is why our knowledge of rotation periods for various latitudes is less accurate than for Jupiter. The rotation period is 10h 14m at the equator, and longer for higher latitudes. There is no sudden change of rotation period as with System I and System II of Jupiter, merely a gradual lengthening of rotation period towards the poles.

Great White Spots appear in Saturn's equatorial zone roughly every 30 years when the planet's north pole is tilted at its maximum towards the Sun. These spots consist of highly reflective clouds of ammonia crystals. Great White Spots were seen in 1933, 1960 and 1990; subsequently, smaller spots were seen for several years into the 1990s. The spots in 1933 and 1990 rapidly grew to fill the entire equatorial zone, taking several months to dissipate.

Hydrogen is the main atmospheric component, with meth-

TABLE 27. *Oppositions of Saturn 2005–2020.*

DATE	DISTANCE FROM EARTH (10^6 km)	APPARENT EQUATORIAL DIAMETER (ARCSEC)	MAGNITUDE
2005 January 13	1208	20.6	−0.4
2006 January 27	1216	20.5	−0.2
2007 February 10	1227	20.3	0.0
2008 February 24	1240	20.1	+0.2
2009 March 8	1256	19.8	+0.5
2010 March 22	1272	19.6	+0.5
2011 April 4	1288	19.3	+0.4
2012 April 15	1304	19.1	+0.2
2013 April 28	1319	18.9	+0.1
2014 May 10	1332	18.7	+0.1
2015 May 23	1342	18.5	0.0
2016 June 3	1349	18.4	0.0
2017 June 15	1353	18.4	0.0
2018 June 27	1354	18.4	0.0
2019 July 9	1351	18.4	+0.1
2020 July 20	1346	18.5	+0.1

ane, ammonia, phosphine, acetylene and ethane also present. The Voyager probes revealed an extensive haze some 70 km thick that affects the visibility of the belts and zones. Wind velocities in the equatorial region are the highest found for any planet. Like Jupiter, Saturn emits more energy than it receives from the Sun; its effective temperature is 96.5 K, nearly 20 K higher than would be expected if it did not give out heat of its own.

Saturn's Rings

The ring system that encircles Saturn's equator can be seen through a small telescope with a magnification of ×50 or so, but larger apertures and magnifications are required to reveal the details of the three major rings: A (the outer), B (the brightest) and C (the inner, also known as the *crêpe ring* as it is partially transparent). Figure 28 shows the main components of Saturn's rings; data for all the known rings are given in Table 22 (p. 71). Ring C is usually most noticeable where it crosses in front of the brighter globe and is seen in silhouette, mimicking the appearance of a dusky belt. The ring system shows itself at various angles, from edge-on to a maximum of 27°.0. The dates in the 21st century on which Saturn's rings are tilted at their maximum towards us are given in Table 28. On average, the rings take about 7¼ years to go from fully open to edge-on. The large brightness variation of the planet at opposition, from mag. 0.8 to mag. −0.5, is mainly a result of the varying presentation of the ring system.

Twice in the course of Saturn's orbit round the Sun, at intervals of 13¾ and 15¾ years alternately, the Earth passes through the plane of the rings (the intervals are unequal because of the

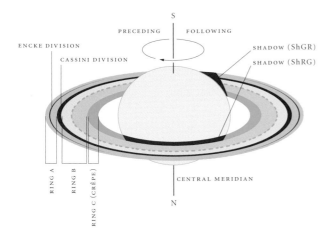

FIGURE 28. *Nomenclature for the main components of Saturn's ring system. Here, the rings are shown tilted at their maximum angle towards the Earth. Ring B is usually the brightest, while ring C is the faintest and is semi-transparent. The Cassini Division between rings A and B is visible at all but the most oblique of tilt angles, while the narrower Encke Division in ring A is difficult even when the rings are wide open. The shadow of the rings on the globe (ShRG) can be seen to the south, north or, as here, through ring C, depending on Saturn's tilt angle and lighting conditions. The shadow of the globe on the rings (ShGR) appears on the preceding side before opposition and on the following side after opposition, as above. The nomenclature of Saturn's globe is similar to that for Jupiter (see Figure 27).*

TABLE 28. *Widest presentations of Saturn's rings in the 21st century. Positive is northern hemisphere towards the Earth, negative southern.*

DATE	ANGLE
2017 October 16	+27°.0
2032 May 12	−27°.0
2046 November 15	+26°.9
2062 March 31	−27°.0
2076 October 9	+27°.0
2091 May 4	−27°.0

Source: Jean Meeus, More Mathematical Astronomy Morsels *(Willmann-Bell, 2002).*

TABLE 29. *Passages of the Earth through the plane of Saturn's rings in the 21st century.*

DATE	DIRECTION
2009 September 4	South to north
2025 March 23	North to south
2038 October 15	South to north
2039 April 1	North to south
2039 July 9	South to north
2054 May 5	North to south
2054 August 31	South to north
2055 February 1	North to south
2068 August 25	South to north
2084 March 14	North to south
2097 October 5	South to north
2098 April 26	North to south
2098 June 18	South to north

Source: Richard E. Schmidt, Sky & Telescope, *Vol. 58, p. 500 (1979).*

eccentricity of Saturn's orbit). On each occasion the Earth can pass through the ring plane up to three times, as on 1995 May 22 (north to south), 1995 August 11 (south to north) and 1996 February 12 (north to south). Subsequent dates of the Earth's passage through the ring plane are given in Table 29. When edge-on, the rings are often invisible in small telescopes as they are only a few hundred metres thick at the most. But they can present a truly magnificent sight when tilted towards us by a degree or so; the satellites then appear like brightly lit droplets on a spider's web.

All too rarely do the rings pass in front of a bright star, although such an event took place in 1989 when the star 28 Sagittarii was occulted first by the rings and then by Saturn's largest satellite, Titan. The event provided information on the structure of the rings and on Titan's atmosphere. Before the arrival of the two Voyager space probes at Saturn in 1980 and 1981, such events were a major source of information on ring structure.

Within ring A is a division known as the *Encke Division* (discovered by Johann Encke), the reality of which was once in dispute but which was shown to be real by the Voyager probes. Separating rings A and B is the *Cassini Division* (discovered by Giovanni Cassini) which, in good seeing, can be observed with a 75-mm telescope if the rings are fairly well open. This conspicuous division, 4800 km wide, is not totally empty but contains numerous thread-like ringlets. Ring C is the faintest ring normally visible from the Earth. Yet another ring, D, observable from the Earth with large telescopes and under excellent seeing conditions, is found between the inner part of ring C and the planet's globe. The Voyager probes revealed at least three rings outside ring A: in order of increasing distance from Saturn, rings F, G and E.

The rings are not solid but are composed of myriads of tiny particles; the 'divisions' mark thinly populated zones in the rings. Many divisions have been seen and have been shown to occur at specific distances from the planet where ring particles would be drawn away by the regularly repeated gravitational pull of some of the planet's satellites, similar to the Kirkwood gaps in the asteroid belt.

Observing Saturn

All amateur observers can play a part in determining the longitudes of surface features. As with Jupiter, a spot's longitude is determined by accurately recording the time at which it transits the central meridian, the imaginary line which bisects the visible disk perpendicular to the belts. With practice, an observer can achieve an accuracy of a minute or two. Longitude tables are available which give the longitude of the central meridian at 0h UT daily. System I (analogous to that on Jupiter) covers the whole of the two equatorial belts and the Equatorial Zone, and has a rotation period of 10h

14m (844°.3 per day). System II, which applies to the rest of the planet, has a period of 10h 38m 25s (812° per day), although there is a gradual lengthening of the rotation period towards the poles.

Re-observation of a given region is possible at the 5th, 7th, 12th and 14th rotations, when a given longitude will be in the same position on the disk within about 3 hours of the time of the initial observation. It is important to ensure that features timed crossing the central meridian are fully identified in the observing log, for this will greatly increase the value of the observations when they come to be compared and collated with those of other observers.

Disk drawings are important, although less so than the determination of longitudes by the transit method. Drawings provide a record of the changing appearance of the surface of the planet throughout each apparition.

As the apparent shape of Saturn is constantly changing with the changing tilt of the ring system, it is not possible to use a single standard blank. The appropriate outline, depending on the value of B (the Saturnicentric latitude of the Earth relative to the ring plane), is usually traced from a set of standard outlines (obtainable from national observing groups). All observations should be fully documented: the date, time (UT) of the observation, instrument, power used and seeing conditions are all essential to the permanent record. Detailed notes of all important features should be added to the record at the time of observation.

Another useful kind of observation, provided it is carried out systematically and regularly, is the estimation of the relative intensity of various parts of the globe and rings. This is done by assessing the intensities on a numerical scale. European astronomers use the convention that 1 is the brightness of ring B (the bright reference) and 10 the absolute blackness of a deep shadow or dark background sky. The scale adopted in the U.S.A. runs the opposite way, from 8 as the brightest (ring B) to 0 for the background sky. Fractions of ½ may be used to denote intermediate intensities, but intensities quoted to a tenth are quite meaningless.

Colour filters are often used; recommended are Wratten 25 (red) and Wratten 44A (blue), or Wratten 47 (blue) for larger apertures (around 300 mm). Different intensities in red, blue and 'white' light can indicate colour tints, and the belts and zones may appear to be of different widths. A curious effect known as the bicoloured aspect of the rings has been noted, in which one side of the rings, when viewed through a given filter, appears to have a different intensity from that of the opposite side. The effect, when present, is usually seen in ring A and is revealed by comparing the brightness of the two ring arms first through a red filter and then through a blue filter. Sometimes a difference is noticed with one filter alone. The reality of this effect has been confirmed with CCDs and webcams.

The Satellites of Saturn

There are ten major satellites, the largest of which is Titan (mag. 8.3 at opposition). A number of smaller satellites have been found by space probes and Earth-based observations, bringing the total to over 30. Most of the major satellites orbit close to the plane of the planet's equator, but Iapetus has a more inclined orbit, and Phoebe's orbit is retrograde. They all display variations in brightness, the greatest variation being shown by Iapetus, which is about two magnitudes brighter at western elongation than at eastern elongation. Iapetus was shown by the Voyager probes to have one hemisphere of bright reflective material, and the other dark.

Titan shows variations in both brightness and colour: it may appear white, yellowish, pink or even red. This is because Titan has an atmosphere of nitrogen in which various organic compounds have been found. It has a surface temperature of 93 K, some 7 K above the equivalent black-body value, so a small greenhouse effect seems to be operating.

The high tilt of the planet's equatorial plane means that satellite phenomena cannot be observed all the time, as is possible with Jupiter, but only during four or five apparitions centred on each passage of the Earth through the ring plane. In each sidereal revolution of the planet there are thus two periods of about ten years when no satellite phenomena can be observed.

Only when the rings are edge-on or the ring angle is very small are the satellites occulted or eclipsed. Titan appears as a dark body against the globe of Saturn when in transit, appearing almost as dark as its shadow, a phenomenon visible in small telescopes. It is interesting to plot the movements of the satellites, but little can be done in the way of useful observation save for estimating their magnitudes; this is difficult to do reliably without photometric equipment, partly because of the glare from Saturn itself. The timing of satellite phenomena which occur at certain apparitions is very useful.

Uranus, Neptune and Pluto

Uranus

At mag. +5.5, Uranus is just visible to the naked eye, but it is far from conspicuous. Good binoculars and small telescopes show a tiny, blue-green disk of somewhat fuzzy appearance, with an apparent angular diameter of about 4 arcsec at best.

Uranus is so remote that its apparent brightness varies by only about 20% from opposition to conjunction, and there is no perceptible difference in its general appearance at opposition and quadrature. However, variations in brightness of around 0.5 mag. that are attributable to changes in the planet's atmosphere have been detected, and can be monitored with binoculars by using the methods of the variable-star observer. More precise measurements can be made with CCD photometry and have confirmed the reality of the fadings reported by visual observers. Such fadings normally last no more than a few weeks although durations of up to 6 months have been known.

Unlike the other planets, Uranus spins on its side – that is, its axis of rotation is almost in the plane of its orbit. The pole of counterclockwise rotation, which, in contrast to the Earth, is the south pole, is tilted 98° to the planet's orbital plane around the Sun. Uranus is therefore seen in a unique series of orientations as it orbits the Sun: pole-on, as around 1985–86; on its side, as around 2007–08; or somewhere in between.

Visually its most striking feature is its very blandness. Lightish bands flanked by narrow dusky belts have been reported when the planet is side-on to us, as have large-scale brightenings near the limb. Much of this detail is at the limits of vision and must be considered suspect, but similar features have been recorded by the Hubble Space Telescope.

Uranus has over 20 known moons. Even the brightest require large telescopes to be seen visually, but can be imaged by CCD through small telescopes if steps are taken to suppress the glare from the planet, either by keeping exposures short or by using an occulting bar. There are in addition 11 faint, narrow rings, which have been imaged by spacecraft. The rings and the innermost moons move in virtually circular orbits in the planet's equatorial plane, so whatever event brought about the current axial tilt of Uranus did not disrupt the satellite and ring system.

Neptune

Neptune is an eighth-magnitude object, visible in a small telescope or binoculars. Its apparent angular diameter is about 2.3 arcsec and shows a bluish disk in telescopes of moderate aperture. No detail can be seen except in very powerful instruments, which reveal traces of belts, reminiscent of those on the other gas giants, bordering a bright equatorial zone. Changes in brightness have been reported as the planet rotates, but there is uncertainty over the amplitude of variation. In addition, Neptune's brightness has been suggested to vary with solar activity, becoming fainter at solar maximum, but this effect awaits confirmation.

Neptune is so remote that little more than visual or CCD photometry can be carried out with small instruments. Its path can be followed using a finder chart as is published in the *Handbook of the British Astronomical Association* or astronomy magazines; alternatively, a finder chart can be plotted from the coordinates in an annual ephemeris or from a computer planetarium-type program.

Neptune has one large satellite, Triton, mag. 13.5, which has been imaged by amateur CCD observers. Its motion is retrograde and its orbit is inclined by 23° to the planet's equator. Nereid, the largest of the known outer moons, is small and very faint, and revolves in an extremely eccentric orbit. Its faintness (mag. 18.7) puts it beyond the reach of all but powerful instruments. There are several other satellites but they are all beyond the range of normal-sized instruments.

Pluto

At visual magnitude 14.5 and photographic magnitude 15.4, Pluto requires a telescope of aperture at least 250 mm to be seen. It will, of course, register on CCD frames and photographic film provided the exposure is long enough, but a good star chart – or a second image taken several hours later to confirm motion – is necessary for it to be identified.

Pluto is the smallest by far of the major planets, with a diameter of 2390 km – smaller than our own Moon. Its orbital eccentricity and inclination are greater than those of any other major planet. It is also unusual in that it crosses inside the orbit of Neptune when it is near perihelion, as it was from 1979 to 1999. These characteristics, allied to the discovery of other trans-Neptunian objects (see p. 86), has prompted debate on whether it should still be considered a major planet.

A satellite, Charon, was discovered in 1978. Charon's maximum separation from Pluto is just 0.8 arcsec and so it can be viewed only under very good seeing conditions through a large instrument, and then only as an elongation of the image of Pluto itself. Charon orbits the planet in 6.39 d at a distance of 19 600 km. Its diameter is half that of Pluto, which makes it by far the largest satellite relative to its primary in the Solar System. This is the only planet–satellite pair with both synchronous rotation and synchronous revolution.

TRANS-NEPTUNIAN OBJECTS

Since 1992 several hundred *trans-Neptunian objects* (TNOs) with a semi-major axis greater than 30 AU have been found, making up the Kuiper Belt (see below). Many of these have orbits similar to that of Pluto and have been termed *Plutinos* ('little Plutos'). The largest TNO, Quaoar, is about 1200 km in diameter, the same as Charon, and the brightest is of 21st magnitude. Several Kuiper Belt objects have been found to be double, like Pluto and Charon. Hence Pluto is now widely regarded as simply the largest member of the Kuiper Belt.

Minor Planets

These bodies, often called asteroids, are small worlds orbiting the Sun. Even the largest, (1) Ceres, is less than 1000 km in diameter, and most are under 10 km in diameter. The only one that can become bright enough to be visible to the naked eye (mag. 5.5 at best) is (4) Vesta, which is some 580 km in diameter and has a high albedo. The first five minor planets to be discovered are listed in Table 30.

There are estimated to be around one million minor planets larger than 1 km in diameter in orbits nearer the Sun than Jupiter. Most of them keep to a broad region between the orbits of Mars and Jupiter known as the *asteroid belt* or *main belt*.

The most interesting minor planets are those that do not orbit within the main belt (see Table 31). For example, (433) Eros, discovered in 1898, may approach the Earth to within 24 million km, as it did in 1975. Other minor planets have passed closer to us than the Moon and there are now dedicated surveys to identify those that may present a hazard to the Earth.

Over 1300 minor planets are known to cross the Earth's orbit. One, 2000 BD$_{19}$, passes closer to the Sun (less than 0.1 AU) than any other known object except for the occasional comet. Since 1992 hundreds of icy minor planets have been found outside the orbit of Neptune, indicating the existence of a second region known as the Kuiper Belt or the *Edgeworth–Kuiper Belt*. This region is thought to be the reservoir for most periodic comets. Several fairly large asteroids, termed the Centaurs, lie in orbits between Jupiter and Neptune. The first of these to be discovered, (2060) Chiron, showed cometary activity as it approached perihelion.

Two groups of minor planets travel in a similar orbit to Jupiter, lying at mean points some 60° ahead of and behind Jupiter. These are named after the heroes of the Trojan war. The first Trojan asteroid discovered, (588) Achilles, was found in 1906. Examples of Trojans in the orbits of Mars and Neptune have now also been found.

Some minor planets in the outer parts of the main belt have orbital periods that are exact fractions (¾ and ⅔) of the orbital period of Jupiter. Generally, however, there are obvious breaks (the *Kirkwood gaps*) in the distribution of minor planets at exact fractions of Jupiter's period (e.g. ½, ⅓, ⅗), which otherwise extends fairly uniformly from about 2.0 to 3.3 AU from the Sun.

The irregular shapes and rough surfaces of minor planets are evident in the detectable variation in reflected sunlight seen from the Earth. This variation, caused mainly by rotation, is usually repeated after several hours. Most of the known rotation periods lie between 5 and 8 hours, although a few minor planets have rotation periods as short as 1 hour or in excess of 24 hours. Spectrophotometry, in which a minor planet is observed at selected regions of the spectrum by the use of filters, yields much additional information on albedo, surface composition and diameter. Radar observations have provided additional information, including the discovery that some asteroids are double, and spacecraft images have revealed that some have satellites.

Minor planets are classified according to spectrum and albedo. About 75% of them are C-class, with albedos as low as 3%, perhaps containing a fair amount of carbonaceous material. Another 15%, of S-class, have moderate albedos (average 16%) and more likely contain silicaceous material. Metals are also present, and in some instances may result in very high albedos, for example 40% in the case of (44) Nysa.

Most minor planets are thought to be debris left over from the early days of the Solar System, the nearby presence of Jupiter having prevented the formation of a larger planet in the region. Subsequent

TABLE 30. *The first five minor planets discovered.*

NUMBER AND NAME	YEAR OF DISCOVERY	SIDEREAL PERIOD (y)	MEAN DIST. FROM SUN (AU)	ORBITAL INCLINATION (DEGREES)	ORBITAL ECCENTRICITY	ALBEDO	DIAMETER (km)	VISUAL MAGNITUDE AT BRIGHTEST
(1) Ceres	1801	4.60	2.77	10.6	0.079	0.11	948	6.7
(2) Pallas	1802	4.62	2.77	34.8	0.230	0.16	532	6.7
(3) Juno	1804	4.36	2.67	13.0	0.259	0.24	234	7.4
(4) Vesta	1807	3.63	2.36	7.1	0.089	0.42	530	5.5
(5) Astraea	1845	4.13	2.57	5.4	0.193	0.23	119	8.8

TABLE 31. *Some interesting and unusual minor planets.*

NUMBER AND NAME	YEAR OF DISCOVERY	SIDEREAL PERIOD (y)	DISTANCE FROM SUN (AU)		i (DEG.)	e	H^a	NOTES
			MINIMUM	MAXIMUM				
Hermes	1937	2.10?	0.617	2.662	6.2	0.624	18.0	The only named but unnumbered asteroid
(279) Thule	1888	8.86	4.23	4.33	2.3	0.012	8.57	At the outer edge of the main asteroid belt
(433) Eros	1898	1.76	1.133	1.783	10.8	0.223	11.16	Regular Earth approaches. Highly elongated shape, 33 km long, 13 km wide
(588) Achilles	1906	11.8	4.422	5.956	10.3	0.148	8.67	First known Jupiter Trojan
(944) Hidalgo	1920	13.8	1.950	9.544	42.6	0.661	10.77	First known aphelion beyond Jupiter
(1221) Amor	1932	2.66	1.085	2.754	11.9	0.435	17.7	First known asteroid to cross the orbit of Mars
(1566) Icarus	1949	1.12	0.187	1.969	22.9	0.827	16.9	Frequent close approaches to Earth
(1862) Apollo	1932	1.78	0.647	2.296	6.4	0.560	16.25	First known asteroid to cross Earth's orbit
(2060) Chiron	1977	50.33	8.433	18.831	6.9	0.381	6.5	First known Centaur-group asteroid
(2062) Aten	1976	0.95	0.790	1.143	18.9	0.183	16.80	First known asteroid with smaller semi-major axis than Earth
(2212) Hephaistos	1978	3.19	0.360	3.973	11.8	0.834	13.87	Possible common parent with Comet Encke
(3200) Phaethon	1983	1.43	0.140	2.403	22.2	0.890	14.6	Parent of Geminid meteor stream
(3753) Cruithne	1986	0.997	0.484	1.511	19.8	0.515	15.1	In 1:1 orbital resonance with the Earth
(4015) Wilson–Harrington	1979	4.29	0.997	4.285	2.8	0.623	15.99	Seen as a comet in 1949
(4179) Toutatis	1989	3.98	0.920	4.102	0.5	0.634	15.30	Close approaches to Earth. Double structure
(5261) Eureka	1990	1.88	1.425	1.622	20.3	0.065	16.1	Mars Trojan
(15760) 1992 QB$_1$	1992	290.5	40.902	46.825	2.2	0.068	7.2	First known Kuiper Belt object
(15789) 1993 SC	1993	247.8	32.225	46.671	5.2	0.183	6.9	First confirmed Plutino
1996 PW	1996	4460	2.544	540	29.8	0.991	14.0	Asteroid in a cometary-shaped orbit
2000 BD$_{19}$	2000	0.82	0.092	1.661	25.7	0.895	17.2	Smallest known perihelion
(50000) Quaoar	2002	284.4	41.732	44.765	8.0	0.035	2.6	Largest Kuiper Belt object, diameter 1200 km

[a] *H is the absolute magnitude, i.e. how bright the object would appear if it were 1 AU from both the Sun and Earth and fully illuminated. Conversion to size depends on the albedo, which usually is not accurately known.*

evolution has been influenced by collisional break-up. Evidence for this idea is provided by the existence of groups of minor planets (the *Hirayama families*) with similar orbits and also similar compositions. The three principal families are associated with (221) Eos, (158) Koronis and (24) Themis.

Terminology and Naming

Minor planets are discovered on CCD images or photographs, as their images trail in relation to the stars during a time exposure. Search programmes at professional observatories account for the majority of newly found minor planets, although so many are detected that, in most cases, only unusually quick (and hence nearby) or slow (and hence remote) examples are investigated fully. On receipt of astrometric observations, the International Astronomical Union's Minor Planet Center (located at the Smithsonian Astrophysical Observatory, Cambridge, Massachusetts) assigns provisional designations. Some new main-belt objects are being identified from CCD images taken by amateurs.

Provisional designations are assigned according to the date of discovery. The designation consists of the year, followed by a letter from A to Y indicating the half-month and then a second letter to show the order of discovery within that half-month (I is omitted in

both cases). Thus 2003 DE is the fifth minor planet (E) discovered in the second half of February (D) in 2003. If there are more than 25 discoveries in any one half-month period, the second letter is recycled and a subscript numeral is added indicate the number of times the second letter has been repeated in that half-month period. Many of the objects assigned provisional designations turn out to be repeat observations of known minor planets, and it is a challenging task to identify observations of the same object in different years.

Official numbers are assigned to minor planets once reliable orbits have been computed for them, usually after the object has been well observed during at least three oppositions (although if pre-discovery photographic records can be found this process can be shortened). When the objects are numbered they can be named by their discoverers. As of mid-2003 there were more than 65 000 numbered minor planets, over 10 000 of them named, with the number of discoveries doubling about every two years.

Numbered minor planets are referred to by their number (usually given in parentheses, as here) and name, for example (439) Ohio, (1677) Tycho Brahe and (3869) Norton. Names are submitted by discoverers to the Minor Planet Center for approval by a committee, and new numbers and names are published in the Minor Planet Circulars, which also contain astrometric observations, orbits and ephemerides.

Observing Minor Planets

With a small telescope or even binoculars it is a simple job to pick out the brighter minor planets, particularly Vesta. The unexpected appearance of a bright asteroid in a familiar star field can give rise to a false alarm of a nova. Positions of the brightest minor planets are published in almanacs such as the *Handbook* of the British Astronomical Association or can be generated by a computer program.

Identification is mainly a matter of finding the field and spotting the interloper. Fainter minor planets can be confirmed by demonstrating their motion; this is done by drawing the stars in the field and comparing this record with the view in the telescope a couple of hours later, when the minor planet will be seen to have moved. Once identified, the minor planet can be followed nightly against the background stars. Images taken using a CCD can be 'blinked' on a screen to show up the motion.

Relatively simple observations can prove interesting and instructive. The brightness of a minor planet may be estimated by making a comparison with stars of known magnitude, in the same way as for variable stars. Minor planets change in brightness as they rotate, which in some cases can be followed visually: (216) Kleopatra varies by 1 magnitude, for example, and the highly elongated (433) Eros by up to 1.5 magnitudes.

Even if the brightness does not appear to change over the course of one night, it will do so over the course of an opposition, increasing as the Earth approaches and decreasing as it draws away again. A plot of a minor planet's brightness against time can reveal whether its orbit is near-circular (the light curve will be symmetrical) or eccentric (if the light curve is asymmetrical).

Brightness estimates can be adjusted to a standard distance of 1 AU from both the Sun and the Earth, and to 0° phase angle (i.e. full illumination), to give values of 'absolute magnitude'. From an analysis of the change in absolute magnitude with phase angle it is possible to derive the albedo, diameter and several other important physical properties. The accuracy of these results will be improved if several observers pool their observations.

Accurate positional measurements obtained from CCD images or photographs are used to refine the orbital elements, and are published in the *Minor Planet Circulars*. Photoelectric photometry is used by some amateurs to produce accurate light curves, from which the rotation periods and physical properties can be estimated.

Comets

According to the *Catalogue of Cometary Orbits* (15th edition), 1642 comets had been observed sufficiently well for their orbits to be calculated as of early 2003. Of these, 83% are *long-period comets*, taking hundreds, thousands or even millions of years to orbit the Sun. Many such comets are 'new', approaching the Sun for the first time from a region known as the *Oort Cloud* that surrounds the Sun at a distance of about 20 000 to 100 000 AU. By contrast, 'old' comets are those that have been perturbed by the major planets, principally Jupiter and Saturn, on numerous previous trips to the inner Solar System so that their orbits are now much smaller than originally. A small minority, such as Comet Bowell (C/1980 E1), which passed very close to Jupiter in late 1980, are ejected permanently from the Solar System.

Periodic comets, also known as *short-period comets*, are those with orbital periods of less than 200 years (a somewhat arbitrary division). Most of these may have come from the Kuiper Belt, beyond Neptune. A typical short-period comet has a period of from 5 to 9 years. Most are too faint for observation with small telescopes, but some can be observed occasionally with binoculars. Of the more than 270 known short-period comets only Halley's Comet consistently attains naked-eye brightness. However, even a very bright comet may escape detection if it approaches us from the direction of the Sun. As most comets have highly elongated orbits (i.e. high orbital eccentricity), they are observable with small telescopes only during a small portion of their orbit near perihelion.

As many as 30 comets may be seen by Earth-based observers each year, of which perhaps a dozen are known comets returning while the rest are new discoveries. Amateur discoveries are now severely restricted by the existence of dedicated professional surveys, but a handful of new comets are still found by amateurs each year. In addition, in recent years a spacecraft called the Solar and Heliospheric Observatory (SOHO), which observes the Sun and its surroundings, has recorded up to a hundred comets each year passing close to the Sun that would otherwise have escaped detection from Earth. Most of these comets have been first spotted by amateurs accessing SOHO images on the Internet.

A new comet is generally seen first as a faint, small, misty object, superficially rather like a nebula (with or without central condensation) but with definite individual motion. At the heart of a comet is a tiny irregular *nucleus* a few kilometres across that consists of frozen ices and dust particles. As the comet approaches the inner Solar System the ices sublime and some of the dust grains are released, creating an atmosphere of gas and dust particles that we observe as the comet's fuzzy head or *coma*. The higher density of particles near the nucleus gives rise to a brighter central condensation that obscures the nucleus from view. Sometimes this condensation is star-like in appearance, and is referred to as the *nuclear condensation* or *false nucleus*.

Many comets have both *dust tails* and *gas tails* formed by the vaporization of ice from the nucleus. The gas (or plasma) tail is formed when ions in the coma are accelerated by the solar wind along lines of magnetic force. Gas tails are usually straight and point directly away from the Sun. The dust tail is caused by the pressure of sunlight on dust grains. When radiation meets small particles with a diameter about one-third the wavelength of the incident radiation, the resulting pressure can be 20 or 30 times greater than the gravitational attraction of the Sun. Hence the dust particles are pushed out into orbits of their own, which lag behind the motion of the comet so that the dust tail may be curved or fan-shaped.

The comet's loss of material into the coma and tail increases as it nears the Sun, and decreases again on receding from the Sun. The release and behaviour of this material varies from comet to comet and thus makes the prediction of cometary brightness very tricky. An often-used equation for total visual magnitude (m_1) of the coma is known as the *power-law formula*. While it has some value in giving an idea of how a comet's brightness will progress (and so is used extensively in published ephemerides), it is usually applicable over only a small fraction of the comet's orbit. This formula is

$$m_1 = H + 5\log \Delta + 2.5n \log r$$

where n is the power-law exponent, and Δ and r are the comet's geocentric and heliocentric distances in astronomical units. For newly discovered comets, n is usually assumed to be 4 for the purpose of calculating ephemerides; for so-called *nuclear magnitudes* it is usually assumed that $n = 2$. H in the formula is the 'absolute magnitude' of the comet, which would be its brightness if it were exactly 1 AU from both the Sun and the Earth.

Comets are named after their discoverer (or occasionally the observatory or artificial satellite from which they were discovered; or in a few cases the people who computed their orbits, such as Halley, Encke or Crommelin); a maximum of three names per comet are allowed. In the system introduced in 1995, comets are designated by the year and month of discovery. Each month is divided into two periods, consisting of days 1–15 and 16–end, which are assigned an upper-case letter from A (January 1–15) to Y (December 16–31), the letter I being omitted. Additionally, a numeral shows the order of discovery in that half-month.

A prefix P/ or C/ denotes whether the comet is periodic or not. For example, Comet Hale–Bopp was discovered on 1995 July 23, and was the first to be discovered in the period July 16–31 (period O), so its designation is C/1995 O1 (Hale–Bopp). Periodic comets that have been seen to return at least once are given a number that indicates the order in which their periodicity was established, such

as 1P/Halley or 109P/Swift–Tuttle. Some periodic comets can now be observed around their entire orbits.

Comets that have become lost or no longer exist are assigned the prefix D/ (for defunct), such as D/1993 F2 (Shoemaker–Levy 9), which broke into more than 20 parts and crashed into Jupiter in 1994 July. The prefix X/ is used where no reliable orbit could be calculated, and A/ may be applied where a comet is subsequently deemed to be asteroidal or is no longer active (with no trace of a coma).

The release of gas and dust from some comets may be suddenly enhanced, leading to an outburst and temporary brightening; one comet noted for its occasional brightenings is 29P/Schwassmann–Wachmann 1. Some comets have been observed to split into two or more parts, such as 3D/Biela, which split and returned as two comets in 1846 and 1852, and C/1975 V1 (West), which split into four major parts in 1976. Much of the dust released eventually spreads right around the comet's orbit. When the Earth intersects this stream of dust a meteor shower may be observed. Many showers are associated with known periodic comets, but the Geminids in December follow the orbit of minor planet (3200) Phaethon.

The orbit of a comet is conveniently defined by the time of perihelion passage (T), perihelion distance (q), eccentricity (e), and three angles which define the orientation of the orbit in space: the argument of perihelion (ω), the longitude of the ascending node (Ω) and the orbital inclination (i). Selected orbital elements of some comets are given in Table 32. Since the orbit is continually being perturbed by the gravitational attractions of the planets, it is also necessary to state the epoch to which these orbital elements apply. In addition, the jet-like release of material from the nucleus results in 'non-gravitational forces' that have to be taken into account in making accurate predictions of a comet's return.

Observing Comets

The four areas where amateur astronomers can contribute most are in seeking new comets; measuring the positions of known comets (astrometry); estimating the brightness of comets; and recording details of physical structure, either visually, photographically or with electronic devices.

For comet-seeking, a moderate-aperture instrument and low magnification are preferable. A good choice of visual instrument is a large pair of binoculars of 100 to 150 mm aperture and a magnification of ×20 to ×30. Most visual comet discoveries are made when the comet is within 100° of the Sun, either in the eastern sky before dawn or in the western sky after dusk, because comets do not usually brighten above 10th or 11th magnitude until they are relatively close to the Sun.

TABLE 32. *Orbital elements of some comets. The orbital elements of a comet, especially the period, can change considerably over the course of time as a result of planetary perturbations.*

COMET	ORBITAL PERIOD (y)	INCLINATION (DEGREES)	PERIHELION DISTANCE (AU)	ECCENTRICITY	NOTES
2P/Encke	3.3	12	0.34	0.85	Parent of the Taurid meteors
26P/Grigg–Skjellerup	5.1	21	1.00	0.66	
96P/Machholz 1	5.2	60	0.12	0.96	Parent of the Quadrantid meteors?
10P/Tempel 2	5.5	12	1.48	0.52	
22P/Kopff	6.5	5	1.58	0.54	
21P/Giacobini–Zinner	6.6	32	1.03	0.71	Parent of the Giacobinid meteors
3D/Biela	6.6	13	0.86	0.76	Parent of the Andromedid meteors. Disintegrated
36P/Whipple	8.5	10	3.09	0.26	
29P/Schwassmann–Wachmann 1	14.7	9	5.72	0.04	
55P/Tempel–Tuttle	33.2	163	0.98	0.91	Parent of the Leonid meteors
1P/Halley	76.0	162	0.59	0.97	Parent of the Eta Aquarid and Orionid meteors
109P/Swift–Tuttle	135	113	0.96	0.96	Parent of the Perseid meteors
153P/Ikeya–Zhang	364	28	0.51	0.99	Longest-period comet seen more than once
C/1861 G1 (Thatcher)	415	80	0.92	0.98	Parent of the Lyrid meteors
C/1956 R1 (Arend–Roland)	—	120	0.32	1.00	
C/1965 S1 (Ikeya–Seki)	880	142	0.008	1.00	Member of the Kreutz sungrazer group, many comets that are parts of what was once a much larger object
C/1969 Y1 (Bennett)	—	90	0.54	1.00	
C/1973 E1 (Kohoutek)	—	14	0.14	1.00	
C/1974 V2 (Bennett)	—	135	0.86	1.00	Disappeared (disintegrated?) just before perihelion
C/1975 V1 (West)	—	43	0.20	1.00	
C/1980 E1 (Bowell)	—	2	3.36	1.06	
C/1983 H1 (IRAS–Araki–Alcock)	—	73	0.99	0.99	Passed 0.03 AU from Earth on 1983 May 11
C/1995 O1 (Hale–Bopp)	—	89	0.91	1.00	
C/1996 B2 (Hyakutake)	—	125	0.23	1.00	Passed 0.10 AU from Earth on 1996 March 25

Source: Brian G. Marsden and Gareth V. Williams, Catalogue of Cometary Orbits, *15th edition (2003).*

Visual seekers should have a careful sweeping routine in which adjacent sweeps overlap slightly and in which the setting or rising regions are accounted for. If an unknown nebulous object is found, the comet hunter should consult one or more detailed star charts and cometary ephemerides to see if the object can be identified as a galaxy or nebula, or as a known comet. A faint close grouping of stars can often be mistaken for a diffuse object; one should also be mindful of ghost images. To check a possible visual discovery, different eyepieces (and preferably also different instruments) should be used to look at the object; in photographic observing, images on two or more photographs should be obtained. A comet will normally betray its nature by its motion over a period of an hour or two. Possible discoveries should be reported to the nearest national observing organization (for addresses see the Appendix), which will forward appropriate information to the Central Bureau for Astronomical Telegrams at the Smithsonian Astrophysical Observatory in the U.S.A. for possible further verification and official announcement of the discovery.

When reporting a discovery, the following information must be given: dates and times (UT) of observations; observing location and equipment used; full name and address, telephone number and email address of discoverer; the object's appearance (including brightness, and amount of condensation, diffuseness and tail, if applicable), and measured positions (with equinox stated) at the times provided.

Photographing comets can yield useful results, such as depicting the changing tail structure in a bright comet. Astrometry is an area in which amateurs have long contributed useful information, particularly on newly discovered comets which are in need of preliminary orbit computations. For astrometry, an accuracy approaching 1 arcsec is required, for which a telescope or camera with a focal length of 1–2 m can give sufficient image scale without too small a field. Most positions of comets are now obtained with CCD devices and software that allows quick and accurate reduction of the comet's position. This process can be quicker than photography and does not require a mechanical measuring machine.

Visual observations of cometary structure are best made at the telescope with a drawing pencil on white paper. Care should be taken to reproduce accurately the relative sizes and intensities of any structure that is visible. Two directions, for example north and east, should also be carefully indicated on the drawing, together with a small scale-bar giving the scale in minutes of arc.

The total brightness of the coma can be useful – though difficult to measure, because the visual brightness varies tremendously with atmospheric conditions and with even small changes in aperture and magnification. The total brightness of the coma is usually measured by defocusing comparison stars and comparing them with the comet. The three most common methods are as follows:

(a) The *Bobrovnikoff method*, whereby the comet and comparison stars are defocused by the same amount, until the stars' sizes approximately equal the size of the defocused comet. The estimate is then made directly by comparing the known brightnesses of the comparison stars with that of the defocused comet. This method should be used only for comets with intense condensations, i.e. generally the brighter objects, where the apparent coma is usually very small.

(b) The *in-out* or *Sidgwick method*, whereby the memorized image of the in-focus comet is compared with equal-sized, defocused images of comparison stars. There are problems with this method when the coma size is large and the comet has a bright condensation.

(c) A method developed by C. S. Morris in which the comet is defocused just enough to give it an approximately uniform surface brightness. This image is memorized and compared with comparison stars which are defocused to this same size (note that the stars must be placed further out of focus than the comet to obtain equal image sizes).

In each case at least three different comparison stars should be used. It is also essential to record the instrument (including aperture, type, *f*-number and magnification) used for the magnitude estimate, as well as the apparent coma diameter and the degree of condensation (usually estimated by giving it a number on a scale of 0, totally diffuse, to 9, completely stellar).

Meteors

A meteor, known popularly as a *shooting star* or *falling star*, is caused when a particle of interplanetary dust (a *meteoroid*) enters the Earth's atmosphere at high speed. Collision of the meteoroid with air molecules produces frictional heating, which normally vaporizes the particle completely. The vaporized atoms from the meteoroid collide with more air molecules, and the energy of the collisions strips electrons from the atoms and molecules, a process called ionization. This ionization forms a long train of positively charged ions and negatively charged free electrons behind the meteoroid.

A typical meteor trail occurs at heights of 80 to 100 km above the Earth's surface, and lasts on average for only a fraction of a second, although some can persist for several seconds. During this short time the ions and electrons recombine, giving off light as they do so. It is this brief streak of light that we call a meteor. Most dust particles that cause meteors come from comets.

Meteors range in brightness from faint telescopic objects of brief duration to bright fireballs lasting several seconds. A few meteors per hour can be seen on any clear, moonless night. These are called *sporadic* meteors, and they enter the atmosphere at random.

Meteor Showers

At certain times each year the Earth encounters streams of cometary dust which produce meteor showers. Members of a meteor shower seem to emanate from a particular point on the celestial sphere known as a *radiant*. About ten major showers occur each year, with rates of between about 10 and 100 meteors per hour, plus a large number of showers of feebler activity. Nearly all showers are named after the constellation in which the radiant lies, for example the Perseids seem to radiate from Perseus, the Geminids from Gemini.

There are considerable differences from one shower to another. In some the meteors move swiftly, in others they move rather slowly; some have a higher proportion of fireballs than others. Some showers produce a higher proportion of meteors with persistent trains than others.

The intensity of a shower may vary considerably with time. The peak activity of the Quadrantids, for instance, is well defined and lasts only a few hours. On the other hand a shower such as the Orionids has no sharp peak, but its maximum is rather flat, lasting two or three days.

The activity of a particular shower can vary from year to year. Many showers, such as the Taurids and Geminids, show much the same level of activity each year. By contrast, the activity of the Leonids is modest in most years, but at 33-year intervals there can be extraordinary displays with rates of thousands of meteors per hour for short periods. This last happened in 1966, and rates of up to 4000 an hour were seen from 1999 to 2002. The activity of a shower is measured by its *zenithal hourly rate* (ZHR), which is the number of meteors that would be seen by a single observer under a perfectly clear sky with the radiant overhead.

Table 33 gives details of the major night-time streams. The dates for the limits of activity and the peak ZHR should be used as only a rough guide, for the activity of showers can vary with time.

The radiant positions are quoted for the date of maximum activity but, in practice, radiants move slightly from night to night so for other dates the radiant daily motion (where listed in Table 33) should be added or subtracted.

TABLE 33. *Principal meteor showers.*

SHOWER	DATES		APPROXIMATE ZHR AT MAXIMUM	RADIANT				PARENT COMET (WHERE KNOWN)	NOTES
	NORMAL LIMITS	MAXIMUM		POSITION AT MAXIMUM (2000.0)		DAILY MOTION (WHERE KNOWN)			
				RA (h m)	DEC. (°)	RA (°)	DEC. (°)		
Quadrantids[a]	Jan. 1–6	Jan. 3	100	15 30 (232)	+50			96P/Machholz (?)	Medium speed; blue
Corona Australids	Mar. 14–18	Mar. 16	5	16 24 (246)	−48				
Lyrids	Apr. 19–25	Apr. 22	10	18 10 (272)	+32	+1.1	0.0	C/1861 G1 (Thatcher)	Swift; brilliant
Eta Aquarids	May 1–10	May 6	35	22 23 (336)	−01	+0.9	+0.4	1P/Halley	Very swift; persistent trains
Ophiuchids	June 17–26	June 20	5	17 23 (261)	−20				
Capricornids	July 10–Aug. 15	July 25	5	21 03 (316)	−15				Yellow; very slow
Delta Aquarids	July 15–Aug. 15	July 29	20	22 39 (340)	−17	+0.8	+0.18		Double radiant; rather faint meteors
		Aug. 7	10	23 07 (347)	+02	+1.0	+0.2		
Piscis Australids	July 15–Aug. 20	July 31	5	22 43 (341)	−30				
Alpha Capricornids	July 15–Aug. 25	Aug. 2	5	20 39 (310)	−10	+0.9	+0.3	45P/Honda–Mrkos–Pajdušáková	Bright yellow; slow
Iota Aquarids	July 15–Aug. 25	Aug. 6	8	22 15 (334)	−15	+1.07	+0.18		Double radiant; rather faint meteors
				22 07 (332)	−06	+0.13	+0.13		
Perseids	July 23–Aug. 20	Aug. 12	80	03 08 (047)	+58	+1.35	+0.12	109P/Swift–Tuttle	Swift; fragmenting; many bright events and trains
Giacobinids	Oct. 6–10	Oct. 8	<5	17 25 (261)	+57			21P/Giacobini–Zinner	Only active (if at all) when parent comet is near perihelion
Orionids	Oct. 16–27	Oct. 20–22	25	06 27 (097)	+15	+1.23	+0.13	1P/Halley	Very swift; with trains
Taurids	Oct. 20–Nov. 30	Nov. 3	10	03 47 (057)	+14	+0.79	+0.15	2P/Encke	Double radiant; very slow; flat maximum
				03 47 (057)	+22	+0.76	+0.10		
Leonids	Nov. 15–20	Nov. 17	15	10 11 (153)	+22	+0.70	−0.42	55P/Tempel–Tuttle	Bright with trains; exceptional activity every 33 years
Geminids	Dec. 7–15	Dec. 13	100	07 28 (112)	+32	+1.1	−0.07	Minor planet (3200) Phaethon	Medium speed; bright; few trains
Ursids	Dec. 17–25	Dec. 22	10	14 27 (217)	+78	+0.88	−0.45	8P/Tuttle	Usually weak

[a] Named after the former constellation Quadrans Muralis; the radiant is in Boötes.

Leap year adjustments cause small variations in the given dates. For example, the best night for Perseid activity in a particular year may be August 11/12 or 12/13. It is always wise to refer to a current handbook or almanac for topical detail of the year's showers, and for other showers that may be particularly active during the year.

Naked-eye Observations

The purpose of visual meteor work is to determine rates and brightness distribution. This is accomplished in watches, normally lasting for one hour, one or more of which are carried out by an observer on a given night.

For each watch the following basic details should be noted: observer's name; correspondence address; observing site, with latitude and longitude; date; time (UT) of start and end of watch to the nearest minute; average stellar limiting magnitude during the watch; and average percentage cloud cover in area of sky being watched.

For each meteor seen during the watch, the observer should note the following details: time (UT) to the nearest minute; meteor brightness to the nearest whole magnitude; whether the meteor was a sporadic, or a member of a shower active on that night; the duration of the associated persistent train, if present, and its behaviour with time; any other details on meteor speed, duration, colour, explosions, etc. Plots of the tracks of meteors are not usually required, except for fireball events or if a radiant position is needed for a suspected new shower.

All observations should be communicated to the meteor section of a national organization (for addresses see the Appendix).

Telescopic Observations

For this work one requires a telescope or a pair of binoculars with as wide a field as possible (and therefore a low power). The telescopic observer should supply much the same information as a visual observer. Details of the instrument used, the eyepiece field of view and the celestial coordinates of the centre of the field are required. The observer should also attempt to plot the paths of meteors seen against the field star background. However, telescopic meteor work demands great patience, as the observed meteor rates are usually considerably lower than for the naked-eye observer.

Photographic Observations

Ordinary cameras with lenses faster than $f/4.5$ and fast film of ISO 400 and above are suitable for meteor work.

Observers should make timed exposures of about 5 to 30 minutes, depending on their equipment and the observing conditions.

The start and end of each exposure should be noted in UT to the nearest 0.1 minute, as should the time, approximate position and nature of any bright meteor thought to have passed through the camera's field of view. Typically, only meteors of magnitude 0 or brighter will be recorded photographically; to increase the number of meteors recorded the observer should use a battery of several cameras rather than just one.

The addition of rotating shutters for velocity measurements, plus cameras at other sites for triangulation, will enable observers to calculate real paths and orbital data for photographed events.

Fireball Observations

If an object entering the atmosphere is large enough, it will produce a brilliant fireball (mag. -5 or brighter) with an apparent diameter similar to that of the Sun or the Moon. Sometimes the object itself will not be completely vaporized before reaching the ground. These surviving rocks, called *meteorites*, are of great importance and it is advantageous to examine them as quickly as possible after the fall. Organizations exist to receive fireball reports so that the fall area can be located quickly. Few major fireballs produce meteorites, but if a sonic boom is heard a meteorite is quite likely to fall.

There is a close similarity between natural fireballs and those produced by the re-entry of artificial satellites, although the latter usually move much more slowly than the natural objects and may take up to a minute to cross the sky.

Because no one can predict the arrival of a fireball, the witnessing of such an event is a matter of luck and reports tend to come from casual observers. Any observer of such an event should contact their local national society or the International Meteor Organization (for addresses see the Appendix). Information on the following is required:

(a) Name and address of observer, observing site, date and time (UT).

(b) Track – if possible, the track through the star background. If this is not possible, then the direction and elevation angle when first seen, when highest in the sky and when last seen. If the object passes virtually overhead, then whether to the left-hand side or right-hand side as the observer faces the direction from which it came.

(c) The magnitude, size, shape and colour, details of tail and presence or absence of dust trail, and fragmentation and extinction.

(d) Sonic effects – estimate of time interval between the passage of the object (known in this case as a *bolide*) and the sonic boom.

Calculation of Meteor Rates

The observed rates of both shower and sporadic meteors depend on the limiting magnitude, cloud cover and the observer's perception. Moonlight significantly reduces the number of meteors visible, and for five days either side of the full moon only bright meteors will be seen. Shower rates also depend on the elevation of the shower radiant above the horizon – the lower the radiant, the fewer shower meteors will be seen.

The activity of a shower is expressed in terms of its ZHR, which is the number of meteors that would be seen by a single observer if the radiant were at the zenith in a clear sky with a limiting magnitude of 6.5. The *sporadic hourly rate* (SHR) is defined as the hourly rate of sporadic meteors seen by a single observer under clear skies. Hence:

$$\text{ZHR} = (N_{sh}/t) \times R \times C_{sh} \times F$$

$$\text{SHR} = (N_{sp}/t) \times C_{sp} \times F$$

where N_{sh} and N_{sp} are respectively the number of shower and sporadic meteors seen during the watch, t is the watch duration in hours, R is the correction factor for radiant elevation at midwatch, C_{sh} and C_{sp} are the correction factors for limiting magnitude for shower and sporadic meteors respectively, and F is the cloud correction factor. The various correction factors are further explained below.

Radiant elevation correction factor (R)

$$R = 1/\sin a$$

where a is the radiant elevation in degrees. This is the simplest formula, though others have been proposed; the higher the elevation, the closer the agreement between the various formulae. The values in Table 34 have been calculated from the above formula. As can be seen, for radiant altitudes lower than 30° the observed rate will be far less than the theoretical ZHR.

Sky limiting magnitude correction factor (C)

$$C = r^{6.5-LM}$$

where r is the *population index* (the ratio between the true number of meteors in adjacent magnitude classes) and LM is the limiting magnitude. The value for the population index varies among showers, and is usually higher for sporadics than for shower meteors, so no precise figures can be quoted. Observations show that for major showers values of r between 2.2 and 2.5 are reasonable, while for sporadic meteors a value from 2.5 to 3.0 is usually adopted.

For a perfectly clear sky $LM = 6.5$, and hence the correction factor is 1. The clearer the sky the better, since C is then closer to 1 and relatively insensitive to the value of r. The poorer the sky, the more uncertain are the derived rates.

TABLE 34. *The radiant elevation correction factor R for various altitudes of the radiant.*

ALTITUDE (DEGREES)	R	ALTITUDE (DEGREES)	R
5	11.5	35	1.7
10	5.8	40	1.6
15	3.9	45	1.4
20	2.9	50	1.3
25	2.4	65	1.1
30	2.0		

Cloud correction factor (F)

$$F = 100/(100 - K)$$

where K is the percentage cloud in the area of sky under watch, averaged over the watch period. For a clear sky $F = 1$.

COMMENTS The above computations are relevant only to the results of a single observer. Correction factors have been derived for two or more observers, but rather than have the uncertainty of yet another correction factor introduced into the computations, observers should keep individual records when observing with others.

A final correction factor, ignored above, is that of contamination of shower rates by sporadic meteors. Since sporadic meteors have random directions, a fraction of them will have paths that align by chance with a shower radiant and will be wrongly recorded as shower members. The effect is small, however, and becomes significant only when considering showers with low rates.

Eclipses

Solar 'eclipses' are actually occultations of the Sun by the Moon and can occur only at new moon. A solar eclipse can be seen from only a limited area of the Earth's surface. Lunar eclipses occur when the Moon passes into the shadow of the Earth; this can happen only at full moon. Lunar eclipses are seen over rather more than half the Earth, wherever the Moon is above the horizon.

Each year at least two solar eclipses occur, and occasionally as many as five. Some years there are no lunar eclipses at all but there can be as many as three, either total or partial (but not counting penumbral eclipses, when the Moon enters only the faint outer penumbra of the Earth's shadow). Solar eclipses outnumber lunar eclipses in the ratio of nearly 5 to 3. Between 2001 and 2100 there are 224 solar eclipses, total and partial. As seen from any one place, though, lunar eclipses are about twice as common as solar ones.

On average, four eclipses occur each year: two pairs of solar and lunar eclipses, separated by about six months. The maximum number of eclipses in a year is seven: either five solar and two lunar or four solar and three lunar, depending on the precise configuration of the Sun and the Moon. In years with seven eclipses, the solar eclipses are always partial; the lunar eclipses can be total or partial. The last year with seven eclipses was 1982 (when all three lunar eclipses were total). Seven eclipses will occur again in 2094 and 2159, four of the Sun and three of the Moon each time.

Other examples of eclipses are those of satellites when they pass through the shadow of their parent planet, or through the shadow of another satellite; and the 'eclipses' (again, actually occultations) of the components of an eclipsing binary star system.

Predictions of eclipses each year are published in almanacs such as the *Handbook* of the British Astronomical Association and the *Observer's Handbook* of the Royal Astronomical Society of Canada. The *Fifty Year Canon of Solar Eclipses* by Fred Espenak (NASA, 1987; available from Sky Publishing Corp.) gives predictions and maps of solar eclipses from 1986 to 2035, and lists the general characteristics of every solar eclipse from 1901 to 2100. The same author's *Fifty Year Canon of Lunar Eclipses* does the same for the Moon. Eclipses of the Sun and Moon from 2004 to 2018 are listed in Table 35.

ECLIPSE MAGNITUDES   The amount by which the Sun or Moon is eclipsed is expressed by a figure termed the *magnitude*. For a solar eclipse, this refers to the fraction of the Sun's diameter that is covered by the Moon, measured along the common diameter. For a lunar eclipse, the magnitude is the fraction of the Moon's diameter that is covered by the shadow of the Earth. At a total eclipse, solar or lunar, the magnitude is 1.00 (100%) or greater.

Solar Eclipses

Solar eclipses may be *total*, *partial* or *annular*, depending on the relative positions and distances apart of the Sun, Moon and Earth at the time. As much as three hours may elapse between first and last contact, but totality may last from an instant, through an average of three or four minutes, to a maximum of 7m 31s as seen from a point on the equator with the Moon at perigee; the longest eclipse in the next few centuries (Table 36) is that of 2186 July 16, predicted to last 7m 29s.

The width of the Moon's umbra at the Earth can reach 273 km, but is on average less than 160 km. However, the width of the eclipse path can be considerably greater than this if the shadow falls obliquely on the Earth, as it can near the poles. For example, the path of the total eclipse of 2033 March 30, visible from within the Arctic Circle, will have the extraordinary width of 777 km. The zone within which a partial eclipse is seen is at least 3200 km either side of the limit of totality. Total eclipses are rare events for an observer at any one point on the Earth's surface.

An annular eclipse occurs when the Moon is near apogee. At these times it does not appear large enough to cover the Sun completely, and the dark disk of the Moon is surrounded by a bright ring (*annulus*) of sunlight. The maximum shadow width at the Earth during an annular eclipse is 313 km, but again the actual path width can be considerably greater if the shadow falls obliquely – for example nearly 4500 km at the annular eclipse of 2003 May 31. Annularity can be instantaneous (Baily's beads around the entire disk, as on 1966 May 20). It has a maximum duration of 12m 30s viewed at the equator with the Moon at apogee, although this theoretical maximum is rarely approached; according to calculations by Jean Meeus the longest annular eclipse between the years 0 and AD 3000 was 12m 23s in AD 150.

Partial eclipses are of little importance, but annular eclipses provide a sensitive way of checking for possible changes in the Sun's diameter. Most important of all are total eclipses, which provide the only opportunity of seeing the Sun's chromosphere, prominences and corona without the need for special instruments. When the Sun is completely hidden the sky becomes dark like deep twilight, and the bright planets and stars can be seen with the naked eye. The Moon appears as an intensely black disk surrounded by the bright inner corona, which farther out changes into the delicate brush-strokes and plumes of the pearly outer corona. Some red-tinted prominences may also be seen.

BAILY'S BEADS are sometimes seen for a few seconds at a total eclipse immediately before and after totality, or during a brief annular eclipse, when the thin crescent of the Sun is broken up into a series of bright moving points by mountain peaks along the limb of the Moon, giving the appearance of a string of shining beads. The name derives from the English astronomer Francis Baily, who first described and discussed the phenomenon after the eclipse of 1836. After only a few seconds this display narrows down to just one bead which, together with the appearance of the inner corona around the entire disk, produces the dazzling *diamond ring* effect.

SHADOW BANDS may also occur, shortly before and after totality. They are bands of shadow 100–150 mm wide and up to a metre apart crossing any light-coloured surface, rather like the projection of surface ripples onto the bottom of a bowl of water, and a very clear, transparent sky is needed for them to occur. They are probably caused by irregular refraction of light from the thin crescent Sun in the Earth's atmosphere.

TABLE 35. *Eclipses from 2004 to 2018 (excluding partial and penumbral eclipses of the Moon).*

DATE	BODY ECLIPSED	TYPE OF ECLIPSE	MAXIMUM DURATION m:s	AREA OF VISIBILITY
2004 Apr. 19	Sun	Partial		S Africa, S of South Atlantic Ocean
May 4	Moon	Total		Antarctica, Australia, New Guinea, Indonesia, Philippines, Asia except NE, Africa, Europe except extreme N, extreme E South America
Oct. 14	Sun	Partial		E Asia, Hawaii, W Alaska
Oct. 28	Moon	Total		W Asia, Africa except extreme E, Europe, Iceland, Greenland, Arctic regions, the Americas
2005 Apr. 8	Sun	Annular/Total	0:42	Southern Pacific Ocean, part of Antarctica, S U.S.A., Central America, extreme W South America
Oct. 3	Sun	Annular	4:31	Africa except S tip, Europe, W Asia, Arabia, W India
2006 Mar. 29	Sun	Total	4:07	Africa except SE, Europe, W Asia, W India
Sept. 22	Sun	Annular	7:09	E South America, Atlantic Ocean, part of Antarctica, W and S Africa, S Madagascar
2007 Mar. 3/4	Moon	Total		W and central Asia, India, Africa, Europe, part of Antarctica, Iceland, Greenland, NE North America, Caribbean Sea, South America except S tip
Mar. 19	Sun	Partial		E Asia, W Japan, W Alaska
Aug. 28	Moon	Total		North America except NE, Canada, W South America, Pacific Ocean, part of Antarctica, Siberia, Japan, Australasia, E Asia
Sept. 11	Sun	Partial		S South America, SW Atlantic Ocean, part of Antarctica
2008 Feb. 7	Sun	Annular	2:08	Antarctica, SE Australia, New Zealand
Feb. 21	Moon	Total		W Asia, E Arabia, Africa, Europe, Iceland, Greenland, Arctic, Atlantic Ocean, the Americas except SW Alaska
Aug. 1	Sun	Total	2:27	Greenland, Iceland, Arctic Ocean, N Europe, Asia except extreme E and SE, Arabia, India
2009 Jan. 26	Sun	Annular	7:54	S Africa, Madagascar, part of Antarctica, Indian Ocean, S India, SE Asia, W Indonesia, S and W Australia
July 22	Sun	Total	6:39	China, Mongolia, SE Asia, Japan, Indonesia, W Pacific Ocean
2010 Jan. 15	Sun	Annular	11:08	E Africa, Madagascar, Arabia, Indian Ocean, S central Asia, India, SE Asia
July 11	Sun	Total	5:20	Pacific Ocean, extreme W central South America
Dec. 21	Moon	Total		N Europe, N Asia, Iceland, Greenland, Arctic, the Americas, Pacific Ocean, New Zealand, Japan
2011 Jan. 4	Sun	Partial		N Africa, E Europe, Arabia, E and central Asia, NW India
June 1	Sun	Partial		NE Asia, Arctic Ocean, N Alaska, N Canada, Greenland, Iceland
June 15	Moon	Total		Australasia, central Asia, India, Africa, Arabia, S Europe, S Atlantic Ocean
July 1	Sun	Partial		S Indian Ocean
Nov. 25	Sun	Partial		Antarctica, S Indian Ocean, S Tasman Sea
Dec. 10	Moon	Total		NW North America, Hawaii, Australasia, Asia, Arabia, E Europe, Arctic, Scandinavia, Iceland, Greenland
2012 May 20	Sun	Annular	5:46	NE Asia, N Philippines, N Pacific Ocean, NW North America, N Greenland, Arctic
Nov. 13	Sun	Total	4:02	E Australia, New Zealand, S Pacific Ocean, part of Antarctica
2013 May 10	Sun	Annular	6:04	Indonesia, E Australia, Pacific Ocean, New Zealand except S, Hawaii
Nov. 3	Sun	Annular/Total	1:40	Atlantic Ocean, S Europe, Africa except extreme E and S tip
2014 Apr. 15	Moon	Total		W Atlantic Ocean, SW Greenland, the Americas, Pacific Ocean, E Siberia, New Zealand, E Australia, part of Antarctica
Apr. 29	Sun	Annular	0:00	S Indian Ocean, W Australia, part of Antarctica
Oct. 8	Moon	Total		Part of Antarctica, W South America, Central America, North America except NE Canada, Arctic, Pacific Ocean, Australasia, Japan, E Asia
Oct. 23	Sun	Partial		N Pacific Ocean, E Siberia, W half of North America, Mexico
2015 Mar. 20	Sun	Total	2:47	E Greenland, Iceland, NE Atlantic Ocean, Europe, NW Africa, W Asia
Apr. 4	Moon	Total		W North America, Pacific Ocean, Antarctica, Australasia, E Asia
Sept. 13	Sun	Partial		Southern Africa, S Madagascar, S Indian Ocean, part of Antarctica
Sept. 28	Moon	Total		Antarctica, Africa except extreme E, Europe, Arctic, Atlantic Ocean, Iceland, Greenland, the Americas except Alaska, part of Antarctica
2016 Mar. 9	Sun	Total	4:10	SE Asia, Indonesia, Australia except SE, Japan, W Pacific Ocean
Sept. 1	Sun	Annular	3:06	Africa except extreme N, SW Arabia, Madagascar, Indian Ocean
2017 Feb. 26	Sun	Annular	0:44	S half of South America, most of Antarctica, S Atlantic Ocean, W Africa except N part
Aug. 21	Sun	Total	2:40	NE Pacific Ocean, Arctic Ocean, Northern and Central America, Caribbean, N half of South America, Greenland, Iceland, extreme NW Europe
2018 Jan. 31	Moon	Total		N Greenland, W North America, Mexico, Pacific Ocean, Australasia, Asia, India, N Scandinavia
Feb. 15	Sun	Partial		S Pacific Ocean, most of Antarctica, S half of South America
July 13	Sun	Partial		S tip of Australia, Tasmania, Tasman Sea
July 27	Moon	Total		Australasia except N New Zealand, SE Asia, S Japan, Indian Ocean, Africa, Europe, E and S Atlantic Ocean, E South America, part of Antarctica
Aug. 11	Sun	Partial		NE Canada, Greenland, Arctic Ocean, Iceland, Scandinavia, N and central Asia

Source: HM Nautical Almanac Office.

TABLE 36. *Forthcoming total solar eclipses lasting longer than 7 minutes.*

DATE	MAX. DURATION	DATE	MAX. DURATION
2150 June 25	7m 14s	2222 Aug. 8	7m 05s
2168 July 5	7m 26s	2504 June 14	7m 10s
2186 July 16	7m 29s	2522 June 25	7m 12s
2204 July 27	7m 22s	2540 July 5	7m 03s

Source: Jean Meeus, More Mathematical Astronomy Morsels *(Willmann-Bell, 2002).*

Lunar Eclipses

Lunar eclipses last longer than solar ones, up to 4 hours from first to last contact of the umbra, and over 6 hours including the penumbral stage. The maximum duration of totality is 1h 47m, as occurred on 2000 July 16. There is a noticeable variation from one eclipse to another in the brightness of the eclipsed Moon. Usually the Moon does not vanish completely, as light is refracted onto its surface through the Earth's atmosphere. The eclipsed Moon commonly appears a coppery red colour, but its exact colour and brightness depend on the conditions of cloud, haze and high-altitude dust in the Earth's atmosphere at the time. During a lunar eclipse, a rapid cooling of the lunar surface occurs as the Earth's shadow sweeps across it; detailed measurements have shown that certain so-called hot spots such as the crater Tycho cool less quickly than their surroundings. In a penumbral eclipse the Moon enters only the light outer part of the Earth's shadow (the penumbra), and often no dimming is noticeable to the eye at all.

DANJON SCALE   The Danjon scale, named after the French astronomer André Danjon, is used to describe the brightness of the Moon during a total lunar eclipse as seen with the naked eye or binoculars:

$L = 0$ Very dark eclipse; Moon almost invisible

$L = 1$ Dark eclipse, grey or brownish coloration, difficult to discern details

$L = 2$ Deep red or rust-coloured eclipse. The central shadow is dark while the outer edge of the umbra is bright.

$L = 3$ Brick red. Umbral shadow has a bright rim which may appear yellowish.

$L = 4$ Very bright copper-red or orange eclipse. Umbral shadow has a very bright rim, sometimes bluish in colour.

Occultations

An occultation occurs when any object obstructs the view of another, more distant object. Hence a total eclipse of the Sun is strictly an occultation, and eclipsing binary stars are really occulting binaries. Planets and asteroids occult stars, and mutual occultations occur among the satellites of Jupiter. However, the most frequent and readily observed types of occultation are those of stars by the Moon.

Lunar Occultations

As it moves in its orbit around the Earth, the Moon regularly occults background stars. Because the Moon has no atmosphere, the disappearance and reappearance of a star is instantaneous, so occultations can be timed very accurately by amateurs with only modest equipment. These timings reveal slight changes in the orbit of the Moon and the gradual slowing of the rotation of the Earth due to tidal friction. The motion of the Moon is used in the determination of Terrestrial Time (TT), which gives a smooth measure of time free from the short-term irregular variations in the Earth's rotation. In addition, occultation timings sometimes reveal errors in the accepted positions and proper motions of stars, and a stepwise fading to disappearance rather than an instantaneous vanishing of the star can indicate the existence of previously unrecognized double or multiple stars.

Since the Moon moves through about 1 second of arc in 2 seconds of time, a timing accuracy of only 0.2 s is needed to give the Moon's position to 0.1 arcsec, corresponding to about 200 m in the Moon's orbit. *Grazing occultations*, when the star skims the upper or lower limb of the Moon, are particularly valuable for determining the precise position of the Moon in its orbit.

OBSERVING LUNAR OCCULTATIONS   A small telescope will permit observation of many occultations of stars of mag. 6.5 and brighter. Disappearances and reappearances are much easier to observe at the dark limb than at the bright limb of the Moon, and disappearances are easier to time accurately than reappearances since the star is in view right until the moment of occultation. The Moon approaches the star in a direction approximately at right angles to the line joining the cusps, or horns, of the Moon. The rate of approach of the Moon to a star is about half a degree (its own apparent diameter) every hour, or the apparent diameter of the crater Copernicus every 1½ minutes.

One simple method of timing occultations requires a good-quality stopwatch, which is started when the star is seen to disappear or reappear and is stopped as soon as possible on a time signal.

The telephone time service is very convenient, and short-wave radio time services are available in most parts of the world. The latter system allows occultations occurring in rapid succession to be recorded, and is essential for the observation of grazing occultations when many events may be seen as the star skims the rugged profile of the lunar limb. The usual method is to leave a tape recorder running with the time signals in the background and to call out the disappearances and reappearances as they are seen. Increasing use is being made of lightweight video cameras attached to the telescope with a visual time-display recorded on the tape. Both these methods allow the results to be analysed at leisure with high accuracy. For all occultations, the observer's location and altitude must be accurately established from a map.

The world centre for predictions and the subsequent reduction of reports is the International Lunar Occultation Centre (ILOC) in Japan. There is also the International Occultation Timing Association (IOTA) in the U.S.A. (see the list of Useful Addresses in the Appendix). Many national astronomical societies coordinate observations among their own members and publish predictions.

Other Occultations

Occasionally stars are occulted by asteroids. Predictions of these events are distributed to members by the International Occultation Timing Association, together with the track across the Earth from which the occultation may be seen. The observing technique is similar to that for lunar occultations, except that the occulting body will not usually be visible since the occulted star is normally much brighter than the asteroid. Accurate timing by several observers of the duration of such an occultation yields a cross-section of the asteroid, allowing its size and shape to be derived and its orbit refined.

From time to time, planets and their satellites occult stars. Two types of event are possible, depending on whether the planet or satellite has an atmosphere. If there is no atmosphere, the light from the star is cut off instantaneously (or in two steps if it is a double), as in a lunar occultation. If the occulted star is large, high-speed photoelectric photometry will show diffraction fringes from which the apparent angular diameter of the star can be derived. When the occulting object has an atmosphere, the star undergoes rapid brightness variations superimposed on a general fading. If the observer is near the centre of the track a central 'flash' may be detected where the light is focused by the body's atmosphere, as happened in 1989 when Titan occulted the star 28 Sagittarii. Timings and photoelectric photometry traces of the variations are valuable for modelling the atmospheric structure of the occulting planet or satellite.

MUTUAL SATELLITE PHENOMENA The satellites of Jupiter and Saturn can occult and eclipse one another when their orbital orientation is favourable. This happens for a short period when the Earth passes through their orbital plane, which occurs every 6 years for Jupiter and alternately every 13 and 15 years for Saturn (because of its appreciably eccentric orbit). Such events give the most precise positions for the satellites, which is valuable for refining their orbits. Predictions of these events, which typically last for 5 minutes, are found in almanacs such as the *Handbook* of the British Astronomical Association. The most favourable can be followed visually, and in some cases the brightness change can reach 1 magnitude. The remainder require the use of either a CCD camera or a photoelectric photometer.

Aurorae, Noctilucent Clouds and the Zodiacal Light

Aurorae

Aurorae are glows in the upper atmosphere of the Earth caused when fast-moving atomic particles (protons and electrons) arrive in the solar wind. Collisions of electrons with atoms and molecules of atmospheric nitrogen and oxygen at heights above 100 km lead to the emission of light, seen as an aurora. Aurorae are commonest at times of high sunspot activity, when solar flares or coronal mass ejections inject 'pockets' of energetic particles into the solar wind. Persistent particle streams emitted from coronal holes can also give rise to aurorae (often recurring at 27-day intervals, corresponding to the rotation period of the Sun), but these tend to be less active and extensive than the displays that follow solar flares.

Aurorae are most commonly seen at high latitudes close to the *auroral ovals*, rings of permanent auroral activity 4000 to 5000 km in diameter that surround the Earth's north and south magnetic poles. Disturbances of the Earth's magnetic field following violent solar activity cause the auroral ovals to expand towards the equator. At such times observers at lower latitudes such as those of the British Isles, central North America and Australasia will see auroral displays.

Mid-latitude aurorae, while never exactly the same from one display to the next, do follow a fairly typical pattern. A display may begin as a fairly weak *glow* on the poleward horizon, aptly described by the name *aurora borealis* ('northern dawn'); the southern-hemisphere equivalent is similarly known as *aurora australis*. This glow may remain relatively static or it may eventually fade away, and simply represents the uppermost parts of a display that is more impressive at higher latitudes.

Under more highly disturbed conditions, the glow may brighten and rise higher into the sky, taking on the form of an *arc* whose lower edge is usually more sharply defined than the upper edge. Folding of an arc produces a ribbon-like *band*. In active displays, vertical *rays* resembling searchlight beams develop along the length of arcs or bands. Rays frequently drift either eastwards or westwards. Sometimes, if a display lies far from the observer, isolated rays extending over the poleward horizon may be the only activity seen.

During extreme disturbances, and rather rarely at temperate latitudes, auroral activity can extend beyond the zenith. In such displays may be seen a *corona*, in which the rays appear to converge on a small area of sky as a result of perspective. Discrete *patches* of auroral light are also seen in some displays, and occasionally they can comprise entire displays.

The brightness of aurorae is rated on a scale from i (the faintest) to iv (the brightest). Aurorae at temperate latitudes range in brightness from weak glows comparable to the Milky Way (brightness i), via moderate displays comparable to moonlit cirrus cloud (brightness ii) to prominent displays comparable to moonlit cumulus cloud (brightness iii). At high latitudes, intense aurorae can cast shadows (brightness iv). Changes in brightness are commonly seen in aurorae. These range from slow *pulsing* over the course of several minutes to extremely rapid *flaming* with a period of seconds. In flaming activity, waves of brightening sweep upwards from the horizon through the aurora, often in the declining phase of a display.

OBSERVING AURORAE   Aurorae are diffuse light sources and are therefore best observed with the naked eye. Accurate measurements of the altitude and azimuth of features in a display, made with a simple sighting device (an *alidade*), can be of considerable value in assessing the geographical extent of the aurora at a given time, particularly if several sets of measurements from widely separated observers are available.

Two altitude measurements are of use: the altitude, in degrees, of the highest point on the base of an arc or band (symbol *h*), and the uppermost extent of the display, also in degrees (symbol ↗). An indication of the auroral forms present at the time of observation is useful, along with brightness estimates. Auroral features may be described as *quiet* or *active*. Thus an observer's record of a display might be as follows:

2000 (UT)   Quiet auroral glow, brightness i,
↗ 10°, 330–020 azimuth

2025 (UT)   Active rayed arc, brightness iii, *h*12° ↗ 40°,
320–040 azimuth

Standard reporting codes recommended by various observing organizations are convenient for providing more concise descriptions of auroral displays.

Photography allows rapid and accurate recording of aurorae. There are few hard-and-fast rules, but the best results seem to be obtained using ISO 400 colour film. Exposures at *f*/2 for 30 to 60 seconds should record weak, static displays, while bright active aurorae should be exposed for only 5 to 10 seconds so as to avoid loss of detail caused by the aurora's motion.

Faint aurorae will normally appear colourless, but more active displays can show pronounced green or red colours. The green colour, which to the naked-eye observer is usually predominant, corresponds to a wavelength of 558 nm; this is caused by excited oxygen atoms, as is the 630 nm red emission. Other colours correspond to emissions from excited nitrogen atoms and molecules. Photographs often enhance auroral greens and reds as a result of the particular colour sensitivities of some emulsions.

Reports of auroral activity are collected by a number of bodies, notably the Aurora Sections of the British Astronomical Association and Royal Astronomical Society of New Zealand.

Noctilucent Clouds

Upwelling of cold air from the polar regions of the lower atmosphere during the summer months in each hemisphere carries traces of water vapour to great heights. Condensation of this water vapour, possibly around nuclei provided by meteoric or volcanic debris, produces very tenuous cloud formations, mainly at latitudes of 60° to 80°. The thin sheets of these noctilucent clouds, forming at heights of 80 to 85 km, lie considerably higher in the atmosphere than the cirrus clouds (maximum height 15 km) to which they bear a superficial resemblance.

At their great height, noctilucent clouds remain sunlit (above the Earth's shadow) long after clouds in the lower atmosphere are in darkness, and are clearly visible in the twilit summer night sky at high temperate latitudes while the Sun is between 6° and 16° below the observer's horizon. Noctilucent clouds can therefore be distinguished by the 'night-shining' nature from which they take their name: clouds in the lower atmosphere are often seen dark in silhouette against noctilucent cloud displays.

Noctilucent clouds often show a distinctive silvery-blue colour, shading to gold towards the sunward horizon. Displays are usually most extensive in the early evening (fading around midnight when the Sun is farthest below the observer's horizon) or just before dawn. Like aurorae, noctilucent clouds show a small range of characteristic structures, classified as follows:

Type I *Veil* – A structureless sheet sometimes seen on its own but usually as a background to other forms.

Type II *Bands* – Long horizontal streaks, either sharp or blurred in appearance, and generally parallel to the horizon.

Type III *Billows* – Short interwoven lines or rippled structures, often compared to the wavy patterns left on a sandy beach at low tide.

Type IV *Whirls* – Large-scale curves or loops.

Noctilucent clouds are most frequently seen in the summers around solar minimum. It seems that heating of the Earth's upper atmosphere by short-wavelength emission from solar flares inhibits noctilucent cloud formation, which is why aurorae and noctilucent clouds are seldom seen together. Observations indicate the existence of a complicated weather system at such altitudes, with winds carrying noctilucent clouds westwards at up to 400 km h^{-1}.

The British Isles, Scandinavia and Canada are well placed for the observation of noctilucent clouds, having long hours of twilight during summer nights and lying reasonably close to the latitudes at which the clouds form. Southern-hemisphere land-masses are less favourably placed, though observations can be made from parts of the South Atlantic and Antarctica.

As with aurorae, measurements of the extent in altitude and azimuth of noctilucent cloud displays are useful. Simple brightness estimates based on the following numbering convention are also valuable: 1 = faint, 2 = prominent, 3 = intensely bright. Since noctilucent clouds seldom show rapid changes of structure, observations need only be taken at 15-minute intervals (ideally, on the hour, quarter past, half past and quarter to). Rough sketches of a display's appearance are of use in conjunction with measurements. Photography of noctilucent clouds is quite straightforward, and attractive results can be obtained with colour film. Typical exposures at *f*/2.8 on ISO 400 film are 2 to 4 seconds.

Reports of noctilucent cloud sightings, and of nights for which the observer can confidently say that *no* noctilucent cloud was present, are welcomed by the organizations that collect auroral observations.

The Zodiacal Light

A faint, hazy, conical beam of light, about 15°–20° wide at the base, is sometimes seen in the west after sunset or in the east before sunrise. The main axis of this beam lies approximately along the ecliptic for 90° or more from the horizon, a little south (in southern latitudes, north) of where the Sun is below the horizon. In its brightest parts, it is two or three times as luminous as the Milky Way, but towards its extreme limits it is always exceedingly faint. Its brightness seems to vary from time to time, possibly in response to fluctuations in solar activity, being brighter at solar minimum. It is brighter when observed from the tropics than from temperate latitudes, partly because the main axis of the cone is more or less at right angles to the horizon, and partly because of the short twilight periods.

From mid-northern latitudes, the zodiacal light is best seen near the vernal equinox in the evening or near the autumnal equinox in the morning (vice versa in the southern hemisphere). Table 30 gives the approximate dates and hours when the ecliptic is most nearly vertical during the short observing season. The position of the foot of the zodiacal light on the horizon for 3 or 4 hours after (or before) the hours mentioned is easily found, as its movement in azimuth may be taken as about 6° per hour over that period; similarly, the decrease per hour in inclination after (or before) greatest verticality is, roughly, 2°.

It is now generally accepted that this light is sunlight that has been scattered by dust particles (micrometeoroids) lying in the plane of the ecliptic and orbiting the Sun. The spectrum of the zodiacal light is essentially the same as that of normal sunlight. The particle sizes lie in the range 1–350 μm, and the zodiacal light is essentially an extension of the F corona (or dust corona) of the Sun.

THE GEGENSCHEIN This phenomenon, also known as the *counterglow*, is a very faint patch of light at or near the antisolar point, i.e. the point on the ecliptic directly opposite the position of the Sun at the time. It is normally elliptical, typically 10° × 20°, although in the tropics it may be seen to extend over 30°. It can normally be seen only on very clear moonless nights, the best times being when the ecliptic is highest above the horizon (December

TABLE 37. *Local times, for northern and southern observers, at which the ecliptic is most nearly vertical and the zodiacal light is best observed.*

N	Feb. 5	Feb. 12	Feb. 20	Feb. 27	Mar. 7	Mar. 14	Mar. 22
	21.00	20.30	20.00	19.30	19.00	18.30	18.00
S	Aug. 6	Aug. 13	Aug. 21	Aug. 29	Sept. 6	Sept. 13	Sept. 21
N	Sept. 22	Sept. 29	Oct. 7	Oct. 14	Oct. 22	Oct. 30	Nov. 7
	06.00	05.30	05.00	04.30	04.00	03.30	03.00
S	Mar. 23	Mar. 31	Apr. 8	Apr. 15	Apr. 23	Apr. 30	May 8

and January for northern observers, June and July for southern observers).

It is thought that the gegenschein is caused by the scattering of sunlight by dust particles orbiting the Sun. The gegenschein is sometimes seen joined to the zodiacal light by a parallel beam of light called the *zodiacal band*.

Artificial Satellites

An artificial satellite appears as a star-like point of light drifting slowly across the sky, possibly flashing as it tumbles or slowly fading out as it moves into the Earth's shadow and is eclipsed. Many satellites have bright surfaces that reflect sunlight well, so they are easily visible at heights of several hundreds or even thousands of kilometres.

Over 8500 artificial objects are being tracked in orbit by military radar installations. They include working satellites, dead satellites, discarded rocket stages, fragments from break-ups and many other pieces of space junk. About 10% of all satellites are above the horizon at any moment. Some satellites rival the brightest stars and are clearly visible to the naked eye, notably the International Space Station, while many more can be seen with binoculars. Visual observations by amateurs can be of considerable value in following the ever-changing orbits of artificial satellites.

The best time to see satellites is shortly after sunset or before sunrise, when they are illuminated by the Sun's rays against a dark or fairly dark background. Bright satellites can usually be detected by the time the Sun has dropped 6° below the horizon (civil twilight), but fainter satellites may require the Sun to be 9° or even 12° down before they can be seen. The length of time for which satellites can be seen during these two periods depends on the height of the satellite and the time of year, these periods being limited by the position of the Earth's shadow. During the summer months the two visibility periods merge together for observers at middle and high latitudes, so that satellites can be seen throughout the night.

A satellite will be invisible if it is eclipsed in the Earth's shadow. Eclipses frequently occur while the satellite is crossing the sky. In the pre-dawn sky the situation is reversed, and satellites can often appear high in the sky as they emerge from eclipse.

The speed at which a satellite moves depends on its height. The lowest ones, which orbit a few hundred kilometres up, move the quickest (in accordance with Kepler's laws), having periods of about 90 minutes. When passing overhead, these satellites will move at a rate of almost 2° (or four Moon diameters) per second, taking two or three minutes to cross the sky from horizon to horizon, far slower than a meteor. Precise determination of the angular velocity of an unexpected satellite is of considerable help in identifying it.

BRIGHTNESS   The brightness of a satellite depends on many factors, including its size and the material of which its surface is made. One that is black or covered in dark solar cells will be fainter than a similar white one. A given satellite will be brighter when nearby (overhead) than when farther away (near the horizon).

As with the Moon, a satellite's brightness varies with *phase angle*, which is the angle between the Sun and the satellite as seen by the observer. When the phase angle is small (i.e. when the Sun and the satellite are almost in line), most of the sunlit side of the satellite is facing away from the observer and it will appear dark. Conversely, a large phase angle (near 180°) means that the satellite appears nearly fully illuminated and will therefore be much brighter.

Variations in brightness will occur if a satellite is not spherical. Large flat surfaces can cause momentary bright reflections, appearing as sudden flashes in the sky; these can be seen on any clear night. The most extreme examples of reflections are caused by the Iridium series of communications satellites, which can become as bright as mag. −8 for several seconds as sunlight catches their flat antenna panels. Such extended brightenings are termed 'flares'.

ORBITAL CHANGES   A number of forces act on an orbiting satellite, causing its orbit to change. It is well known that air-drag causes a satellite to spiral slowly back to Earth, making its orbit smaller and its period shorter until it finally re-enters the atmosphere and burns up. But other forces are in action, including the gravitational effects of the Earth's equatorial bulge and the pressure of sunlight.

As a result of the changing orientation of the orbit and the movement of the Earth around the Sun, a satellite will not be visible every night, but will have periods of visibility which can range from a few days for low, non-polar satellites to several months for satellites in moderately high-altitude polar orbits with inclinations greater than 90°.

PREDICTIONS   The commonest form of predictions for artificial satellites are referred to as 'look-angle' predictions. These give the apparent track across the sky as seen by an observer at a known position on the Earth's surface. Look-angle data can readily be computed from the satellite's orbital elements. Predictions for the brightest satellites, including Iridium flares, can be obtained for any location from the Heavens Above website:

http://www.heavens-above.com/

OBSERVATIONS   From the predicted path of a satellite, plotted on a star atlas or shown on a computer-generated chart, background stars may be selected against which to determine the satellite's position at a given time. This is usually done by timing the passage of the satellite across an imaginary line joining two stars, simultaneously estimating the position of the crossing point. For example, you might estimate that a satellite passed two-fifths of the way between stars A and B. The smaller the separation between the two stars, the more accurate the estimated position; if the reference stars are several degrees apart the resulting observation will have low accuracy. The observations can be converted into celestial coordinates either by measurement on large-scale charts or by a planetarium-style computer program.

Observers of artificial satellites usually prefer binoculars with specifications such as 7×50 or 11×80, or even larger. Some experienced observers have made observations of geostationary satellites, but much skill is required, especially in identifying the objects seen. For accurate timings a stopwatch is required, as is a source of precise time signals – either the telephone time service or a radio broadcast. Timings are made by starting the watch as the satellite crosses the line between two stars, and stopping it against a time signal. Two timed positions on one transit are preferable to only one.

The times and positions, and possibly an estimate of the brightness of the satellite, are sent to a prediction centre, where they will be combined with data from other observers to produce new orbital elements for future predictions, and to study the perturbations affecting the orbit.

DESIGNATIONS There are three systems for designating the various objects in orbit. The first is a simple catalogue number assigned in sequence by the United States Air Force Space Command as each new object is detected and tracked in orbit. The second is a popular name assigned by the owner of the satellite such as Cosmos 1500, Rosat, Iridium 96 or Hubble Space Telescope.

The third system is the international designation, which before 1963 made use of Greek letters but now consists of three parts: the year of launch, the number of the launch in that year and a letter (or letters) indicating the various objects resulting from that launch. Frequently the payload is assigned the letter A, the orbiting rocket stage is labelled B and other items, such as discarded panels, are designated C, D and so on. The letters I and O are omitted to avoid confusion with numbers, so the system provides letters for 24 separate objects. If this is insufficient, e.g. for an explosion in orbit which produces hundreds of objects, double letters are used starting with AA to AZ, followed by BA to BZ, and so on, until ZZ is reached, which allows for up to 600 objects.

CHAPTER FOUR

Stars, Nebulae and Galaxies

The Stars

Constellations and Nomenclature

THE CONSTELLATIONS

In total there are 88 constellations, listed in Table 38, covering the entire sky. Nowadays constellations are regarded as being fixed areas of sky rather than star patterns as originally envisaged by the ancient Greeks, but they remain convenient guides to the location of celestial objects.

The constellations that we know today have grown from a list of 48 published around AD 150 by the Greek astronomer Ptolemy in a book called the *Almagest*. At that time, constellations were regarded as star patterns with no definite boundaries. Many of those star figures, particularly the 12 constellations of the zodiac, were apparently invented by the Babylonians before 2000 BC.

Ptolemy's list remained essentially unchanged until the end of the sixteenth century when two Dutch navigators, Pieter Dirkszoon Keyser and Frederick de Houtman, added 12 new constellations in the south polar region of the sky. A century later seven new constellations were introduced into the northern sky by the Polish astronomer Johannes Hevelius, tucked into gaps between Ptolemy's figures. In the eighteenth century the French astronomer Nicolas Louis de Lacaille placed 14 more constellations in the southern hemisphere, and split up Ptolemy's large and unwieldy Argo Navis, the ship of the Argonauts, into Carina (the keel), Puppis (the poop) and Vela (the sails).

CONSTELLATION BOUNDARIES

The modern list of 88 constellations was adopted in 1922 by the newly formed International Astronomical Union. Even so, there were still no generally accepted constellation boundaries – charts such as those in the early editions of *Norton's Star Atlas* simply showed dotted lines meandering vaguely between the stars. On behalf of the IAU the Belgian astronomer Eugène Delporte drew up constellation boundaries along arcs of right ascension and declination for the year 1875 (that

date was chosen because the American astronomer B. A. Gould had already devised boundaries for the southern constellations for that epoch). Delporte's boundaries were published in 1930. They are fixed with respect to the stars (ignoring the stars' proper motions), but the effect of precession means that they are gradually departing from the lines of right ascension and declination along which they were originally drawn, as can be seen on the charts in this Atlas.

STANDARD NAMES AND ABBREVIATIONS OF THE CONSTELLATIONS

When the IAU adopted its official list of 88 constellations in 1922, it drew up a list of three-letter abbreviations for each constellation, as given in Table 38. Hence a star such as Alpha (α) Ursae Majoris can for brevity be referred to as α UMa. Note that the genitive case of a constellation's name is used when referring to a star within it (hence Alpha Ursae Majoris means 'alpha of Ursa Major'). The genitive case for each constellation is also given in Table 38.

STAR NOMENCLATURE

Many of the brightest stars have proper names (Table 39). Some of these names are Greek or Roman in origin, such as Sirius and Spica, but many are Arabic, such as Aldebaran. Astronomers use proper names sparingly, preferring instead the system of Greek letters introduced in 1603 by the German astronomer Johann Bayer, and hence known as *Bayer letters*. (See Table 40 for the Greek alphabet.)

Bayer labelled the brightest stars in a constellation with the letters alpha (α), beta (β) and so on, usually (but not always) in order of brightness. Stars are also given numbers, e.g. 61 Cygni, known as *Flamsteed numbers*. These come from a catalogue published in 1725 by the English Astronomer Royal John Flamsteed, in which he listed the stars of each constellation in order of right ascension; however, the Flamsteed numbers were not actually assigned by him but were added later by other astronomers. A star can therefore have several aliases – Betelgeuse is also Alpha (α) Orionis and 58 Orionis, for example. Fainter stars are known by their numbers in other catalogues. Variable stars have a nomenclature of their own (see pp. 122–123). Some of the best-known star catalogues are listed in Table 41, and star atlases in Table 42.

(text continued on p.107)

TABLE 38. *The constellations.*

NAME	GENITIVE	ABBREVIATION	AREA (SQUARE DEGREES)	AREA OF SKY (%)	ORDER OF SIZE
Andromeda	Andromedae	And	722	1.750	19
Antlia	Antliae	Ant	239	0.579	62
Apus	Apodis	Aps	206	0.499	67
Aquarius	Aquarii	Aqr	980	2.376	10
Aquila	Aquilae	Aql	652	1.580	22
Ara	Arae	Ara	237	0.575	63
Aries	Arietis	Ari	441	1.069	39
Auriga	Aurigae	Aur	657	1.593	21
Boötes	Boötis	Boo	907	2.199	13
Caelum	Caeli	Cae	125	0.303	81
Camelopardalis	Camelopardalis	Cam	757	1.835	18
Cancer	Cancri	Cnc	506	1.227	31
Canes Venatici	Canum Venaticorum	CVn	465	1.127	38
Canis Major	Canis Majoris	CMa	380	0.921	43
Canis Minor	Canis Minoris	CMi	183	0.444	71
Capricornus	Capricorni	Cap	414	1.004	40
Carina	Carinae	Car	494	1.197	34
Cassiopeia	Cassiopeiae	Cas	598	1.450	25
Centaurus	Centauri	Cen	1060	2.570	9
Cepheus	Cephei	Cep	588	1.425	27
Cetus	Ceti	Cet	1231	2.984	4
Chamaeleon	Chamaeleontis	Cha	132	0.320	79
Circinus	Circini	Cir	93	0.225	85
Columba	Columbae	Col	270	0.654	54
Coma Berenices	Comae Berenices	Com	386	0.936	42
Corona Australis	Coronae Australis	CrA	128	0.310	80
Corona Borealis	Coronae Borealis	CrB	179	0.434	73
Corvus	Corvi	Crv	184	0.446	70
Crater	Crateris	Crt	282	0.684	53
Crux	Crucis	Cru	68	0.165	88
Cygnus	Cygni	Cyg	804	1.949	16
Delphinus	Delphini	Del	189	0.458	69
Dorado	Doradus	Dor	179	0.434	72
Draco	Draconis	Dra	1083	2.625	8
Equuleus	Equulei	Equ	72	0.175	87
Eridanus	Eridani	Eri	1138	2.759	6
Fornax	Fornacis	For	398	0.965	41
Gemini	Geminorum	Gem	514	1.246	30
Grus	Gruis	Gru	366	0.887	45
Hercules	Herculis	Her	1225	2.969	5
Horologium	Horologii	Hor	249	0.604	58
Hydra	Hydrae	Hya	1303	3.159	1
Hydrus	Hydri	Hyi	243	0.589	61
Indus	Indi	Ind	294	0.713	49
Lacerta	Lacertae	Lac	201	0.487	68
Leo	Leonis	Leo	947	2.296	12
Leo Minor	Leonis Minoris	LMi	232	0.562	64
Lepus	Leporis	Lep	290	0.703	51
Libra	Librae	Lib	538	1.304	29
Lupus	Lupi	Lup	334	0.810	46
Lynx	Lyncis	Lyn	545	1.321	28
Lyra	Lyrae	Lyr	286	0.693	52
Mensa	Mensae	Men	153	0.371	75
Microscopium	Microscopii	Mic	210	0.509	66
Monoceros	Monocerotis	Mon	482	1.168	35
Musca	Muscae	Mus	138	0.335	77
Norma	Normae	Nor	165	0.400	74
Octans	Octantis	Oct	291	0.705	50
Ophiuchus	Ophiuchi	Oph	948	2.298	11
Orion	Orionis	Ori	594	1.440	26
Pavo	Pavonis	Pav	378	0.916	44
Pegasus	Pegasi	Peg	1121	2.717	7
Perseus	Persei	Per	615	1.491	24
Phoenix	Phoenicis	Phe	469	1.137	37
Pictor	Pictoris	Pic	247	0.599	59
Pisces	Piscium	Psc	889	2.155	14
Piscis Austrinus	Piscis Austrini	PsA	245	0.594	60
Puppis	Puppis	Pup	673	1.631	20
Pyxis	Pyxidis	Pyx	221	0.536	65
Reticulum	Reticuli	Ret	114	0.276	82
Sagitta	Sagittae	Sge	80	0.194	86
Sagittarius	Sagittarii	Sgr	867	2.102	15
Scorpius	Scorpii	Sco	497	1.205	33
Sculptor	Sculptoris	Scl	475	1.151	36
Scutum	Scuti	Sct	109	0.264	84
Serpens	Serpentis	Ser	637	1.544	23
Sextans	Sextantis	Sex	314	0.761	47
Taurus	Tauri	Tau	797	1.932	17
Telescopium	Telescopii	Tel	252	0.611	57
Triangulum	Trianguli	Tri	132	0.320	78
Triangulum Australe	Trianguli Australis	TrA	110	0.267	83
Tucana	Tucanae	Tuc	295	0.715	48
Ursa Major	Ursae Majoris	UMa	1280	3.103	3
Ursa Minor	Ursae Minoris	UMi	256	0.621	56
Vela	Velorum	Vel	500	1.212	32
Virgo	Virginis	Vir	1294	3.137	2
Volans	Volantis	Vol	141	0.342	76
Vulpecula	Vulpeculae	Vul	268	0.650	55

TABLE 39. *Proper names of stars. A selection of the most frequently encountered proper names, with the Bayer or Flamsteed designations of the stars to which they apply. Note that some stars have more than one name: for example, Alpha (α) Andromedae is known as either Alpheratz or Sirrah, both derived from Arabic. Alternative spellings of many names may be encountered in different sources.*

PROPER NAME	DESIGNATION	PROPER NAME	DESIGNATION	PROPER NAME	DESIGNATION
Acamar	θ Eridani	Celaeno	16 Tauri	Nihal	β Leporis
Achernar	α Eridani	Chara	β Canum Venaticorum	Nunki	σ Sagittarii
Acrab	β Scorpii	Cor Caroli	α Canum Venaticorum	Peacock	α Pavonis
Acrux	α Crucis	Cursa	β Eridani	Phact	α Columbae
Acubens	α Cancri	Dabih	β Capricorni	Phecda	γ Ursae Majoris
Adhara	ε Canis Majoris	Deneb	α Cygni	Pherkad	γ Ursae Minoris
Agena	β Centauri	Deneb Algedi	δ Capricorni	Pleione	28 Tauri
Albireo	β Cygni	Deneb Kaitos	β Ceti	Polaris	α Ursae Minoris
Alcor	80 Ursae Majoris	Denebola	β Leonis	Pollux	β Geminorum
Alcyone	η Tauri	Diphda	β Ceti	Porrima	γ Virginis
Aldebaran	α Tauri	Dschubba	δ Scorpii	Procyon	α Canis Minoris
Alderamin	α Cephei	Dubhe	α Ursae Majoris	Propus	η Geminorum
Alfirk	β Cephei	Electra	17 Tauri	Pulcherrima	ε Boötis
Algedi	α Capricorni	Elnath	β Tauri	Rasalgethi	α Herculis
Algenib	γ Pegasi	Eltanin	γ Draconis	Rasalhague	α Ophiuchi
Algieba	γ Leonis	Enif	ε Pegasi	Rastaban	β Draconis
Algol	β Persei	Errai	γ Cephei	Regulus	α Leonis
Alhena	γ Geminorum	Etamin	γ Draconis	Rigel	β Orionis
Alioth	ε Ursae Majoris	Fomalhaut	α Piscis Austrini	Rigil Kentaurus	α Centauri
Alkaid	η Ursae Majoris	Gacrux	γ Crucis	Ruchbah	δ Cassiopeiae
Alkalurops	μ Boötis	Gemma	α Coronae Borealis	Rukbat	α Sagittarii
Almaak *or* Almach	γ Andromedae	Giedi	α Capricorni	Sabik	η Ophiuchi
Alnair	α Gruis	Girtab	θ Scorpii	Sadachbia	γ Andromedae
Alnasl	γ Sagittarii	Gomeisa	β Canis Minoris	Sadalmelik	α Aquarii
Alnath	β Tauri	Graffias	β Scorpii	Sadalsuud	β Aquarii
Alnilam	ε Orionis	Hadar	β Centauri	Sadr	γ Cygni
Alnitak	ζ Orionis	Hamal	α Arietis	Saiph	κ Orionis
Alphard	α Hydrae	Homam	ζ Pegasi	Scheat	β Pegasi
Alphecca *or* Alphekka	α Coronae Borealis	Izar	ε Boötis	Seginus	γ Boötis
Alpheratz	α Andromedae	Kitalpha	α Equulei	Shaula	λ Scorpii
Alrami	α Sagittarii	Kocab *or* Kochab	β Ursae Minoris	Schedar *or* Shedar	α Cassiopeiae
Alrescha	α Piscium	Kornephoros	β Herculis	Sheliak	β Lyrae
Alshain	β Aquilae	Lesath	υ Scorpii	Sheratan	β Arietis
Altair	α Aquilae	Maia	20 Tauri	Sirius	α Canis Majoris
Alya	θ Serpentis	Markab	α Pegasi	Sirrah	α Andromedae
Ankaa	α Phoenicis	Megrez	δ Ursae Majoris	Spica	α Virginis
Antares	α Scorpii	Menkalinan	β Aurigae	Tarazed	γ Aquilae
Arcturus	α Boötis	Menkar	α Ceti	Taygeta	19 Tauri
Arkab	β Sagittarii	Merak	β Ursae Majoris	Thuban	α Draconis
Arneb	α Leporis	Merope	23 Tauri	Toliman	α Centauri
Asellus Australis	δ Cancri	Mesarthim	γ Arietis	Unukalhai	α Serpentis
Asellus Borealis	γ Cancri	Miaplacidus	β Carinae	Vega	α Lyrae
Asterope	21 Tauri	Mimosa	β Crucis	Vindemiatrix	ε Virginis
Atlas	27 Tauri	Mintaka	δ Orionis	Wasat	δ Geminorum
Atria	α Trianguli Australis	Mira	o Ceti	Wezen	δ Canis Majoris
Becrux	β Crucis	Mirach	β Andromedae	Yed Posterior	ε Ophiuchi
Bellatrix	γ Orionis	Mirfak *or* Mirphak	α Persei	Yed Prior	δ Ophiuchi
Benetnasch	η Ursae Majoris	Mirzam	β Canis Majoris	Yildun	δ Ursae Minoris
Betelgeuse	α Orionis	Mizar	ζ Ursae Majoris	Zaurak	γ Eridani
Canopus	α Carinae	Mothallah	α Trianguli	Zavijava	β Virginis
Capella	α Aurigae	Muliphein	γ Canis Majoris	Zosma	δ Leonis
Caph	β Cassiopeiae	Naos	ζ Puppis	Zubenelgenubi	α Librae
Castor	α Geminorum	Nashira	γ Capricorni	Zubeneschamali	β Librae
Cebalrai	β Ophiuchi	Nekkar	β Boötis		

TABLE 40. *The Greek alphabet.*

A	α	Alpha
B	β	Beta
Γ	γ	Gamma
Δ	δ	Delta
E	ε	Epsilon
Z	ζ	Zeta
H	η	Eta
Θ	θ	Theta
I	ι	Iota
K	κ	Kappa
Λ	λ	Lambda
M	μ	Mu
N	ν	Nu
Ξ	ξ	Xi
O	o	Omicron
Π	π	Pi
P	ρ	Rho
Σ	σ	Sigma
T	τ	Tau
Y	υ	Upsilon
Φ	φ	Phi
X	χ	Chi
Ψ	ψ	Psi
Ω	ω	Omega

TABLE 41. *Star catalogues.*

TITLE	EPOCH	AUTHOR(S)	DATE	COVERAGE[a]	APPROXIMATE LIMITING MAGNITUDE	NO. OF STARS LISTED
Positional and general						
Astronomische Gesellschaft Katalog (AGK3)	1950.0	O. Heckmann and W. Dieckvoss	1975	+90° to −2°	12	183 000
Bonner Durchmusterung (BD)	1855.0	F. W. A. Argelander	1859–62[b]	+90° to −2°	9.5	324 000
Bonner Durchmusterung (BD) extension	1855.0	E. Schönfeld	1886	−2° to −23°	9.5	133 000
Bright Star Catalogue (BS)[c]	1900.0 & 2000.0	D. Hoffleit and W. H. Warren Jr	1991 (5th edn)	w.s.	6.5	9110
Supplement to the Bright Star Catalogue	1900.0 & 2000.0	D. Hoffleit *et al.*	1983	w.s.	7.1	2603
Cape Photographic Durchmusterung (CPD)	1875.0	D. Gill and J. C. Kapteyn	1895–1900	−18° to −90°	10	455 000
Catalog of 3539 Zodiacal Stars (ZC)	1950.0	J. Robertson	1940	z.b.		3539
Cordoba Durchmusterung (CoD or CD)	1875.0	J. M. Thome	1892–1932	−22° to −90°	10	614 000
Sixth Catalogue of Fundamental Stars (FK6)	2000.0	R. Wielen *et al.*	1999+ (in progress)	w.s.	9.5	*c.* 5650
General Catalogue of 33 342 Stars (GC)	1950.0	B. Boss	1937	w.s.	7	33 342
Hipparcos Catalogue	1991.25	M. A. C. Perryman *et al.*	1997	w.s.	12.4	118 218
PPM Star Catalogue	2000.0	S. Roeser *et al.*	1991 (north) 1993 (south)	w.s.	12	378 910
Sky Catalogue 2000.0 Vol. 1	2000.0	A. Hirshfeld *et al.*	1991 (2nd edn)	w.s.	8.0	50 071
Smithsonian Astrophysical Observatory Star Catalog (SAO)	1950.0	K. L. Haramundanis	1966	w.s.	9	259 000
Tycho 2 Catalogue	2000.0	E. Høg *et al.*	2000	w.s.	11.5	2 539 913
US Naval Observatory CCD Astrograph Catalog (UCAC)	2000.0	N. Zacharias *et al.*	2005?	w.s.	16[d]	80 000 000
Photometric						
Catalogue of Mean UBV Data on Stars		J. C. and M. Mermilliod	1994	w.s.	—	103 000
Spectroscopic						
Henry Draper Catalogue (HD)	1900.0	A. J. Cannon and E. C. Pickering	1918–24	w.s.	8.5	225 300
Henry Draper Extension (HDE)	1900.0	A. J. Cannon	1925–36	w.s.	11	47 000
Henry Draper Extension 2 (HDE)	1900.0	A. J. Cannon and M. W. Mayall	1949	w.s.	11	86 000
Michigan Catalogue of Two-Dimensional Spectral Types for the HD stars	1900.0	N. Houk *et al.*	1975+ (in progress)	w.s.	11	225 300
Double stars						
New General Catalogue of Double Stars (ADS)	1900.0 & 2000.0	R. G. Aitken	1932	+90° to −30°	—	17 180
Catalogue of the Components of Double and Multiple Stars (CCDM)	2000.0	J. Dommanget and O. Nys	1994	w.s.	—	34 031
Sixth Catalog of Orbits of Visual Binary Stars	2000.0	W. I. Hartkopf and B. D. Mason	ongoing	w.s.	—	1692
Washington Visual Double Star Catalog (WDS)	2000.0	B. D. Mason *et al.*	ongoing	w.s.	—	98 084
Tycho Double Star Catalogue (TDSC)	2000.0	C. Fabricius *et al.*	2002	w.s.	—	32 631
Variable stars						
General Catalogue of Variable Stars (GCVS)	1950.0 & 2000.0	P. N. Kholopov *et al.*	1985–88 (4th edn)	w.s.	—	28 484
New Catalogue of Suspected Variable Stars (NSV)	1950.0	B. V. Kukarkin *et al.*	1982	w.s.	—	14 811
New Catalogue of Suspected Variable Stars. Supplement	1950.0 & 2000.0	E. V. Kazarovets	1999	w.s.	—	11 206

[a] w.s. whole sky; z.b. zodiacal band.
[b] Reprinted 1903.
[c] The star numbers of the Bright Star Catalogue (BS) are identical to those of the Harvard Revised Photometry (HR).
[d] Contains no stars brighter than mag. 7.5.
Many of the above catalogues can be accessed online via the Centre de Données astronomiques de Strasbourg (CDS): http://cdsweb.u-strasbg.fr/

TABLE 42. *Star atlases.*

TITLE	EPOCH	AUTHOR(S)	DATE	COVERAGE[a]	APPROXIMATE LIMITING MAGNITUDE	SCALE (mm PER DEGREE)
Visual						
Millennium Star Atlas	2000.0	R. W. Sinnott and M. Perryman	1997	w.s.	11	36
Norton's Star Atlas	2000.0	Ian Ridpath *et al.*	2003 (20th edn)	w.s.	6.5	3.3
Sky Atlas 2000.0	2000.0	W. Tirion and R.W. Sinnott	1998 (2nd edn)	w.s.	8.5	8.2
SAO Star Atlas	1950.0	Smithsonian Institution	1969	w.s.	9	8.6
Uranometria 2000.0	2000.0	W. Tirion *et al.*	2001 (2nd edn)	w.s.	9.75	18.5
AAVSO Variable Star Atlas	1950.0	C. E. Scovil	1980	w.s.	9	15
Photographic						
True Visual Magnitude Photographic Star Atlas	1950.0	C. Papadopoulos	1979–80	w.s.	14	30
Atlas Stellarum	1950.0	H. Vehrenberg	1977	w.s.	14	30
Palomar Observatory Sky Survey (POSS I)	1950.0	National Geographic Society/ Palomar Observatory	1959–63	+90° to −30°	20 (red) 21 (blue)	54
Second Palomar Observatory Sky Survey (POSS II)	1950.0	Palomar Observatory	1991–2002	+90° to 0°	20.8 (red) 22.5 (blue)	54
Southern Sky Atlas	1950.0	ESO/SERC	1975–91	−20° to −90°	22 (red) 23 (blue)	53 (ESO) 54 (SERC)

[a] w.s. whole sky.

Radiation, Magnitude and Luminosity

RADIATION

Virtually all the information we have about most celestial objects comes from analysis of the energy they radiate: radio waves, heat, light, X-rays and gamma rays. These are all forms of *electromagnetic radiation* – energy propagated through space in the form of waves.

The *electromagnetic spectrum* is the complete range of wavelengths of electromagnetic radiation, from the very longest (radio waves) to the very shortest (gamma rays). The radiation is commonly classified somewhat arbitrarily into different ranges of wavelength (or of frequency), as shown in Figure 29. The Earth's atmosphere is opaque to radiation of most wavelengths. Observations in the 'windows' shown in Figure 29, to which the atmosphere is more or less transparent, can be made from ground-based observatories, but radiation from celestial objects at other wavelengths can be studied only from space.

BRIGHTNESS AND MAGNITUDE

The apparent brightness of a celestial object is related to the amount of radiation from it that is received by the eye (or measuring instrument). Brightnesses are usually expressed on a scale of magnitudes, as explained below. They may be determined accurately by the use of light-sensitive instruments called photometers. Separate measures of magnitudes are made in different ranges of wavelength; among these, the V band corresponds quite closely to the response of the human eye. On all the star charts in this Atlas, stars are plotted according to their V magnitude.

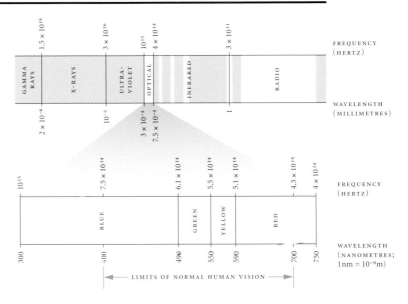

FIGURE 29. *The electromagnetic spectrum. The unshaded regions are 'windows' – wavelength bands to which the full depth of the Earth's atmosphere is transparent.*

In Ptolemy's star catalogue (second century AD), which is based on an earlier one by Hipparchus (second century BC), the naked-eye stars were classed into six grades of brightness or *magnitude*. The brightest stars were said to be 1st magnitude, those less bright 2nd magnitude, and so on. The faintest stars that could just be clearly seen with the naked eye were called 6th magnitude. The scale was later extended for the telescopic stars to magnitudes 7, 8, and so on.

When instrumental methods of measuring the relative brightnesses of stars were developed in the nineteenth century, it was found that two stars differing by one magnitude, as estimated by earlier

TABLE 43. *Ratio of brightness and combined magnitude. For two stars differing in magnitude by a given amount ('Diff.'), the table gives the ratio of their brightnesses and the amount ('Comb.') by which their combined magnitude exceeds that of the brighter star. Thus a double star whose components have magnitudes of 6.00 and 6.50 (Diff. − 0.5) has a combined magnitude of 6.00 − 0.53 = 5.47. The formulae used are:*

$$\text{Ratio} = \text{antilog}(0.4 \times \text{Diff.})$$

$$\text{Comb.} = 2.5 \times \log[1 + \text{antilog}(-0.4 \times \text{Diff.})]$$

DIFF.	RATIO	COMB.	DIFF.	RATIO	COMB.	DIFF.	RATIO	COMB.
0.00	1.00	0.75	1.20	3.02	0.31	4.00	39.81	0.03
0.10	1.10	0.70	1.30	3.31	0.29	4.50	63.10	0.02
0.20	1.20	0.66	1.40	3.63	0.26	5.00	100.00	0.01
0.30	1.32	0.61	1.50	3.98	0.24	5.50	158.49	0.01
0.40	1.45	0.57	1.60	4.37	0.22	6.00	251.19	—
0.50	1.58	0.53	1.70	4.79	0.21	6.50	398.11	—
0.60	1.74	0.49	1.80	5.25	0.19	7.00	630.96	—
0.70	1.91	0.46	1.90	5.75	0.17	7.50	1000.00	—
0.80	2.09	0.42	2.00	6.31	0.16	8.00	1585	—
0.90	2.29	0.39	2.50	10.00	0.10	9.00	3981	—
1.00	2.51	0.36	3.00	15.85	0.07	10.00	10000	—
1.10	2.75	0.34	3.50	25.12	0.04	12.50	100000	—

observers, had a nearly constant brightness ratio of about 2.5, and that an interval of five magnitudes corresponded to a brightness ratio of about 2.5⁵, or nearly 100. That discovery led the way to a more precise definition of a continuous scale of magnitude.

Magnitudes are now recorded to tenths, hundredths or even thousandths of a magnitude. Numerically smaller magnitudes denote greater brightness, so mag. 3.00 is slightly brighter than mag. 3.01, but slightly less bright than mag. 2.99. Some objects are brighter than magnitude 1, so for them the scale is extended backwards to zero and then to negative magnitudes, e.g. Sirius (mag. −1.44), Venus (mag. −4.4 at mean greatest elongation), the full moon (mean mag. −12.7), the Sun (mag. −26.8). Where there is no sign, magnitudes are always understood to be positive.

Modern magnitudes are always calculated for above the Earth's atmosphere. Even for stars at the zenith the atmosphere absorbs or scatters light of different wavelengths to a degree which varies from time to time and from one place to another. For stars at lower altitudes, atmospheric effects further diminish the brightness and must be allowed for when comparing stars at different altitudes.

RELATIONSHIPS FOR CALCULATING MAGNITUDE DIFFERENCES The magnitude scale is defined exactly by specifying that two stars having brightnesses in the ratio 1:100 have a magnitude difference of precisely 5; and by an agreed list of standard stars which effectively define the zero-point of the scale. If two stars have brightnesses B_1 and B_2, their magnitude difference $m_2 - m_1$ is given by

$$m_2 - m_1 = 2.5\log(B_1/B_2)$$

or $\quad B_1/B_2 = \text{antilog}[0.4(m_2 - m_1)] \approx 2.512^{m_2 - m_1}$

The ratios in Table 43 are calculated from this relationship.

COMBINED MAGNITUDE The combined brightness of two stars (e.g. ones that are so close together that they appear as a single star) is obviously the sum of their individual brightnesses. The combined magnitude m of two stars of magnitudes m_1 and m_2 is given by a more complicated formula:

$$m = m_1 \, 2.5\log\{1 + \text{antilog}[-0.4(m_2 - m_1)]\}$$

Table 43 gives the amounts by which the combined magnitude of two stars is brighter (i.e. numerically less) than that of the brighter component. The combined magnitude of three or more stars can be found by repeated application of the formula. The *integrated* (or *total*) *magnitude* of an extended object such as a galaxy or comet is a measure of the brightness it would have if all its light were condensed to a single, star-like point.

APPARENT MAGNITUDE This is the magnitude of a celestial object as directly estimated by the human eye, or determined from a photograph, or measured instrumentally by a photometer, without any correction for the object's distance. It is denoted by the symbol m. In astronomical photometry, measurements of apparent magnitude are made in different wavelength ranges. *Visual magnitudes*, as perceived by the human eye, are denoted by the symbol m_v; *photographic magnitudes*, estimated or measured from photographs, are denoted by m_{pg} if traditional blue-sensitive plates or films are used, or m_{pv} (*photovisual*) if a combination of photographic emulsion and filter is used whose colour response approximates that of the human eye.

For precise work, measurements are made with a CCD or photoelectric photometer. Measurements may be made in various wavelength ranges; the most commonly used is the UBV system, U standing for ultraviolet, B for blue and V for visual. The B magnitude approximates to the older m_{pg} scale, and the V magnitude

TABLE 44. *Effective wavelengths of bands used in photometry. U stands for ultraviolet, B for blue, V for visual and R for red; I to Q are infrared bands.*

BAND	U	B	V	R	I	J
Effective wavelength (nm)	360	440	550	700	900	1250

BAND	H	K	L	M	N	Q
Effective wavelength (µm)	1.62	2.2	3.4	5.0	10.2	19.5

TABLE 45. *Apparent magnitude (m_V), B − V colour index, absolute magnitude (M_V) and spectral type of a few well-known stars.*

STAR	m_V	COLOUR INDEX	M_V	SPECTRAL TYPE
Aldebaran (α Tauri)	0.87	+1.54	−0.6	K5
α Centauri A	−0.01	+0.71	4.3	G2
Vega (α Lyrae)	0.03	0.00	0.6	A0
Spica (α Virginis)	0.98	−0.23	−3.6	B1

TABLE 46. *Distance and magnitude. Increase of distance ('Dist.') for various differences ('Diff.') from 1 to 20 magnitudes, in the absence of interstellar extinction. Thus a mag. 5 star, if it were 100 times farther away, would be 10 magnitudes fainter (or mag. 15). The formula used is:*

$$\text{Dist.} = \text{antilog}(0.2 \times \text{Diff.})$$

DIFF.	DIST.	DIFF.	DIST.	DIFF.	DIST.	DIFF.	DIST.
1	1.585	6	15.85	11	158.5	16	1585
2	2.512	7	25.12	12	251.2	17	2512
3	3.981	8	39.81	13	398.1	18	3981
4	6.310	9	63.10	14	631.0	19	6310
5	10.00	10	100.00	15	1000	20	10000

to the older m_v or m_{pv} scale. In some photometers the wavelength range is extended to include red and infrared wavelengths. The effective wavelengths of the U, B and V bands, and the most commonly used infrared bands, are given in Table 44.

Another important system of photoelectric photometry, introduced by Bengt Strömgren, uses four filters passing narrower wavelength bands than in the UBV system. These bands are u (ultra-violet), v (violet), b (blue) and y (yellow), and are centred on 350, 410, 470 and 550 nm (nanometres) respectively.

As a general rule, photoelectric magnitudes can be assumed to be correct to within one-hundredth of a magnitude. A similar accuracy is obtained with the best ground-based CCD photometry. The Hipparcos satellite, whose results were published in 1997, has set new standards for photometric precision: the magnitudes of over 118 000 stars are included in the *Hipparcos Catalogue* with an accuracy of 0.0015 mag.

COLOUR INDEX   This is the difference between the magnitudes of a star measured in two different wavelength bands, most commonly B and V, or U and B. The B − V colour index of white stars is close to zero, while for the reddest stars it may be over two magnitudes. The B − V colour indices of a few well-known stars are given in Table 45.

ABSOLUTE MAGNITUDE   Apparent magnitude is no criterion of intrinsic luminosity, as many nearby stars appear far brighter than more luminous ones which are at greater distances. Absolute magnitude is the brightness a star would have if it were a standard distance from us: it is found by calculating what the observed magnitude would be if the star were at a distance of 10 parsecs (about 33 light years), equivalent to a parallax (see pp. 112–13) of 0.1 arcsec. This calculation requires a knowledge of the star's distance. Conversely, if the absolute magnitude can be found by some other means, the distance of the star can be found.

Absolute magnitude is of great importance in stellar research as it enables luminosities to be compared. The Sun's visual absolute magnitude is 4.82; Table 45 gives the apparent and absolute magnitudes of a few well-known stars. The absolute magnitudes of dwarf stars fall off by over a magnitude in each successive spectral type, reaching +15.5 in the red dwarf Proxima Centauri. Those of typical giant stars vary by only two magnitudes (from about +1.1 to

−0.8) in the progression from spectral type G to M. The supergiant star Deneb has an absolute magnitude of −8.7, and supernovae at maximum range from about −16.5 to −21. One of the most luminous stars known is S Doradus, at about −9.2, and one of the least luminous is Van Biesbroeck's Star (V1289 Aql), at +18.6.

A star's absolute magnitude M can be calculated from its apparent magnitude m and parallax π in arc seconds, as follows:

$$M = m + 5 + 5\log\pi$$

Table 46 gives the relationship between distance and magnitude.

INTERSTELLAR EXTINCTION AND DISTANCE MODULUS
The formula for absolute magnitude given above is strictly correct only in the absence of absorption or scattering of light by interstellar gas and dust. For distant objects a correction needs to be made. If the amount of extinction A can be estimated, the intrinsic apparent magnitude m_0 can be found from $m_0 = m - A$. The *distance modulus*, $m - M$, is calculated from

$$m - M = 5\log(\text{distance in parsecs}) - 5 + A$$

and the corrected (or 'absolute') distance modulus is then

$$m_0 - M = 5\log(\text{distance in parsecs}) - 5$$

The amount of extinction suffered by a star's light depends on the star's location in the Galaxy. It is greatest in the plane of the Milky Way, where it averages about one magnitude per kiloparsec at visual wavelengths; but in certain regions the extinction can amount to several magnitudes. Extinction diminishes rapidly with distance from the galactic plane. There are formulae that describe statistically the amount of extinction at various galactic latitudes, but to obtain a reliable estimate for a particular star, maps showing interstellar extinction in various directions must be consulted.

INTRINSIC COLOUR INDEX This is obtained by applying corrections for interstellar extinction to the observed colour index. The difference between the observed colour index $(B - V)_{obs}$ and the intrinsic colour index $(B - V)_i$ is called the *colour excess*, denoted by $E(B - V)$. Thus:

$$E(B - V) = (B - V)_{obs} - (B - V)_i = A(B) - A(V)$$

where $A(B)$ and $A(V)$ are the amounts of extinction in the B and V wavelength bands.

BOLOMETRIC MAGNITUDE This is a measure of the total radiation received from a star: ultraviolet, light, heat, radio, and so on. Measurement may be made by a *bolometer*, a detecting device which produces an output signal that depends on the total incident radiation irrespective of wavelength; but only radiation that penetrates the Earth's atmosphere is registered in this way. An alternative is to make an estimate on the basis of separate measurements taken in the various wavelength bands. The amount of energy received outside the Earth's atmosphere from a star of bolometric magnitude 0.00 is equal to 2.48×10^{-8} watts per square metre.

BOLOMETRIC CORRECTION (BC) This is the difference between the bolometric magnitude m_{bol} and the apparent visual magnitude (m_v). Hence the bolometric correction is BC = $m_v - m_{bol}$. The correction is zero for a star with a surface temperature of about 6500 K. The bolometric correction is positive for both hotter and cooler stars because such stars emit more of their radiation outside the visual range, either in the ultraviolet (hotter stars) or in the infrared (cooler stars). Sometimes, though, the bolometric correction is expressed in the opposite way, $m_{bol} - m_v$, so it is then negative, not positive.

POLARIMETRY Light is a wave motion, the waves being transverse (i.e. the wave oscillations are at right angles to the direction in which the light travels). If the direction of the electric field associated with the radiation remains constant, the radiation is said to be *plane-polarized*. The light from some astronomical sources is partly polarized, and the amount and direction of the polarization can be measured with a photoelectric photometer in conjunction with a Polaroid filter. The polarization of starlight can reveal the existence of interstellar dust and strong magnetic fields.

LUMINOSITY

The luminosity of a star is its intrinsic or absolute brightness: it is a measure of the total outflow of radiation from the star. Luminosities may be calculated for any particular wavelength band, or they may be bolometric – covering radiation of all wavelengths.

The symbol is L, and the unit is the watt. For example, the Sun's bolometric luminosity is about 3.8×10^{26} W, which corresponds to radiation emitted at the surface at 6.2×10^7 W m^{-2}; it also corresponds to an apparent bolometric magnitude $m_{bol} = -26.85$ and an absolute bolometric magnitude $M_{bol} = +4.75$, slightly brighter than its absolute visual magnitude of $M_V = 4.82$.

The relationship between the luminosity and absolute magnitude of the Sun $(L_\odot, M_\odot)$ and those of another star (L, M) is given by

$$\log(L/L_\odot) = 0.4(M_\odot - M)$$

Hence if the absolute magnitude of a star can be determined, a rearrangement of the above equation will enable its luminosity to be found. The star Sirius (Alpha (α) Canis Majoris) has an apparent visual magnitude of -1.44, and a parallax of 0.379 arcsec, yielding an absolute visual magnitude of $+1.45$. The bolometric correction for a main-sequence star of spectral type A0, such as Sirius, is -0.30, so its absolute bolometric magnitude is 1.15. Thus Sirius is intrinsically brighter than the Sun by 3.37 magnitudes visually, or 3.60 magnitudes bolometrically. By applying the above relationship we find that the visual luminosity of Sirius is 22.3 times that of the Sun, and its bolometric luminosity 27.5 times that of the Sun.

MASS–LUMINOSITY RELATIONSHIP For stars on the main sequence, luminosity increases proportionally with mass. The general relationship between mass, M, and luminosity, L, can be expressed in the form

$$L = M^k$$

where M and L are in solar units. This is termed the mass–luminosity relationship. The exact value of the factor k varies along the main sequence but may be taken to have an average value of 3.3. For example, a star twice as massive as the Sun will have a luminosity approximately $2^{3.3} = 10$ times greater, and so on. The mass–luminosity relationship may be used to estimate the mass of single stars, i.e. those not forming part of a binary or multiple system. Supergiants, giants and white dwarfs do not follow the relationship as they are respectively over-luminous and under-luminous for their masses by comparison with stars on the main sequence.

Distances, Motions and Physical Parameters

THE NEAREST AND BRIGHTEST STARS

The 28 nearest stars (including the Sun) are listed in Table 47, and the 26 stars of greatest visual apparent magnitude in Table 48. For each star the position, apparent and absolute magnitude, spectral classification, parallax and distance are given.

TABLE 47. *The nearest stars.*

STAR	RA 2000.0 h m	DEC. 2000.0 ° ′	APPARENT MAGNITUDE	SPECTRAL TYPE	PARALLAX ″	DISTANCE (l.y.)	ABSOLUTE MAGNITUDE
Sun	—	—	−26.78	G2V	—	—	4.82
Proxima Centauri (V645 Cen)	14 29.8	−62 41	11.01 (var.)	M5Ve	0.77233	4.22	15.45
Alpha Centauri A	14 39.7	−60 50	−0.01	G2V	0.74212	4.39	4.34
B			1.35	K1V			5.70
Barnard's Star	17 57.8	+04 40	9.54	M4V	0.54901	5.94	13.24
CN Leo (Wolf 359)	10 56.5	+07 01	13.44 (var.)	M6Ve	0.419	7.8	16.55
Lalande 21185 (HD 95735)	11 03.3	+35 59	7.49	M2V	0.39240	8.31	10.46
Sirius A	06 45.2	−16 43	−1.44	A0m	0.37921	8.60	1.45
B			8.44	DA2			11.34
UV Ceti A	01 39.0	−17 57	12.54 (var.)	M5.5Ve	0.374	8.7	15.40
B			12.99 (var.)	M6Ve			15.85
V1216 Sgr (Ross 154)	18 49.8	−23 50	10.37	M3.5Ve	0.33648	9.69	13.00
Ross 248	23 41.9	+44 10	12.29	M5.5Ve	0.316	10.3	14.79
Epsilon Eridani	03 32.9	−09 27	3.72	K2V	0.31075	10.50	6.18
HD 217987 (CoD −36° 15693)	23 05.8	−35 51	7.35	M2/M3V	0.30390	10.73	9.76
FI Vir (Ross 128)	11 47.7	+00 48	11.12 (var.)	M4.5V	0.29958	10.89	13.50
EZ Aqr (L 789–6 ABC)	22 38.6	−15 18	13.33 (var.)	M5Ve	0.290	11.2	15.64
61 Cyg A (V1803 Cyg)	21 06.8	+38 44	5.20 (var.)	K5V	0.28713	11.36	7.49
Procyon A	07 39.3	+05 14	0.40	F5IV–V	0.28593	11.41	2.68
B			10.7	DF			13.0
61 Cyg B	21 06.9	+38 44	6.05 (var.)	K7V	0.28542	11.43	8.33
HD 173740 (BD +59° 1915 B)	18 42.8	+59 37	9.70	M4V	0.28448	11.47	11.97
HD 173739 (BD +59° 1915 A)	18 42.8	+59 38	8.94	M3.5V	0.28028	11.64	11.18
GX And (BD +43° 44 A)	00 18.3	+44 01	8.09 (var.)	M1V	0.28027	·11.64	10.33
(BD +43° 44 B)			11.10	M4V			13.35
DX Cnc (G51–15)	08 29.8	+26 47	14.78 (var.)	M6.5Ve	0.276	11.8	16.98
Epsilon Indi	22 03.3	−56 47	4.69	K5V	0.27576	11.83	6.89
Tau Ceti	01 44.1	−15 56	3.49	G8V	0.27417	11.90	5.68

Sources: Data for stars to 11th magnitude from the Hipparcos Catalogue; fainter stars from the Research Consortium on Nearby Stars (RECONS)
http://www.chara.gsu.edu/RECONS/TOP100.htm

TABLE 48. *The brightest stars.*

STAR	NAME	RA 2000.0 h m	DEC. 2000.0 ° ′	APPARENT MAGNITUDE	SPECTRAL TYPE	PARALLAX ″	DISTANCE (l.y.)	ABSOLUTE MAGNITUDE[a]
	Sun	—	—	−26.78	G2V	—	—	4.82
α CMa	Sirius	06 45.2	−16 43	−1.44	A0m	0.37921	8.60	1.45
α Car	Canopus	06 24.0	−52 42	−0.62	A9II	0.01043	313	−5.53
α Cen	Rigil Kentaurus	14 39.7	−60 50	−0.28[b]	G2V + K1V	0.74212	4.39	4.07[b]
α Boo	Arcturus	14 15.7	+19 11	−0.05	K1.5III	0.08885	36.71	−0.31
α Lyr	Vega	18 36.9	+38 47	0.03 (var.)	A0V	0.12892	25.30	0.58
α Aur	Capella	05 16.7	+46 00	0.08	G6III + G2III	0.07729	42.20	−0.48
β Ori	Rigel	05 14.5	−08 12	0.18 (var.)	B8Ia	0.00422	773	−6.69
α CMi	Procyon	07 39.3	+05 14	0.40	F5IV–V	0.28593	11.41	2.68
α Eri	Achernar	01 37.7	−57 14	0.45	B3Vnp	0.02268	144	−2.77
α Ori	Betelgeuse	05 55.2	+07 24	0.45 (var.)	M1–M2Ia–Iab	0.00763	427	−5.14
β Cen	Hadar	14 03.8	−60 22	0.61 (var.)	B1III	0.00621	525	−5.42
α Aql	Altair	19 50.8	+08 52	0.76 (var.)	A7V	0.19444	16.77	2.20
α Cru	Acrux	12 26.6	−63 06	0.77[b]	B0.5IV + B1V	0.01017	321	−4.19[b]
α Tau	Aldebaran	04 35.9	+16 31	0.87	K5+III	0.05009	65.11	−0.63
α Vir	Spica	13 25.2	−11 10	0.98 (var.)	B1V	0.01244	262	−3.55
α Sco	Antares	16 29.4	−26 26	1.05 (var.)	M1.5Iab–Ib + B2.5V	0.00540	604	−5.29
β Gem	Pollux	07 45.3	+28 02	1.16	K0III	0.09674	33.72	1.09
α PsA	Fomalhaut	22 57.6	−29 37	1.16	A3V	0.13008	25.07	1.73
β Cru	Mimosa	12 47.7	−59 41	1.25 (var.)	B0.5III	0.00925	353	−3.92
α Cyg	Deneb	20 41.4	+45 17	1.25 (var.)	A2Ia	0.00101	3230	−8.73
α Leo	Regulus	10 08.4	+11 58	1.36	B7V	0.04209	77.49	−0.52
ε CMa	Adhara	06 58.6	−28 58	1.50	B2II	0.00757	431	−4.10
α Gem	Castor	07 34.6	+31 53	1.58[b]	A1m	0.06327	51.55	0.59[b]
γ Cru	Gacrux	12 31.2	−57 07	1.59	M3.5III	0.03709	87.94	−0.56
λ Sco	Shaula	17 33.6	−37 06	1.62 (var.)	B1.5IV + B	0.00464	703	−5.05

[a] Ignoring interstellar absorption

[b] Combined magnitude of double star. For the individual components of α Cen, see Table 47 above.

Source: Hipparcos Catalogue; spectral types from The Astronomical Almanac.

TABLE 49. *Distances in parsecs (pc) and light years (l.y.) equivalent to any parallax (π in arcsec). For parallaxes of 0.0001, 0.0002, etc., move the parsec or light year decimal point one place to the right.*

π	pc	l.y.	π	pc	l.y.	π	pc	l.y.	π	pc	l.y.	π	pc	l.y.	π	pc	l.y.
0.001	1000	3262.0	0.021	47.62	155.3	0.041	24.39	79.55	0.061	16.39	53.47	0.081	12.35	40.27	0.12	8.33	27.18
0.002	500.0	1631.0	0.022	45.45	148.3	0.042	23.81	77.66	0.062	16.13	52.61	0.082	12.20	39.78	0.14	7.14	23.30
0.003	333.3	1087.0	0.023	43.48	141.8	0.043	23.26	75.85	0.063	15.87	51.77	0.083	12.05	39.30	0.16	6.25	20.39
0.004	250.0	815.4	0.024	41.67	135.9	0.044	22.73	74.13	0.064	15.63	50.96	0.084	11.90	38.83	0.18	5.56	18.12
0.005	200.0	652.3	0.025	40.00	130.5	0.045	22.22	72.48	0.065	15.38	50.18	0.085	11.76	38.37	0.20	5.00	16.31
0.006	166.7	543.6	0.026	38.46	125.4	0.046	21.74	70.90	0.066	15.15	49.42	0.086	11.63	37.93	0.22	4.55	14.83
0.007	142.9	465.9	0.027	37.04	120.8	0.047	21.28	69.40	0.067	14.93	48.68	0.087	11.49	37.49	0.24	4.17	13.59
0.008	125.0	407.7	0.028	35.71	116.5	0.048	20.83	67.95	0.068	14.71	47.96	0.088	11.36	37.06	0.25	4.00	13.05
0.009	111.1	362.4	0.029	34.48	112.5	0.049	20.41	66.56	0.069	14.49	47.27	0.089	11.24	36.65	0.26	3.85	12.54
0.010	100.0	326.2	0.030	33.33	108.7	0.050	20.00	65.23	0.070	14.29	46.59	0.090	11.11	36.24	0.28	3.57	11.65
0.011	90.91	296.5	0.031	32.36	105.2	0.051	19.61	63.95	0.071	14.08	45.94	0.091	10.99	35.84	0.30	3.33	10.87
0.012	83.33	271.8	0.032	31.25	101.9	0.052	19.23	62.72	0.072	13.89	45.30	0.092	10.87	35.45	0.35	2.86	9.319
0.013	76.92	250.9	0.033	30.30	98.84	0.053	18.87	61.54	0.073	13.70	44.68	0.093	10.75	35.07	0.40	2.50	8.154
0.014	71.43	233.0	0.034	29.41	95.93	0.054	18.52	60.40	0.074	13.51	44.08	0.094	10.64	34.70	0.45	2.22	7.248
0.015	66.67	217.4	0.035	28.57	93.19	0.055	18.18	59.30	0.075	13.33	43.49	0.095	10.53	34.33	0.50	2.00	6.523
0.016	62.50	203.9	0.036	27.78	90.60	0.056	17.86	58.24	0.076	13.16	42.92	0.096	10.42	33.98	0.55	1.82	5.930
0.017	58.82	191.9	0.037	27.03	88.15	0.057	17.54	57.22	0.077	12.99	42.36	0.097	10.31	33.62	0.60	1.67	5.436
0.018	55.56	181.2	0.038	26.32	85.83	0.058	17.24	56.23	0.078	12.82	41.82	0.098	10.20	33.28	0.65	1.54	5.018
0.019	52.63	171.7	0.039	25.64	83.63	0.059	16.95	55.28	0.079	12.66	41.29	0.099	10.10	32.95	0.70	1.43	4.659
0.020	50.00	163.1	0.040	25.00	81.54	0.060	16.67	54.36	0.080	12.50	40.77	0.100	10.00	32.62	0.75	1.33	4.349

STELLAR DISTANCES

LIGHT YEAR (l.y.) The light year is a unit of distance frequently used for stars and galaxies. It is the distance covered in one calendar year by a beam of light, which travels at a speed of $299792.458\,\mathrm{km\,s^{-1}}$. One light year is 9.46×10^{12} kilometres, equivalent to 63 240 astronomical units or 0.3066 parsecs. The star nearest the Sun, Proxima Centauri, is 4.22 light years away. Smaller units such as the light month, light week, light day, light minute and light second, which are sometimes encountered, are measures of the distances covered by a beam of light in those lengths of time; they are used for instance for distances on the scale of the Solar System. In such units the Moon is about 1.3 light seconds away, and the Sun is 8.3 light minutes away.

PARSEC (pc) One parsec is the distance at which a star or other object would have an annual parallax (see below) of 1 second of arc. One parsec is 30.857×10^{12} kilometres, equal to 206 265 astronomical units or 3.2616 light years. No star is known with a parallax this large, the greatest measured parallax being that of Proxima Centauri, 0.772 arcsec, corresponding to a distance of 1.3 parsecs. Commonly used multiples of the parsec are the *kiloparsec* (1000 pc, abbreviated kpc) and *megaparsec* (1 000 000 pc, abbreviated Mpc).

TRIGONOMETRIC PARALLAX is the angular difference in position of an object when seen from two different places. In Figure 30, which shows a relatively nearby star seen against a background of distant stars, the difference in position of the star when seen by a hypothetical observer located on the Sun and an observer on

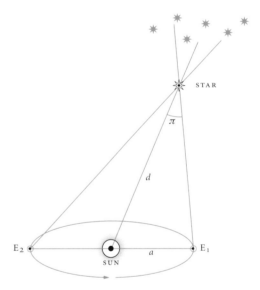

FIGURE 30. *Trigonometric parallax. The distance of a nearby star can be measured by noting its change in position relative to background stars as the Earth moves in its orbit.*

the Earth at E_1 is called the *instantaneous parallax*. Its value can be determined in principle by making observations of the star when the Earth is at E_1 and again six months later when the Earth is at the opposite side of its orbit, at E_2.

In practice, during the course of a year the motion of the Earth around the Sun makes the star appear to trace out an ellipse on the celestial sphere. If the star is near the ecliptic, the ellipse will be highly flattened; if the star is near either ecliptic pole it will be almost a circle (the effect of the slight ellipticity of the Earth's orbit can be neglected). The value of the semi-major axis of the ellipse in which the star appears to move is known as the *annual parallax* (π) of the star. This is the maximum displacement of the star from its mean position (as a result of parallax) and corresponds to the configuration where the angle star–Sun–Earth in Figure 30 is 90°. If a is the Sun–Earth distance and d is the distance of the star, then if π is expressed in radians, we have $\pi = a/d$. If a is known and π is measured, then d can be obtained. Table 49 is a conversion table for parallax into distances in light years and parsecs.

The value of a is the astronomical unit (AU). In principle it can be measured by means of *planetary parallax*, the difference in the position of the Sun or other object in the Solar System as measured by two observers located at different points on the Earth. The effect of planetary parallax is greatest for an object on the horizon, and it is then known as *horizontal parallax*. The (horizontal) solar parallax is 8.79 arcsec. It is the angle subtended by the equatorial radius of the Earth at the Sun's mean distance of 1 AU, so from a knowledge of the Earth's size we can calculate the size of the astronomical unit. In practice, more accurate methods are used, including determination of the distances of objects in the Solar System by radar.

Parallax depends on distance, the nearest objects having the largest parallaxes. The largest parallax known for an object outside the Solar System is for the star Proxima Centauri, 0.772 arcsec. Most stellar parallaxes are very much smaller than this: only about 1000 stars are known to have parallaxes greater than 0.05 arcsec, and only about 3000 greater than 0.04 arcsec (equivalent to a distance of 25 parsecs).

As a distance-finding method, trigonometric parallax can be employed with Earth-based telescopes only to a range of a few hundred light years at most because of the difficulties of measuring very small angles. However, the Hipparcos satellite has obtained highly accurate parallax measurements out to about 500 light years. Other forms of parallax can be obtained, several of them utilizing the distance modulus (see p. 109) for stars at large distances. Some of these forms are described below.

SPECTROSCOPIC PARALLAX Good estimates of the true absolute magnitudes of many stars can be obtained from examination of their spectra (see p. 115). The absolute magnitude can be compared with the observed apparent magnitude, and, after correction for interstellar extinction, a distance or parallax derived.

Absolute magnitudes can be estimated in other ways for certain types of star. In particular the Cepheid variables (see p. 128) show a well-defined relationship between absolute magnitude and period of light variation (the period–luminosity law), so measurement of the period provides a value of absolute magnitude for calculation of the distance modulus. Because of their great luminosity, Cepheids can be used as 'standard candles' for measurements out to very large distances.

DYNAMICAL PARALLAX For a binary star (see p. 120) whose orbit is well known, the distance can be estimated by assuming initially that the combined mass is two solar masses (see p. 114), then using Newton's generalization of Kepler's third law, which relates the period of revolution of the pair of stars to the linear dimensions of the orbit. The linear dimensions are compared with the observed angular dimensions of the stars' orbit to give an initial estimate of the distance. As the apparent magnitudes of the components are known, their absolute magnitudes may then be calculated and the mass–luminosity relationship (see p. 110) used to improve the estimate of their masses. The sequence of calculations is repeated until the difference between successive mass estimates is sufficiently small. Fortunately an error in the estimated mass of the system does not produce a large error in the value of the dynamical parallax.

SECULAR PARALLAX The nearby stars (those within about 100 pc of the Sun) together define a *local standard of rest* relative to which their mean motion is zero. These stars are together revolving around the galactic centre with a velocity of about 250 km s^{-1}, and this motion is at present carrying them towards a point in the constellation Cygnus. Relative to this moving group of nearby stars, the Sun has its own velocity of about 19.5 km s^{-1} towards a point in the constellation Hercules. This motion provides a base-line for parallax measurements that is continually increasing. The average distance of a group of stars can thus be derived from observations of their proper motions (see below).

STELLAR MOTIONS

The motion of a star relative to the Sun can be considered to consist of two components: a radial component R (*radial velocity*), i.e. motion in the line of sight, and a transverse component T (*transverse velocity*). If both R and T can be determined, then the star's velocity V is obtained from $V = \sqrt{(R^2 + T^2)}$, and the direction θ of its motion relative to the radial direction from the Sun is found from $\tan \theta = T/R$.

RADIAL VELOCITY can be obtained from the displacement of the lines of a star's spectrum caused by the *Doppler effect*. The velocity is obtained directly, and it is not necessary to know the distance of the star. A positive radial velocity means that the star is receding, while a negative value means that the star is approaching. Radial velocities in excess of ±100 km s^{-1} are rarely found in stars; most values lie between −40 and +40 km s^{-1}. Periodic variations in radial velocity reveal the orbital motion of spectroscopic binaries (double stars too close to be separated visually).

PROPER MOTION The transverse component of stellar motion shows up as a secular change in the position of a star. The observed angular displacement in one year is known as *annual proper motion* and is generally expressed in arc seconds. The largest known value of proper motion (symbol μ) is that of Barnard's Star at 10.4 arcsec per year. To convert proper motion to transverse velocity it is necessary to know the parallax of the star. The transverse velocity V_μ is then given by $V_\mu = 4.74\,\mu/\pi\,\mathrm{km\,s}^{-1}$. The proper motion thus obtained, after allowing for parallax, aberration and so on, gives the transverse component of velocity relative to the Sun; to obtain the star's motion relative to the local standard of rest it is necessary to allow for solar motion (see the section on Secular parallax, p. 113). Because of proper motion the shapes of the constellations are slowly changing, but the effect is imperceptible in a human lifetime.

HIGH-VELOCITY STARS

A number of stars in the neighbourhood of the Sun have velocities relative to the Sun that are extremely high, greater than $200\,\mathrm{km\,s}^{-1}$. The explanation for such apparently high velocities is that most of the stars in the Sun's neighbourhood, and the Sun itself, are moving around the centre of the Galaxy in approximately circular orbits with velocities of about $250\,\mathrm{km\,s}^{-1}$. The high-velocity stars, however, do not share this circular motion, but usually travel around the galactic centre in eccentric orbits. They are generally members of the galactic halo.

RUNAWAY STARS

are stars of spectral type O or early B with unusually high *space velocities* (i.e. velocities relative to the Sun). They are thought to be produced when a supernova explosion disrupts a close binary system. Three of the best known are 53 Arietis, AE Aurigae and Mu (μ) Columbae, which diverge from a comparatively small area in the constellation Orion.

STELLAR MASSES

Only for binary stars (see p. 120) can the masses of stars be obtained directly. If the orbital period P in years, the mean angular separation a and the parallax π are known, then the combined mass of the pair can be obtained in terms of solar masses from the formula

$$(M_1 + M_2)/M_\odot = a^3/\pi^3 P^2$$

If the position of the centre of mass can be obtained, then the ratio of the distances of the two stars from the centre of mass will yield the ratio of the masses; the individual masses can then be found. Only a few tens of stars have accurately known masses.

The masses of stars on the main sequence can be estimated from the mass–luminosity relationship (p. 110) if their absolute magnitudes are known.

STELLAR TEMPERATURES

It is difficult to assign an unambiguous value to the 'temperature' of a star. Several definitions are used, some of which are listed below.

EFFECTIVE TEMPERATURE (T_{eff})

For a star, the effective temperature is the temperature of a *black body* – i.e. a perfect radiator – of the same radius as the star that radiates the same total amount of radiation – i.e. one that has the same bolometric luminosity as the star. For the Sun, the effective temperature is the temperature of the photosphere, about 5800 K.

COLOUR TEMPERATURE (T_c)

is the equivalent black-body temperature that fits the slope of the observed energy distribution measured between two wavelengths. It can be related to colour index. If the B − V colour index is denoted by I, then the colour temperature in kelvin is

$$T_c = 7200/(I + 0.64)$$

The value of the colour temperature determined in this way may differ from the effective temperature, as stars do not radiate exactly as black bodies. For the Sun ($I = 0.63$), the colour temperature is 5700 K.

The central temperatures of stars are much higher than the surface temperatures: the temperature at the centre of the Sun is thought to be about 15 million kelvin. In the tenuous outer atmosphere of the Sun, the corona, the kinetic temperature (i.e. the temperature corresponding to the velocities of atomic particles) is of the order of two million kelvin.

STELLAR DIAMETERS

Only a few dozen stellar diameters were known until recently, because ground-based telescopes do not reveal stellar disks. The first direct measurements of stellar diameters were made by *stellar interferometry*, which is based on applying the principle that there will be interference of light from different parts of an object of finite size. Three major classes of interferometer have been used: the Michelson (phase) interferometer, the Brown–Twiss (intensity) interferometer, and the Labeyrie (speckle) interferometer. These have yielded values for the angular diameters of a few giant stars such as Betelgeuse. More recently, the diameter of Betelgeuse has been measured directly by the Hubble Space Telescope, and found to be 0.05 arcsec. If the Hipparcos distance of 427 l.y. is taken, the diameter comes out at about 10^9 km (6.5 AU), about twice the size of the orbit of Mars.

Where interferometric observations are not possible, diameters can be estimated by observing occultations of stars by the Moon. They can also be calculated for the components of eclipsing binaries, from the observed durations of their eclipses, in conjunction with radial-velocity measurements that yield the velocities of the components in their orbits. In general, however, diameters are inferred from a consideration of effective temperature and luminosity, using *Stefan's law*, which states that the flux of radiation from a black body is proportional to the square of its radius and the fourth power of its temperature. Thus if two stars have the same effective temperature, but differing luminosities, then it follows that the radius of the one with the higher luminosity must be greater than that of the other (see the section on Spectral classification, below).

Typical stellar diameters range from several hundred million kilometres (supergiants), through 1.4 million km for the Sun, down to a few thousand kilometres for some white dwarfs. Neutron stars (see p. 118) are thought to have diameters of only tens of kilometres.

STELLAR DENSITIES

Although stellar radii vary enormously, stellar masses do not vary by such large amounts. Consequently there are large variations in stellar densities. The Sun has a mean density of $1.4 \times 10^3 \, \mathrm{kg \, m^{-3}}$; supergiants may have mean densities of about $10^{-2} \, \mathrm{kg \, m^{-3}}$; white dwarfs have densities in the range $10^8–10^{11} \, \mathrm{kg \, m^{-3}}$; and neutron stars probably have densities of $10^{16}–10^{18} \, \mathrm{kg \, m^{-3}}$.

Spectral Classification

Stars may be classified into various types on the basis of features in their spectra. In the 1860s the Italian astronomer Angelo Secchi made the first attempt to classify the stars by visually observing their spectra, and divided the stars into four groups. Later classifications were based on photographs of spectra and were much more finely divided. The Harvard classification system, first introduced by Edward Pickering in 1890 and later developed by Annie Cannon and Williamina Fleming, was the immediate precursor of the system currently in use. The current system is variously called the MKK (after Morgan, Keenan and Kellman), MK (Morgan, Keenan) or Yerkes system.

The MKK system applies two labels to a spectrum. The first, the *spectral type*, correlates closely with a star's temperature; the second, *luminosity class*, is related to the star's intrinsic brightness. Stars are allocated to the spectral and luminosity classes by comparison with standard stars which define the system.

SPECTRAL TYPE

Over 90% of stars can be allocated to one of seven spectral types. These are designated by letters inherited from the older Harvard system, and in order of decreasing temperature they are:

O   B   A   F   G   K   M

A traditional mnemonic is 'Oh Be A Fine Girl Kiss Me'. In principle each type is potentially divisible into (at least) ten *subclasses*. As defined by Philip Keenan in 1985, there are actually only between four and nine subclasses in each type, but some astronomers have introduced more. The subclasses are indicated by a numerical suffix (e.g. O5, B9.5), with some gaps in the numbering. A star of class A5 has a spectrum roughly half-way between spectra of stars of classes A0 and F0.

The criteria used to place a star accurately into its spectral type are exceedingly complex. The principal features in the spectra of each of the main types, indicated by absorption lines, are as follows:

O   ionized helium (He II)

B   neutral helium; first appearance of hydrogen

A   hydrogen dominant, plus singly ionized metals

F   hydrogen weaker, ionized calcium (Ca II)

G   Ca II prominent, hydrogen very much weaker; neutral metals

K   neutral metals prominent

M   molecular bands, particularly titanium oxide (TiO).

In addition to the above sequence there are various side-branches and additional codes. Type W (the Wolf–Rayet stars) contains hot stars showing broad, intense emission lines, including He II. Type C (formerly split into types R and N) contains the cool carbon stars, where the TiO bands of type M are replaced by bands of cyanogen, carbon monoxide and molecular carbon (C_2). The spectra of type S stars have bands of zirconium oxide. White dwarf stars comprise type D and, although they are not part of the MKK system, the letters P and Q are sometimes used for, respectively, the emission spectra of planetary nebulae and the peculiar spectra of novae. In recent years, types L and T have been added after M to encompass the very lowest-luminosity red dwarfs and the substars known as *brown dwarfs* which have been discovered at infrared wavelengths.

For historical reasons the spectra of hot stars (O, B, A) are often referred to as *early-type*, and those of cool stars (K, M, C, S) as *late-type* spectra; stars of types F and G are sometimes called *intermediate-type*.

LUMINOSITY CLASS

Within a given spectral type a bright star will be larger and have more rarefied outer regions than a faint star. Hence, the more luminous a star, the narrower its spectral lines, since pressure is often one of the principal line-broadening mechanisms. Thus, on the basis of the quality of the spectral lines (together with, in some cases, intensity differences), the stars in a given spectral type may be further separated into luminosity classes. The luminosity class is denoted by a roman numeral between I and VI placed after the spectral type, e.g. F2III. The main luminosity classes are:

I supergiants

II bright giants

III giants

IV subgiants

V main-sequence dwarfs

VI or sd subdwarfs

(see Figure 31, p. 119). Some classes (in particular the supergiants) are subdivided by adding the suffixes a, ab and b. A notation such as III–IV indicates an object intermediate between two classes.

Thus the full MKK classification of a normal star consists of a letter and an arabic numeral to denote the temperature class, and a roman numeral to denote the luminosity class. The full spectral types of some of the brighter stars are:

δ Ori:	O9.5II	α UMi:	F5–8Ib
β Tau:	B7III	ε Vir:	G8IIIab
β Leo:	A3Va	α Hya:	K3II–III
β Cas:	F2III	β Peg:	M2.5II–III

In addition to the standard spectral type notation, lower-case letters may be added after the luminosity class to indicate certain non-standard features in the spectrum, e.g:

e   emission lines (f in some O-type stars)

m   metallic lines

n   nebulous lines

p   peculiar spectrum

q   lines with blueshifted absorption and redshifted emission, indicating the presence of an expanding shell (P Cygni stars)

v   variable spectrum.

Examples of the use of this notation are:

γ Cas:	B0IVnpe	α CMa:	A0m
P Cyg:	B1Iapeq	ζ Pup:	O5Iafn

The full system of classification allows 90–95% of all stellar spectra to be dealt with. The remainder are composite spectra of unresolved double or multiple stars, or stars with major individual peculiarities.

Table 50 gives the distribution of stars in the *Bright Star Catalogue* according to their spectral type. Table 51 gives the spectral type, absolute magnitude M_V, bolometric magnitude M_{bol}, effec-

TABLE 50. *Distribution of stars according to spectral type. The table gives the percentage of the stars in the* Bright Star Catalogue *belonging to each of the main spectral types.*

O	B	A	F	G	K	M	OTHERS
0.5	19	22	14	13	25	6	0.4

tive temperature T_{eff}, mass, diameter, luminosity and mean density for dwarf, giant and supergiant stars. The spectral type, luminosity class and colour index are closely related, and often the B − V colour index may be used in place of spectral type; Table 52 relates the spectral type and luminosity classes of stars to their B − V and U − B colour indices.

Stellar Evolution

THE HERTZSPRUNG–RUSSELL (HR) DIAGRAM

This diagram is a convenient way in which to display the relationship between the spectral type (colour index, temperature) and luminosity (absolute magnitude) of stars. Figure 31 (p. 119) is an HR diagram in which spectral type is plotted along the horizontal axis, with cool stars to the right and hot stars to the left, and absolute magnitude is plotted vertically. Most stars are found to lie in a band running from top left to bottom right, called the *main sequence*. The remainder – supergiants, giants, white dwarfs, and so on – are found in other specific regions. The HR diagram provides a useful visual aid to the understanding of the evolution of stars.

STAR FORMATION Stars are thought to condense out of clouds of gas, principally hydrogen in composition, which heat up as they contract and so move from right to left across the HR diagram. When the temperature in the central regions of the protostar reaches about ten million kelvin, nuclear reactions can take place in which helium is produced by fusion from hydrogen, with the release of large amounts of energy. At this stage the star reaches a stable state as it joins the main sequence at a position determined principally by its mass. The more massive the star, the greater its luminosity, and the farther up will it join the main sequence. Technically, all stars on the main sequence are classified as dwarfs, even though the most massive ones have diameters many times that of the Sun.

There are physical limits on the mass of a star. A contracting gas cloud with a mass less than about $0.08\,M_{\odot}$ does not become hot enough at its centre for nuclear reactions to begin, and so can never be a true star; instead it is classified as a brown dwarf, cooler and fainter even than the red dwarfs at the foot of the main sequence. Stars at the very top of the main sequence have masses of about $120\,M_{\odot}$. A star with a mass above this would have a luminosity so great that radiation pressure would drive it apart.

TABLE 51. *Typical physical parameters for stars of various luminosity classes and spectral types.*

SPECTRAL TYPE	M_V	M_{bol}	T_{eff} (K)	MASS (RELATIVE TO SUN)	DIAMETER (RELATIVE TO SUN)	LUMINOSITY (RELATIVE TO SUN)	MEAN DENSITY (10^3kg m^{-3})
Main sequence (V)							
O5	−5.7	−10.1	44500	60	12	790000	0.05
B0	−4.0	−7.1	30000	17.5	7.4	52000	0.06
B5	−1.2	−2.7	15400	5.9	3.9	830	0.14
A0	+0.6	+0.3	9520	2.9	2.4	54	0.30
A5	+1.9	+1.7	8200	2.0	1.7	14	0.58
F0	+2.7	+2.6	7200	1.6	1.5	6.5	0.67
F5	+3.5	+3.4	6440	1.4	1.3	3.2	0.90
G0	+4.4	+4.2	6030	1.05	1.1	1.5	1.11
G5	+5.1	+4.9	5770	0.92	0.92	0.79	1.67
K0	+5.9	+5.6	5250	0.79	0.85	0.42	1.80
K5	+7.4	+6.7	4350	0.67	0.72	0.15	2.54
M0	+8.8	+7.4	3850	0.51	0.60	0.077	3.33
M5	+12.3	+9.6	3240	0.21	0.27	0.011	15.1
M8	+16.0	+11.9	2640	0.06	0.1	0.0012	84.7
Giants (III)							
B0	−5.1	−8.0	29000	20	15	110000	0.008
B5	−2.2	−3.5	15000	7	8	1800	0.019
A0	+0.0	−0.4	10100	4	5	106	0.045
G0	+1.0	+0.8	5850	1.0	6	34	0.007
G5	+0.9	+0.6	5150	1.1	10	43	0.002
K0	+0.7	+0.2	4750	1.1	15	60	0.0005
K5	−0.2	−1.2	3950	1.2	25	220	0.0001
M0	−0.4	−1.6	3800	1.2	40	330	0.00003
M5	−0.3	−2.8	3330	—	—	930	—
Supergiants (Iab)							
O5	−6.6	−10.5	40300	70	30	1100000	0.0037
B0	−6.4	−8.9	26000	25	30	260000	0.0013
A0	−6.3	−6.7	9730	16	60	35000	0.00010
F0	−6.6	−6.6	7700	12	80	32000	0.000033
G0	−6.4	−6.6	5550	10	120	30000	0.0000082
G5	−6.2	−6.5	4850	12	150	29000	0.0000051
K0	−6.0	−6.5	4420	13	200	29000	0.0000023
K5	−5.8	−6.8	3850	13	400	38000	0.0000006
M0	−5.6	−6.9	3650	13	500	41000	0.0000001

Source: Adapted from Kenneth R. Lang, Astrophysical Data *(Springer-Verlag, 1992), pp. 132–145.*

MAIN-SEQUENCE LIFETIME A star will spend most of its lifetime on the main sequence. Just how long it does spend there is determined by the *Chandrasekhar–Schönberg limit*, which states that the amount of helium in the central core of the star cannot exceed about 12% of the mass of the star. More massive stars consume energy much faster, as shown by the mass–luminosity relationship, so their available hydrogen is consumed more quickly and their main-sequence lifetimes are much shorter. While the Sun is expected to have a main-sequence lifetime of about 10^{10} years, a highly luminous B0 star probably spends only a few million years on the main sequence.

RED GIANTS The main-sequence stage is followed by the onset of hydrogen burning in a shell surrounding the core, and the star begins to evolve fairly rapidly away from the main sequence into the upper right part of the HR diagram. The surface temperature usually decreases, but the radius of the star increases during this stage of its evolution and so too does the luminosity, and the star becomes a red giant.

The subsequent evolution of a red giant depends on its mass, and may be very complicated with several passages back and forth across the HR diagram. In more massive stars the density and temperature of the core reach a flashpoint and a new series of nuclear

TABLE 52. *Relationship between spectral type and colour index.*

| | COLOUR INDEX | | | | | |
| | MAIN SEQUENCE (V) | | GIANTS (III) | | SUPERGIANTS (Iab) | |
SPECTRAL TYPE	B − V	U − B	B − V	U − B	B − V	U − B
O5	−0.33	−1.19	−0.32	−1.18	−0.31	−1.17
B0	−0.30	−1.08	−0.30	−1.08	−0.23	−1.06
B5	−0.17	−0.58	−0.17	−0.58	−0.10	−0.72
A0	−0.02	−0.02	−0.03	−0.07	−0.01	−0.38
A5	+0.15	+0.10	+0.15	+0.11	+0.09	−0.08
F0	+0.30	+0.03	+0.30	+0.08	+0.17	+0.15
F5	+0.44	+0.02	+0.43	+0.09	+0.32	+0.27
G0	+0.58	+0.06	+0.65	+0.21	+0.76	+0.52
G5	+0.68	+0.20	+0.86	+0.56	+1.02	+0.83
K0	+0.81	+0.45	+1.00	+0.84	+1.25	+1.17
K5	+1.15	+0.98	+1.50	+1.81	+1.60	+1.80
M0	+1.40	+1.22	+1.56	+1.87	+1.67	+1.90
M5	+1.64	+1.24	+1.63	+1.58	+1.80	+1.60

Source: Kenneth R. Lang, Astrophysical Data (Springer-Verlag, 1992), pp. 150–153.

reactions begins: first the burning of helium into carbon, and later of carbon into heavier elements. Stars of less than about $0.4\,M_\odot$ (so-called red dwarfs) will not have a helium-burning stage, but their evolution is so slow that probably no such star in our Galaxy has yet had time to complete its main-sequence stage.

Eventually every star must run out of nuclear fuel, at which point it will cease to generate the radiation needed to support its structure. Its core will then contract, although its outer layers may be expelled in the form of a stellar wind (perhaps giving rise to a *planetary nebula*), or in a more violent process.

WHITE DWARFS   After losing its outer layers at the end of its red-giant stage, a relatively low-mass star such as the Sun is thought to evolve into a white dwarf, with a diameter of only about 10000 km and a density of at least $10^8\,\mathrm{kg\,m^{-3}}$. The luminosity is then very low but the surface temperature is high, so such stars are found to the lower left of the HR diagram. There is no energy-producing process operating in white dwarfs, so they will eventually cool down to non-luminous bodies (*black dwarfs*, not to be confused with black holes).

NEUTRON STARS   It seems that a star more massive than $1.4\,M_\odot$ cannot become a white dwarf unless it loses sufficient mass in some way to bring it below that limit. A star or stellar remnant of mass greater than $1.4\,M_\odot$ may collapse to a superdense state in which atoms are broken down and nuclear components combined to form a body composed of neutrons. Such neutron stars, with densities of the order of $10^{18}\,\mathrm{kg\,m^{-3}}$, have been identified with the *pulsars*, discovered in 1967. The outbursts of some supernovae are thought to be triggered by the collapse of a star's central regions to the neutron star stage.

BLACK HOLES   Very massive objects may enter a state of gravitational collapse where no known physical process can halt the contraction. The body will then contract to within a critical radius known as the *Schwarzschild radius*, at which point its gravitational field becomes so strong that no radiation can escape from it. Such an object is known as a black hole.

STELLAR POPULATIONS

Stars in the arms of spiral galaxies are, in general, bluer and richer in heavy elements than stars in the galactic nuclei or in elliptical galaxies. The stars in the spiral arms are called *Population I* stars, and are thought to be younger than stars in the nuclei and in elliptical galaxies, which are called *Population II* stars. In our own Galaxy, Population I stars are found in the flattened disk of the Galaxy, while Population II stars are found in the spherical halo, which includes the globular clusters, and towards the galactic centre. The greater proportion of heavy elements in Population I stars is probably a result of their having formed in part from material which had already been processed in earlier generations of stars and returned to the interstellar medium by, for example, stellar winds or supernova explosions.

Double Stars

Double stars appear to the naked eye as single points of light, but when viewed through a telescope are found to be two stars. The stars may be connected gravitationally (a *binary*), or may simply happen to lie in nearly the same direction (an *optical pair*). Triple stars have three, quadruple stars four and multiple stars many

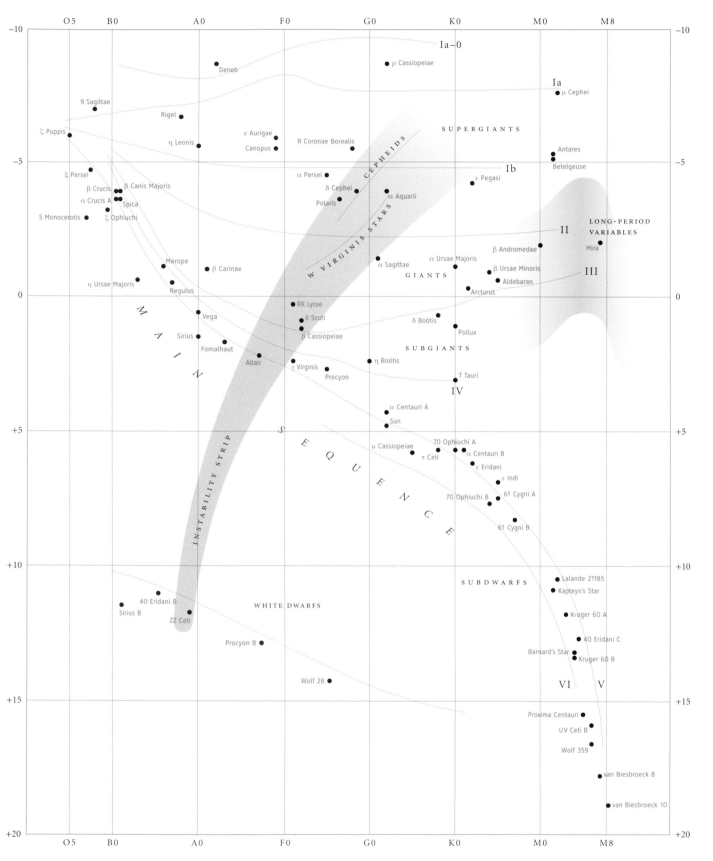

FIGURE 31. *Hertzsprung–Russell (HR) diagram showing stars plotted according to their absolute magnitude and spectral type. Luminosity increases from bottom to top; temperature increases from right to left; colour index increases from left to right. The roman numerals indicate luminosity classes. Many stars are seen to lie on the main sequence; the instability strip is a region in which pulsating variables are found. Adapted from James B. Kaler, Sky & Telescope, Vol. 75, p. 482 (1988).*

Double Star Program

This BASIC program takes as its input the seven orbital elements for a binary system, and calculates for any specified date the position angle and separation to an accuracy sufficient for observational purposes.

```
10 DEF FNC(W) = 1.745329252E-2*W
20 PX = 3.141591 : C = 6.283185
30 INPUT "Period, P (years)          ";P
40 INPUT "Date of periastron, T      ";T
50 input "Semi-major axis, a         ";a1
60 INPUT "Eccentricity, e            ";S
70 INPUT "Inclination, i             ";I
80 INPUT "Arg. of periastron, w      ";W
90 INPUT "PA of ascending node       ";L
100 I = FNC(I) : L = FNC(L) : W = FNC(W)
110 N = C/P
120 INPUT "Date of obs. (year)        ";D
130 MA = N*(D-T)
140 GOSUB 300
150 R = A1-A1*S*COS(EA)
160 Y = SIN(NU+W)*COS(I)
170 X = COS(NU+W)
180 Q = ARCTAN(Y/X)
190 IF X < 0 THEN Q = Q+PX : GOTO 210
200 IF Q < 0 THEN Q = Q+C
210 TH = Q+L : IF TH > C THEN TH = TH-C
220 RH = R*X/COS(Q)
230 PRINT "PA   =   ";INT(TH/FNC(1)*10+0.5)/10;" deg."
240 PRINT "Sep. =   ";INT(RH*100+0.5)/100;" arcsec"
250 INPUT "New date? (Y/N) ";AN$
260 IF AN$ = "Y" THEN GOTO 120
270 INPUT "New binary? (Y/N) ";AN$
280 IF AN$ = "Y" THEN GOTO 10
290 IF AN$ = "N" THEN END
300 M = MA-C*INT(MA/C) : EA = M
310 A = EA-(S*SIN(EA))-M
320 IF ABS(A) < 1E-06 THEN GOTO 350
330 A = A/(1-(S*COS(EA)))
340 EA = EA-A : GOTO 310
350 TU = SQR((1+S)/(1-S))*TAN(EA/2)
360 NU = 2*ARCTAN(TU)
370 RETURN
```

Example. The orbital elements for Xi (ξ) Ursae Majoris are entered, and the program calculates the PA and separation on 2000 Jan. 1 (2000.0) and on 2010 Jul. 1 (2010.5):

Period, P (years)	59.88
Date of periastron, T	1995.07
Semi-major axis, a	2.536
Eccentricity, e	0.398
Inclination, i	122.13
Arg. of periastron, w	127.94
P.A. of ascending node	101.85
Date of obs. (year)	2000
P.A. = 274.4 deg.	
Sep. = 1.8 arcsec	
New date (Y/N) Y	
Date of obs. (year)	2010.5
P.A. = 207.5 deg.	
Sep. = 1.61 arcsec	

components. The brightest star of a multiple is usually designated A, and the companion(s) B, C and so on, as in Sirius A, Sirius B. The fainter star of a pair is sometimes called the *comes* (plural *comites*) or companion.

BINARY STARS

Binary stars are physically related pairs that orbit around a common centre of gravity. A binary star is said to be a *visual binary* if the components may be resolved in the telescope and their orbital motion can be measured over a period of time. If the relative motion of the components is constant and in a straight line, however, they are probably not a binary but an *optical pair*. A *spectroscopic binary* is one that is detected from the periodic doubling or displacement of lines in its spectrum; an *eclipsing binary* is detected by the periodic variations in its magnitude (see the section on Variable stars, p. 122). A binary star may be simultaneously a visual and a spectroscopic binary, or simultaneously a spectroscopic and an eclipsing binary. The orbital periods range from a fraction of a day to many centuries; visual binaries generally have periods of at least two years, while the other two types generally have much shorter periods.

Even if no relative motion is detectable, the physical association of two stars may be suspected if they appear to be at a similar distance and to have the same radial and transverse motion through space. Often, though, the only indication is that a pair of stars have *common proper motion* (c.p.m.).

Multiple stars may also be connected gravitationally; they are usually found to consist of a close binary pair that moves in a larger orbit with another star, which may also be a close binary. Theta[1] (θ[1]) Orionis, known as the Trapezium, contains four bright, well-separated components visible in small telescopes; they are undoubtedly physically connected, but it seems unlikely that they can be moving in stable orbits. Two of the bright components are eclipsing binaries, so the system contains at least six stars, and there are several fainter components which may also be connected. Trapezium-like objects may be thought of as small star clusters. About half the stars in the neighbourhood of the Sun are components of binary or multiple systems.

OBSERVING DOUBLE STARS

Some of the more interesting doubles are indicated in the notes accompanying the star charts. The column 'PA' gives the position angle of the companion relative to the brighter component, and the column 'Dist.' gives the separation of the components in arc seconds.

The closer the two components of a double star, the greater the aperture needed to separate them. A good telescope of aperture

TABLE 53. *Orbital elements of some visual binaries: the orbital period P, date of periastron T, semi-major axis of orbit a, eccentricity of orbit e, inclination of orbit to plane of sky i, arguments of periastron ω and PA of ascending node Ω.*

ADS[a]	STAR	RA 2000.0 h m	DEC. ° ′	MAGNITUDES		P (y)	T	a ″	e	i °	ω °	Ω °
17175	85 Peg	00 02.2	+27 05	5.8	8.9	26.28	1989.4	0.83	0.38	49.0	96.0	290.0
434	λ Cas	00 31.8	+54 31	5.3	5.6	515	1930.0	0.58	0.0	56.5	0.0	163.4
520	β 395	00 37.3	−24 46	6.2	6.6	25.09	1998.86	0.667	0.235	77.6	317.0	291.8
671	η Cas	00 49.1	+57 49	3.5	7.4	480	1889.6	11.994	0.497	34.76	268.59	278.42
755	36 And	00 55.0	+23 38	6.1	6.5	167.71	1956.2	1.002	0.304	45.8	359.1	173.1
1538	Σ 186	01 55.9	+01 51	6.8	6.8	162.02	1893.31	1.033	0.695	72.87	220.89	40.03
1598	48 Cas	02 02.0	+70 54	4.7	6.7	60.55	1965.8	0.628	0.386	19.4	244.0	188.4
1615	α Psc	02 02.0	+02 46	4.1	5.2	933	2098.6	4.0	0.696	120.9	225.4	23.3
1631	10 Ari	02 03.7	+25 56	5.8	7.9	325	1931.6	1.39	0.59	51.0	165.0	20.5
2402	α For	03 12.1	−28 59	4.0	7.2	269	1947	4.0	0.73	81.0	43.0	117.0
2616	7 Tau	03 34.4	+24 28	6.6	6.9	522.16	1911.62	0.625	0.679	157.2	238.1	13.0
2799	OΣ 65	03 50.3	+25 35	5.7	6.5	60.59	1998.32	0.437	0.641	83.9	344.6	26.2
4241	σ Ori	05 38.7	−02 36	3.7	6.3	155.3	1997	0.264	0.051	160.4	18.0	136.0
5423	α CMa	06 45.1	−16 43	−1.4	8.5	50.09	1994.31	7.500	0.592	136.53	147.27	44.57
5400	12 Lyn AB	06 46.2	+59 27	5.4	6.0	706.09	2446.21	1.66	0.03	178.03	244.86	90.0
5514	14 Lyn	06 53.1	+59 27	6.0	6.5	290.45	1963.14	0.621	0.41	68.1	170.7	54.0
6175	α Gem	07 34.6	+31 53	1.9	3.0	444.95	1960.10	6.593	0.323	114.61	253.31	41.46
6420	9 Pup	07 51.8	−13 54	5.6	6.5	22.7	1985.92	0.602	0.741	80.4	73.1	102.9
6650	ζ Cnc AB	08 12.2	+17 39	5.3	6.3	59.56	1989.19	0.862	0.32	167.0	187.0	13.0
6650	ζ Cnc AB–C	08 12.2	+17 39	5.1	6.2	1115	1970	7.7	0.24	146.0	345.5	74.2
6914	β 208	08 39.1	−22 40	5.4	6.8	123.0	1986.6	1.705	0.33	82.9	309.2	31.8
	I 314	08 39.4	−36 36	6.4	7.9	66.5	1992.2	0.527	0.86	102.0	341.0	55.7
	δ Vel	08 44.7	−54 43	var.	5.1	142	2000.8	1.99	0.47	105.2	188.0	163.6
6993	ε Hya	08 46.8	+06 25	3.5	6.7	990	1920	4.66	0.30	39.0	200.0	49.3
7307	Σ 1338	09 21.0	+38 11	6.7	7.1	303.27	2023.25	1.336	0.254	29.9	191.9	137.3
7390	ω Leo	09 28.5	+09 03	5.7	7.3	118.23	1959.40	0.880	0.557	66.05	302.65	325.69
	ψ Vel	09 30.7	−40 28	3.9	5.1	33.95	1969.68	0.862	0.433	58.0	44.3	291.0
7545	φ UMa	09 52.1	+54 04	5.3	5.4	105.4	1987.4	0.349	0.45	24.5	35.0	130.3
7555	γ Sex	09 52.5	−08 06	5.4	6.4	77.55	1957.92	0.383	0.691	145.1	141.5	31.0
7724	γ Leo	10 20.0	+19 51	2.4	3.6	618.56	1743.32	2.505	0.843	36.37	162.54	143.24
7846	β 411	10 36.1	−26 41	6.7	7.8	170.14	1948.39	0.886	0.765	128.2	37.9	145.1
8119	ξ UMa	11 18.2	+31 32	4.3	4.8	59.88	1995.07	2.536	0.398	122.13	127.94	101.85
8148	ι Leo	11 23.9	+10 32	4.1	6.7	186.0	1948.8	1.91	0.53	128.0	325.0	235.0
8197	OΣ 235	11 32.4	+61 05	5.7	7.6	72.7	1981.8	0.79	0.40	46.0	132.0	80.0
8539	Σ 1639	12 24.4	+25 35	6.7	7.8	575.44	1891.75	1.224	0.926	150.4	9.7	140.8
8573	β 28	12 30.1	−13 24	6.5	9.6	151	1944	1.4	0.71	24.0	75.0	94.0
	γ Cen	12 41.5	−48 58	2.8	2.9	84.49	2015.71	0.936	0.791	113.5	187.2	2.4
8630	γ Vir	12 41.7	−01 27	3.5	3.5	168.9	2005.3	3.68	0.89	148.0	257.0	37.0
8695	35 Com	12 53.3	+21 15	5.2	7.1	359	2038.0	1.181	0.145	34.0	30.0	201.4
8974	25 CVn	13 37.5	+36 18	5.0	7.0	228.0	1864.0	1.02	0.80	147.0	159.0	87.0
	α Cen	14 39.6	−60 50	0.0	1.3	79.914	2035.49	17.575	0.518	79.21	231.65	204.85
9343	ζ Boo	14 41.1	+13 44	4.5	4.6	123.44	2021.03	0.595	0.957	142.0	1.47	129.99
9413	ξ Boo	14 51.4	+19 06	4.8	7.0	151.6	1909.3	4.94	0.51	139.0	203.0	347.0
9425	OΣ 288	14 53.4	+15 42	6.9	7.6	313	1824	1.36	0.50	108.5	49.0	12.5
9494	44,i Boo	15 03.8	+47 39	5.2	var.	206.0	2013.0	3.80	0.55	84.0	45.0	57.0
9617	η CrB	15 23.2	+30 17	5.6	6.0	41.585	2016.89	0.868	0.262	59.03	38.42	203.19
9626	μ² Boo	15 24.5	+37 23	7.1	7.6	257.0	1864.2	1.47	0.58	134.0	336.0	174.0
	γ Lup	15 35.1	−41 10	3.5	3.6	190	1885.0	0.655	0.51	95.0	311.5	94.6
9909	ξ Sco AB	16 04.4	−11 22	5.2	4.9	45.68	1997.0	0.663	0.75	33.0	343.0	206.0
9979	σ CrB	16 14.7	+33 52	5.6	6.5	889.0	1826.9	5.93	0.76	31.8	72.2	16.9
10087	λ Oph	16 30.9	+01 59	4.2	5.2	129.0	1939.7	0.91	0.611	23.0	157.5	53.3
10157	ζ Her	16 41.3	+31 36	3.0	5.4	34.45	1967.7	1.33	0.46	131.0	111.0	50.0
10279	20 Dra	16 56.4	+65 02	7.1	7.3	422.22	1838.91	1.044	0.143	96.0	216.2	68.3
10345	μ Dra	17 05.3	+54 28	5.7	5.7	672	1949.0	3.95	0.45	144.7	197.0	282.8
	MlbO 4 AB	17 19.0	−34 59	6.4	7.4	42.15	1975.9	1.81	0.58	128.0	247.0	313.0
10660	26 Dra	17 35.0	+61 52	5.3	8.5	76.1	1947.0	1.53	0.18	104.0	307.0	151.0
11005	τ Oph	18 03.1	−08 11	5.3	5.9	257.0	1829.0	1.40	0.77	52.0	42.0	60.0
11046	70 Oph	18 05.5	+02 30	4.2	6.2	88.38	1984.32	4.554	0.499	121.16	14.0	302.12
	h 5014	18 06.8	−43 25	5.7	5.7	450	1854.7	2.04	0.65	123.1	282.7	85.8
11483	OΣ 358	18 35.9	+16 59	6.9	7.1	380	1816	1.84	0.57	119.0	74.0	30.8

[a] *The ADS number is the number given in the* New General Catalogue of Double Stars, *which covers as far south as −30°.*

TABLE 53 *(continued). Orbital elements of some visual binaries.*

ADS[a]	STAR	RA 2000.0 h m	DEC. ° '	MAGNITUDES		P (y)	T	a "	e	i °	ω °	Ω °
11635	ε^1 LyrAB	18 44.3	+39 40	5.0	6.1	1165.6	1152.4	2.78	0.19	138	165.7	29.0
11635	ε^2 LyrCD	18 44.4	+39 37	5.3	5.4	724.3	2223.9	2.92	0.35	126.1	73.8	26.2
	γ CrA	19 06.4	−37 04	4.5	6.4	121.76	2000.64	1.896	0.320	149.6	349.0	50.3
12880	δ Cyg	19 45.0	+45 08	2.9	6.3	780.3	1880.0	3.0	0.47	151.0	120.2	91.4
14296	λ Cyg	20 47.4	+36 29	4.7	6.3	391.3	1795.0	0.78	0.45	133.8	298.4	138.6
14360	4 Aqr	20 51.4	−05 38	6.4	7.4	194.0	1896.3	0.86	0.489	67.3	46.2	174.5
14499	ε Equ AB	20 59.1	+04 18	6.0	6.3	101.485	2021.855	0.647	0.705	92.17	340.19	105.15
14636	61 Cyg	21 06.9	+38 45	5.4	6.1	659	1697	24.4	0.48	54.0	146.0	176.0
14787	τ Cyg	21 14.8	+38 03	3.8	6.6	49.6	1989.0	0.91	0.24	133.0	118.0	159.0
15270	μ Cyg	21 44.1	+28 45	4.8	6.2	789	1958.0	5.32	0.66	75.5	145.7	110.1
15971	ζ Aqr	22 28.8	−00 01	4.3	4.5	760	1968.0	4.51	0.50	135.9	63.4	304.6
16538	π Cep	23 07.9	+75 23	4.6	6.8	160.0	1933.95	0.84	0.58	28.4	115.3	63.5
16836	72 Peg	23 34.0	+31 20	5.7	6.1	246.17	1856.09	0.447	0.28	35.6	129.0	123.8

[a] The ADS number is the number given in the New General Catalogue of Double Stars, *which covers as far south as −30°.*
Source: Orbital elements from the Sixth Catalog of Orbits of Visual Binary Stars *by William I. Hartkopf and Brian D. Mason*
http://ad.usno.navy.mil/wds/orb6.html

D mm should just enable a pair of stars of sixth magnitude to be distinguished under high power if their separation is $116/D$ arcsec (the Dawes limit; see Table 10, p. 29). If the components are unequal in brightness, or if they are much brighter or fainter than sixth magnitude, a larger aperture will be needed to separate a given pair than this formula would suggest.

For binary stars whose orbits have been calculated, the BASIC computer program on p. 120 may be used to obtain predictions of PA and distance for any required date. The calculation requires seven *orbital elements*:

P orbital period (years)

T date of periastron (in decimal form)

a semi-major axis of orbit (arc seconds)

e eccentricity of orbit

i inclination of orbit to plane of sky (degrees)

ω argument of periastron (degrees)

Ω PA of ascending node (degrees)

In a visual binary system, the companion is said to be at *periastron* when its actual distance (as distinct from its apparent distance) from the main star is a minimum, and at *apastron* when it is a maximum. Together a and e define the size and shape in space of the orbit of the companion relative to the brighter component; these may be very different from the size and shape of the apparent orbit projected onto the celestial sphere. Between them i, ω and Ω define the orbit's orientation. The motion of the companion is said to be *direct* when the position angle is increasing, and *retrograde* when it is decreasing; the inclination i is given as between

0° and 90° for direct motion, and between 90° and 180° for retrograde motion.

Many double and multiple stars present an attractive spectacle in the telescope, especially when the components are of contrasting colours. Useful work can be done by making regular measurements of the PA and distance of binary stars with a micrometer. The masses of stars can be directly determined only by the observation of binary stars. Fewer than 400 visual binaries have well-determined orbits; some orbital elements (not all well-determined) are given in Table 53.

Variable Stars

Stars whose brightness changes with time are known as variable stars. Figure 32 shows some typical graphs of magnitude against time (*light curves*) for different types of variable star. The *amplitude* of a variable star is the difference between its magnitudes at maximum and at minimum. The variations may be periodic, semi-periodic or irregular, with time-scales from a fraction of a second to many centuries. Frequently, other aspects of the star – such as its radial velocity or spectrum – are also found to be variable.

NOMENCLATURE OF VARIABLE STARS   For the variables in each constellation not already assigned a Bayer letter or roman letter, the German astronomer F. W. A. Argelander set aside the capital roman letters from R to Z. After Z, the double forms RR to RZ, SS to SZ, and so on to ZZ were used, which provided for 54 variable stars in any constellation. As that number proved insufficient, AA to AZ, BB to BZ, and so on were also used, J being omitted, extending the capacity to 334 variables per constellation.

The simplest system, by which the variables of each constellation are denoted by the letter V followed by a number, is used from V335, when QZ has been reached. These designations are assigned when the type of variability has been ascertained. Various provisional designations are used for unconfirmed variables, of which the most important are the NSV numbers of the *New Catalogue of Suspected Variable Stars*, e.g. NSV 14811.

Novae are now designated in the same way as other variable stars, but until they receive a final designation they are provisionally referred to by constellation, year and (if necessary) number; e.g. Nova And 1986 is now called OS And; Nova Vul 1984 No. 2 is now QU Vul. Supernovae are given a designation on discovery consisting of the year followed by an upper-case letter from A to Z. After the 26th discovery in any year, double lower-case letters are applied, from aa to az, then ba to bz, and so on.

TYPES OF VARIABLE STAR

Table 54 lists the types currently recognized. They are arranged in six classes. Stars in the eruptive, pulsating, cataclysmic and X-ray classes are sometimes called *intrinsic* variables, as the light changes are due to physical changes in the stars themselves; stars in the rotating and eclipsing classes are *extrinsic* variables, as the light changes are a geometrical effect. Some of the more important types are described below.

EXTRINSIC VARIABLES

Eclipsing variables are binary systems in which the stars' orbital plane lies in our line of sight, so the two stars periodically eclipse each other as seen from Earth. The consequent light variations show two different minima. The deeper minimum, when the star with the greater surface brightness is eclipsed, is called the *primary eclipse*; the shallower minimum is the *secondary eclipse*. If the eclipse is total or annular then the minima may have flat bases. Eclipsing binaries account for nearly a fifth of all known variables. Their periods are subject to slight variation, and useful work can be done in timing the eclipses, either by making visual estimates or, for more accurate results, by photoelectric photometry. They are classified into three types, depending on the shapes of their light curves.

In *Algol-type* eclipsing binaries (type EA), the times when eclipses begin and end can be identified from the light curve, and between consecutive eclipses there is little variation. Algol-type curves may be produced by systems in which the components are sufficiently far apart for them (or at least the one with the higher surface brightness) to retain a normal shape and structure (detached system, Figure 33(*a*), p. 128). The *Roche lobe* is a volume around the star beyond which it cannot expand without losing material in the direction of the other component.

(text continued on p.127)

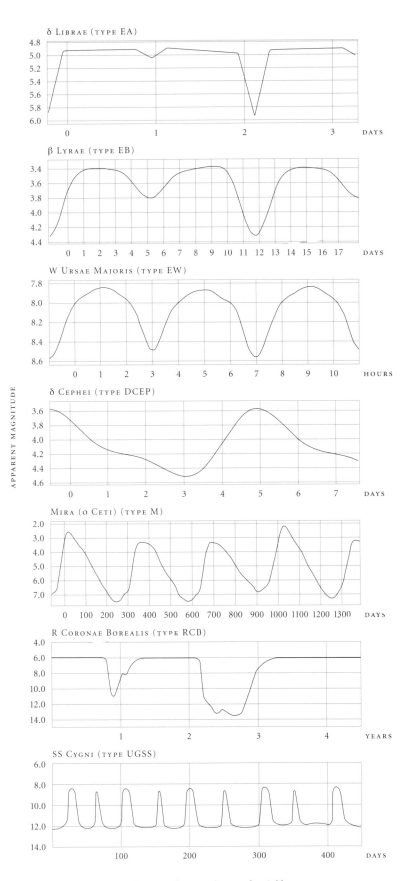

FIGURE 32. *Typical light curves for several types of variable star. Sources:* Hipparcos Catalogue *and AAVSO.*

TABLE 54. *Types of variable star.*

TYPE	ABBREVIATION	AMPLITUDE (MAGNITUDES)	PERIOD[a]	SPECTRUM	DISTRIBUTION[b] (%)	NOTES
Eruptive variables						
FU Ori	FU	6	—	Ae–Ge	0.01	Gradual rise over months to max. lasting many years; also called *fuors*
γ Cas	GCAS	up to 1.5	—	BIII–Ve	0.4	Shell stars; temporary fades
Be	BE	—	—	Be		Similar to GCAS, but small-scale variations, sometimes quasi-periodic; not related to shell events
Irregular	I	—	—	—		Poorly studied stars of unknown spectral type
	IA	—	—	O–A		Poorly studied irregular variables of early spectral type
	IB	—	—	F–M		Poorly studied irregular variables of intermediate or late spectral type
Orion	IN, INS	up to several magnitudes	—	—		Young objects in diffuse nebulae; 'S' is added to denote rapid variation
	INA, INSA	—	—	B–A or Ae		Orion variables of early spectral type; occasional abrupt Algol-like fades; example T Ori
	INB, INSB	—	—	F–M, Fe–Me		Orion variables of intermediate or late spectral type; F-type stars may show Algol-like fades; example BH Cep
T Tau	IT	—	—	Fe–Me		Orion variables with intense emission of Fe I at 404.6 and 413.2 nm: example RW Aur
	INT	—	—	Fe–Me		T Tau stars in diffuse nebulae; example T Tau itself
	IS	0.5–0.10	—	—		Rapid irregular variables not in nebulae
	ISA	—	—	B–A or Ae		Rapid irregular variables of early spectral type
	ISB	—	—	F–M, Fe–Me		Rapid irregular variables of intermediate or late spectral type
All types I					5	
R CrB	RCB	1–9	—	Bpe–R	0.1	Cyclic pulsations and irregular deep fades
RS CVn	RS	0.2	—	—	0.05	Close binaries with chromospheric activity
S Dor	SDOR	1–7	—	Bpeq–Fpeq	0.05	High-luminosity stars, usually in diffuse nebulae and with expanding shells; example P Cyg
LPB	LPB	0.01	>6 h to centuries	B		Long-period B stars, related to S Dor stars, some possibly pulsating
UV Cet	UV	up to 6	—	KVe–MVe	3	Flare stars
	UVN	—	—	Ke–Me	1	Flaring Orion variables; example V389 Ori
Wolf–Rayet	WR	up to 0.1	—	W	0.03	Non-stable mass outflow; example V1042 Cyg
Pulsating variables						
α Cyg	ACYG	0.1	days to weeks	B–AIaeq	0.09	Non-radially pulsating supergiants
β Cep	BCEP	0.01–0.3	0.1–0.6	O8–B6I–V	0.3	Radial or non-radial pulsation
	BCEPS	0.15–0.025	0.02–0.04	B2–B3IV–V		Short-period group of β Cep variables
Cepheids	CEP	up to 2	1–135	F–KIb–II	0.6	Radial pulsation
W Vir	CW	0.3–1.2	0.8–35	—	0.01	Population II Cepheids
	CWA	—	8–35	—	0.4	Long-period W Vir stars; example W Vir itself
	CWB	—	0.8–8	—	0.2	Short-period W Vir stars; example BL Her
δ Cep	DCEP	—	—	—	1	Population I or classical Cepheids
	DCEPS	up to 0.5	up to 7	—	0.2	Short-period group of classical Cepheids; example δ Cep itself
δ Sct	DSCT	0.003–0.9	0.01–0.02	A0–F5III–V	0.3	Radial or non-radial pulsators; Population I
	DSCTC	up to 0.1	—	—	0.4	Low-amplitude group of δ Sct stars; present in open clusters; example EW Aqr
Irregular	L	—	—	—	3	Slow irregular variables
	LB	—	—	K, M, C, S	6	Slow irregular variables of late spectral type; example CO Cyg

[a] *In days unless otherwise stated.*
[b] *The percentages are based on the numbers of each type in the* General Catalogue of Variable Stars. *They do not necessarily reflect the true distribution; for example, brighter stars and stars showing larger variations are more likely to be discovered than fainter or small-amplitude variables.*

TABLE 54 *(continued)*. *Types of variable star.*

TYPE	ABBREVIATION	AMPLITUDE (MAGNITUDES)	PERIOD[a]	SPECTRUM	DISTRIBUTION[b] (%)	NOTES
	LC	1	—	K, M, C, S	0.2	Slow irregular supergiant variables of late spectral type; examples Antares, TZ Cas
Mira	M	2.5–11	80–1000	Me, Ce, Se	21	Long-period variable giants
PV Tel	PVTEL	0.1	0.1 d to 1 y	Bp	0.01	Helium supergiants; example PV Tel itself
RR Lyr	RR	0.2–2	0.2–1.2	A–F	6	Radial pulsators of Population II; formerly called short-period Cepheids or cluster-type variables
	RRAB	0.5–2	0.3–1.2	—	14	Steep ascending branch on light curves; example RR Lyr itself
	RRC	up to 0.8	0.2–0.5	—	1	Nearly symmetrical light curves; example SX UMa
RV Tau	RV	up to 4	30–150	F–M	0.3	Radially pulsating supergiants with alternating primary and secondary minima
	RVA	—	—	—	0.09	RV Tau stars with constant mean magnitude; example R Sct
	RVB	—	—	—	0.05	RV Tau stars with mean magnitude varying up to 2 mags in periods of 600–1500 d; examples RV Tau, DF Cyg
Semi-regular	SR	1–2	20–2000+	—	5	Noticeable periodicity, but with irregularities
	SRA	up to 2.5	35–1200	M, C, S	3	Red giants with persistent periodicity; example Z Aqr
	SRB	—	20–2300	M, C, S	3	Red giants wth poorly defined periodicity; example RR CrB
	SRC	1	30 d to several years	M, C, S	0.2	Red supergiants; examples Betelgeuse, α Her, μ Cep
	SRD	0.1–4	30–1100	F–K	0.3	Giants and supergiants of intermediate spectral type; examples SX Her, ρ Cas
	SRS	up to 0.25?	3–30?	—	—	Short-period semi-regular pulsating red giants, probably high-overtone pulsators; example AU Ari
SX Phe	SXPHE	up to 0.7	0.04–0.08	A2–F5	0.05	Population II subdwarfs resembling δ Sct stars
ZZ Cet	ZZ	0.001–0.2	30–1500 s	—		Non-radially pulsating white dwarfs
	ZZA	—	—	DA		ZZ Cet stars with only hydrogen absorption lines in spectrum; example ZZ Cet itself
	ZZB	—	—	DB		ZZ Cet stars with only helium absorption lines in spectrum; example V777 Her
	ZZO	—	—	DO		Very hot ZZ Cet stars with He II and C IV absorption lines; example GW Vir
All types ZZ					0.08	

The following suffix may be added (e.g. CEP(B)):

| | B | — | — | — | | Beats caused by two simultaneous pulsation modes |

Rotating variables

TYPE	ABBREVIATION	AMPLITUDE (MAGNITUDES)	PERIOD[a]	SPECTRUM	DISTRIBUTION[b] (%)	NOTES
α² CVn	ACV	0.01–0.1	0.5–160	B8p–A7p	0.6	Main-sequence stars with strong magnetic fields and anomalously strong lines of Si, Sr, Cr and rare earth elements
	ACVO	0.01	0.004–0.1	Ap	0.02	α² CVn stars with rapid non-radial pulsations; example DO Eri
BY Dra	BY	up to 0.5	up to 120	G–M, Ge–Me	0.1	Rotating dwarfs with starspots and chromospheric activity
Ellipsoidal	ELL	up to 0.1	—	—	0.2	Close binaries with changing visible surface area, but no eclipses; example b Per
FK Com	FKCOM	*c.* 0.5	up to several days	G–K	0.01	Rapidly rotating giants with non-uniform surface brightness
Pulsars	PSR	up to 0.8	0.001–4 s	—	0.004	Rapidly rotating neutron stars with narrow beams of optical radiation; example CM Tau
	R	0.5–1.0	—	—		Close binaries showing reflection of light of hot component on surface of cool component; brightness varies as system rotates; example KV Vel
SX Ari	SXARI	0.1	1	B0p–B9p	0.06	High-temperature analogues of α² CVn variables, sometimes called *helium variables*

[a] *In days unless otherwise stated.*
[b] *The percentages are based on the numbers of each type in the* General Catalogue of Variable Stars. *They do not necessarily reflect the true distribution; for example, brighter stars and stars showing larger variations are more likely to be discovered than fainter or small-amplitude variables.*

TABLE 54 *(continued). Types of variable star.*

TYPE	ABBREVIATION	AMPLITUDE (MAGNITUDES)	PERIOD[a]	SPECTRUM	DISTRIBUTION[b] (%)	NOTES
Cataclysmic variables (explosive and nova-like)						
AM Her	AM	up to 5	—	—	0.004	Polars; close binaries containing compact object with strong magnetic field; accretion on magnetic poles gives rise to emission of polarized light
Novae	N	7–19	—	—	0.2	Thermonuclear runaway on white dwarf component of close binary
	NA	—	—	—	0.3	Fast novae, fading by 3 mags in 100 d or less; example GK Per
	NAB	—	—	—	0.004	Novae of intermediate speed; example V400 Per
	NB	—	—	—	0.1	Slow novae, fading by 3 mags in 150 d or more; example RR Pic
	NC	up to 10	—	—	0.03	Very slow novae, at max. for more than 10 y; often classed with A And stars; example RR Tel
Nova-like	NL	—	—	—	0.1	Insufficiently studied objects with outbursts like novae, or resembling old novae; example V Sge
	NR	—	10–80 y	—	0.03	Recurrent novae; example R CrB
Supernovae	SN	20+	—	—		Catastrophic explosion of star
	SNI	—	—	—		Type I supernovae: no hydrogen lines; fading at 0.1 mag per day for 20–30 d, then at 0.01 mag per day; example Z Cen
	SNII	—	—	—		Type II supernovae: hydrogen lines present; usually fading at 0.1 mag per day 40–100 d after max.; example SN 1987A
All types SN					0.02	
U Gem	UG	2–9	10 to 1000+	—	0.6	Dwarf novae; pulsed release of gravitational energy from accretion disk around white dwarf component of close binary
SS Cyg	UGSS	2–6	—	—	0.3	Dwarf novae with outbursts lasting several days
SU UMa	UGSU	4–9	—	—	0.08	Dwarf novae with short outbursts like SS Cyg stars, and occasional supermaxima 2 mags brighter and five times longer
Z Cam	UGZ	2–5	10–40	—	0.2	Dwarf novae with cyclic outbursts interrupted by standstills at intermediate magnitudes
Z And	ZAND	up to 4	—	—	0.2	Close binaries consisting of a cool star and a hot one exciting an extended envelope; often called *symbiotic stars*
Eclipsing variables						
Eclipsing	E	—	—	—	3	Binary stars in which one component periodically passes in front of the other
Algol	EA	—	0.2–10000+	—	11	Nearly spherical components, with contact times identifiable from light curve
β Lyr	EB	up to 2	over 1	B–A	2	Ellipsoidal components, with continuous change in brightness
W UMa	EW	up to 1	up to 0.8	F–G	2	Components almost in contact; primary and secondary minima nearly equal
Planetary transits	EP	up to 0.02	days to years	—	—	Stars showing eclipses by their planets; example V376 Peg
The following suffixes may be added (e.g. EA/AR/RS):						
	AR	—	—	—		Detached system of AR Lac type; both components are subgiants, and neither fills its inner equipotential surface (Roche lobe)
	D	—	—	—		Detached system
	DM	—	—	—		Detached main-sequence system
	DS	—	—	—		Detached system with subgiant component
	DW	—	—	—		Resembles contact systems of W UMa type, but components not in contact
	GS	—	—	—		Giant or supergiant component(s)
	K	—	—	—		Contact system, both components filling their Roche lobes

[a] *In days unless otherwise stated.*
[b] *The percentages are based on the numbers of each type in the* General Catalogue of Variable Stars. *They do not necessarily reflect the true distribution; for example, brighter stars and stars showing larger variations are more likely to be discovered than fainter or small-amplitude variables.*

TABLE 54 *(continued)*. *Types of variable star.*

TYPE	ABBREVIATION	AMPLITUDE (MAGNITUDES)	PERIOD[a]	SPECTRUM	DISTRIBUTION[b] (%)	NOTES
	KE	—	—	O–A		Contact system of early spectral type
	KW	—	—	F0–K		Contact system of W UMa type, primary is main-sequence star, and secondary lies below and to left of it in HR diagram
	PN	—	—	—		Nucleus of planetary nebula
	RS	—	—	—		RS CVn system; see eruptive variables above
	SD	—	—	—		Semi-detached system; less massive, subgiant component fills Roche lobe
	WD	—	—	—		White dwarf components
	WR	—	—	—		Wolf–Rayet component(s)
X-ray binaries						
	X	—	—	—		Close binaries containing compact object (white dwarf, neutron star or black hole)
Burster	XB	0.1	—	—		X-ray and optical bursts lasting seconds or minutes; example V801 Ara
	XF	—	—	—		Rapid X-ray and optical fluctuations in fraction of a second; example V1357 Cyg
Irregular	XI	1	—	—		Variations over minutes or hours; example V818 Sco
	XJ	—	—	—		Relativistic jets present; example V1343 Aql
	XND	4–9	—	—		X-ray novae or transients, with a dwarf or subgiant component of spectral type G–M; outbursts lasting up to several months but no envelope ejected; example V616 Mon
	XNG	1–2	—	—		X-ray novae or transients with an early-type giant or supergiant component; example V725 Tau
	XP	up to several magnitudes	1–10	—		X-ray pulsars, with periods of 1 s to 100 m; slower light change caused by rotation of ellipsoidal component; example GP Vel
	XPR	2–3	—	—		X-ray pulsars showing reflection effect – 'normal' component is irradiated by X-rays, and brightness varies as system rotates; example HZ Her
Polars	XPRM	1–5	—	—		X-ray pulsars with strong magnetic field; accretion on magnetic poles gives rise to emission of polarized light; example BL Hyi
	XR. XRM	—	—	—		Resemble types XPR, XPRM, but presumed X-ray pulsar not observed as the X-ray beam is never in the line of sight; example AN UMa
All types X					0.2	
Other types						
	S	—	—	—	0.6	Unstudied stars with rapid light changes
	★	—	—	—	0.2	Unique types of variable not fitting above classification; example VY CMa

[a] *In days unless otherwise stated.*
[b] *The percentages are based on the numbers of each type in the* General Catalogue of Variable Stars. *They do not necessarily reflect the true distribution; for example, brighter stars and stars showing larger variations are more likely to be discovered than fainter or small-amplitude variables.*

In *Beta Lyrae-type* systems (type EB) the brightness varies continuously, and the times when eclipses begin and end cannot be identified from the light curve. They may be detached systems with ellipsoidal components, or ones in which the component of greater surface brightness fills its Roche lobe (semi-detached system, Figure 33(*b*)), and much of the variation is a result of the changing visible area of this star as the system rotates.

In *W Ursae Majoris* systems (type EW) both components almost fill their Roche lobes, or overfill them so that the stars are actually in contact (Figure 33(*c*)). The light curves resemble those of Beta Lyrae systems, but the periods are generally shorter and the primary and secondary minima are of similar depth.

Pulsars (type PSR) are normally radio variables, but two – the pulsar in the Crab Nebula and the Vela Pulsar – have also shown periodic light variations. It is believed that a pulsar is a neutron star (see p. 118) which is left behind after a supernova explosion.

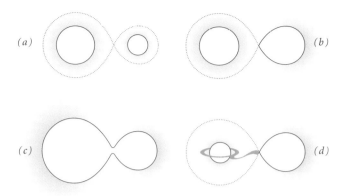

FIGURE 33. *Configurations of interacting binary stars: (a) detached, (b) semi-detached, (c) contact and (d) semi-detached with accretion disk. In each case the dashed line represents a surface of equal gravitational potential in the rotating frame of the binary.*

The variations are caused by the object's rotation combined with a directional form of light emission. The Crab Pulsar (CM Tau) has a period of 0.033 s. The Vela Pulsar (HU Vel) is one of the faintest known variable stars, ranging between B magnitudes 23.2 and 25.2 in a period of 0.089 s.

INTRINSIC VARIABLES

Two-thirds of the known variable stars are *pulsating variables*. They may pulsate radially, remaining spherical in shape, or non-radially, with the shape deviating periodically from a sphere. *Cepheid variables* are bright, radially pulsating stars whose period and mean magnitude are closely related. They are subdivided into *classical Cepheids* (type DCEP) and *W Virginis stars* (type CW); closely related are the *Beta Cephei* or *Beta Canis Majoris stars* (type BCEP), the *Delta Scuti stars* (type DSCT), the *RR Lyrae stars* (type RR) and the *SX Phoenicis stars* (type SXPHE). The cause of the variations appears to be a self-propagating, periodic ionization and recombination of helium in the star's atmosphere which causes a periodic variation in the opacity. The changing opacity in turn produces a change in the star's temperature and radius, and hence in its luminosity.

Mira stars (type M), also known as *long-period variables*, are late-type giants with periods typically of several months to a year or so and amplitudes of several magnitudes; both the periods and the amplitudes of Mira stars are subject to appreciable variation from cycle to cycle, making them objects which amateur observers can usefully study.

The *semi-regular variables* (type SR) sometimes resemble small-amplitude Mira stars, but the variations are often subject to great irregularity, as in the *irregular variables* (type L) and the *RV Tauri stars* (RV). The periods of these objects are generally not well known, and they merit further study, although their typically small amplitudes and redness make them difficult to observe visually.

TABLE 55. *Recurrent novae of mag. 6.5 or brighter.*

NOVA	YEAR OF OUTBURST	MAGNITUDE RANGE	RA 2000.0 h m	DEC. o '
T CrB	1866, 1946	2.0–10.8	15 59.5	+25 55
T Pyx	1890, 1902, 1920, 1944, 1966	6.5–15.3	09 04.7	−32 23
RS Oph	1898, 1933, 1958, 1967, 1985	4.3–12.5	17 50.2	−06 43

Eruptive variables are a very mixed class of stars which vary because of violent processes and flares taking place in their chromospheres and coronae. They constitute about a tenth of the known variables. The *R Coronae Borealis stars* (type RCB) are carbon-rich stars that are particularly worth monitoring for their sudden fades, which are caused by the ejection of clouds of soot.

Cataclysmic variables make up only 2% of known variable stars, but they are one of the most important subjects for research by amateurs. Most are close binaries in which one component is usually a white dwarf surrounded by an accretion disk formed by matter lost by the other component, which is usually a cool star (Figure 33(*d*)). They are liable to undergo occasional outbursts. In the *dwarf novae* (type UG) the outbursts are semi-periodic, pulsed releases of gravitational energy from material in the accretion disk at intervals from days to months. Outbursts in the *novae* (type N) are the result of thermonuclear runaway, the explosive burning of hydrogen to helium on the surface of the white dwarf. It is likely that all novae are *recurrent novae* with very long periods, and that in most cases only one outburst has been observed. Known outbursts of the three brightest recurrent novae are listed in Table 55. Estimates of the frequency of nova outbursts in our Galaxy range from 12 to 100 per year, although only two or three per year are actually seen. The *symbiotic stars* (types NC and ZAND) are also important objects to monitor for possible outbursts, as are the *nova-like variables* (type NL) which may be ex-novae or pre-novae. The outbursts of *supernovae* (type SN) are typically one million times as energetic as those of novae; at maximum they can be as bright as an entire galaxy. *Type Ia supernovae*, which are less common, are exploding white dwarfs in close binaries, while *Type II supernovae* are massive stars whose cores undergo a catastrophic collapse after their nuclear fuel has been exhausted.

Table 56 lists the brightest known supernovae and novae in our Galaxy. Many novae are not found until they are on the decline, but an estimate of their maximum brightness can be made by comparison with the light curves of other novae. Many so-called 'new stars' were recorded in earlier years. Thus the appearance of a new star in about 150 BC is said to have led Hipparchus to draw up his catalogue of stars, but generally the old records are vague and indefinite, and in some cases undoubtedly refer not to novae but to comets.

TABLE 56. *The brightest galactic novae and supernovae.*

YEAR	NOVA	TYPE[a]	GREATEST MAGNITUDE	APPROXIMATE GALACTIC LONG. °	APPROXIMATE GALACTIC LAT. °	RA 2000.0 h m	DEC. 2000.0 ° '
185	Cen	SN	−6	313	0	14 20	−60
1006	Lup	SN	−8	327	+14	15 02.8	−41 57
1054	CM Tau	SN	−6	185	−6	05 34.5	+22 01
1572	B Cas	SNI	−4.0	120	+1	00 25.3	+64 08
1604	V843 Oph	SNI	−3	5	+7	17 30.6	−21 29
1670	CK Vul	NB	2.7	63	+1	19 47.6	+27 19
1866	T CrB[b]	NR	2.0	42	+48	15 59.5	+25 55
1876	Q Cyg	NA	3.0	90	−8	21 41.7	+42 50
1901	GK Per	NA	0.2	151	−10	03 31.2	+43 54
1918	V603 Aql	NA + E	−1.4	33	+1	18 48.9	+00 35
1920	V476 Cyg	NA	2.0	87	+12	19 58.4	+53 37
1925	RR Pic	NB	1.2	272	−26	06 35.6	−62 38
1934	DQ Her	NB + EA	1.3	73	+26	18 07.5	+45 51
1936	CP Lac	NA	2.1	102	−1	22 15.7	+55 37
1936	V630 Sgr	NA	1.6	358	−7	18 08.0	−34 20
1942	CP Pup	NA	0.5	253	−1	08 11.8	−35 21
1960	V446 Her	NA	2.8	45	+5	18 57.4	+13 14
1963	V533 Her	NA	3	69	+24	18 14.3	+41 51
1975	V1500 Cyg	NA	1.7	90	0	21 11.6	+48 09
1999	V382 Vel	NA	2.7	284	+6	10 44.8	−52 26

[a] A key to the abbreviations can be found in Table 54.
[b] See also Table 55.
Source: General Catalogue of Variable Stars.

Most bright novae have appeared in the Milky Way regions, within 10° of the galactic equator. They have also been commoner towards the galactic centre in Sagittarius. Half of those detected have been between galactic longitudes 0° and 90°, in the region from Sagittarius to Cygnus; but it is possible that a similar number have occurred between 270° and 360°, from Vela to Sagittarius, in the less well-observed southern sky. Amateurs have had considerable success in discovering novae, by searching with binoculars and by making photographic patrols.

X-ray binaries resemble cataclysmic variables, except that the compact object may be not a white dwarf but a neutron star, or even a black hole. Only a few dozen have so far been identified as optically variable objects.

Secular variables are stars that are suspected to have faded or brightened slowly and steadily over a long period of time. No case is certain, but the presence of such a variable in the Pleiades (possibly Pleione) could be the reason why they are sometimes called the Seven Sisters even though only six are obvious to the naked eye.

OBSERVING VARIABLE STARS

Visual estimates are made by comparing the variable with comparison stars of constant brightness. Special charts are required. These are issued by organizations such as the American Association of Variable Star Observers (AAVSO) and the variable-star sections of national societies. (For addresses see the Appendix.) At least two comparison stars should be used for each estimate. It is possible to make visual estimates that are accurate to within about 0.1 mag.

A few variables are bright enough to be followed with the naked eye. Binoculars bring many more into range, but most are telescopic objects. Variables with an amplitude of at least 0.4 mag. that reach mag. 6.5 or brighter at maximum are noted in the lists of interesting objects accompanying the star charts. In addition, known variables with an amplitude of at least 0.1 mag. are indicated by a special symbol on the charts.

For really accurate estimates, photoelectric methods are used. CCD photometry of variables is increasingly employed by amateurs, and is capable of reaching very faint magnitudes. However, considerable work is required to transform CCD results to an accepted photometric system so that magnitudes may be compared with those obtained by other methods. Further information is available from the variable-star sections of national societies and an organization called International Amateur–Professional Photoelectric Photometry (see the Appendix).

ATMOSPHERIC EXTINCTION If the comparison stars used for estimating the brightness of a variable star are not at a similar altitude to the variable, as is sometimes unavoidable, the result will be in error as the light from the variable and the comparison stars will have traversed different thicknesses of the Earth's atmosphere.

TABLE 57. *Atmospheric extinction. The dimming of starlight by scattering and absorption increases with the thickness of the Earth's atmoshere through which it passes, i.e. with zenith distance, and must be allowed for when estimating a star's brightness. This table gives values for the approximate extinction in a clear sky by comparison with the value at the zenith.*

Zenith distance	47°	58°	64°	69°	71°	73°	75°	77°	79°	80°	82°	84°	86°	88°	89°
Extinction (mag.)	0.1	0.2	0.3	0.4	0.5	0.6	0.7	0.8	0.9	1.0	1.2	1.5	2.0	2.5	3.0
Altitude	43°	32°	26°	21°	19°	17°	15°	13°	11°	10°	8°	6°	4°	2°	1°

The errors can be balanced to some extent by making estimates against comparison stars both above and below the variable. Alternatively, allowance may be made for the difference in extinction between the variable and each comparison star. The approximate figures in Table 57 may be used, unless haze is present.

Clusters, Nebulae and Galaxies

NOMENCLATURE Brighter star clusters, nebulae and galaxies are often referred to by their number in the catalogue prepared by the French astronomer Charles Messier and published over the period 1771–1784. These so-called *Messier objects* are listed in Table 58. The Messier or M numbers are used on the star charts in this Atlas, other objects being identified by their number in J. L. E. Dreyer's *New General Catalogue of Nebulae and Clusters of Stars* (NGC) published in 1888, or their number in its two supplements, the *Index Catalogue* (1895) and the *Second Index Catalogue* (1908).

STAR CLUSTERS

Many of the stars that we see are scattered randomly along the spiral arms of our Galaxy, but a great number are concentrated in relatively compact groups called star clusters. They fall into two main categories, *open clusters* and *globular clusters*, each type having its own characteristics.

OPEN CLUSTERS Over a thousand open star clusters are known. They are often referred to as *galactic clusters*, as they are found in the plane of the Galaxy. As their name suggests, open clusters are loose collections of stars that have no well-defined shape; their diameters are generally no more than a few tens of light years. The numbers of stars in open clusters can vary considerably. For example, the sparse cluster M18 (NGC 6613) in Sagittarius contains only about a dozen members whereas M11 (NGC 6705) in Scutum contains 500 stars or more, making it one of the richest open star clusters known.

A number of open clusters are visible to the unaided eye, notably the Pleiades (M45) in Taurus. As with other open star clusters, the Pleiades stars were formed together in the same region of space.

The Pleiades cluster is thought to be about 50 million years old, making it comparatively young. Once a cluster forms, gravitational perturbations from the rest of the Galaxy can slowly break up the group, eventually dispersing the individual stars. However, some dense clusters such as NGC 188 in Cepheus and NGC 6791 in Lyra have remained bound together for 5000 million years or more.

ASSOCIATIONS are loose-knit groups of young stars that have recently been born in the spiral arms of the Galaxy. *OB associations* consist of hot, massive stars of spectral types O and B, numbering from 10 to 100, scattered over an area several hundred light years in diameter. OB associations are often centred on an open cluster, as in the case of the Perseus OB1 association, which is centred on the double cluster h and Chi (χ) Persei. The Orion Nebula is the centre of a major OB association. The nearest to us is the Sco–Cen association, about 500 light years away, stretching from Scorpius to Crux. *T associations* are similar groups containing faint, low-mass T Tauri stars and hence are much less prominent than OB associations.

MOVING CLUSTERS As an open cluster travels through space, its member stars move in paths which are more or less parallel to one another, but because of perspective the paths as seen from the Earth appear to converge on (or diverge from) a particular point in the sky known as the convergent point. This fact provides astronomers with a means of determining the distances of clusters, by finding three values for the individual stars in the moving cluster: their radial velocities towards or away from us (measured from the Doppler shift in their spectra), their proper motions and their angular distances from the convergent point. Once these values are known accurately for as many stars as possible, then the distance to the cluster can be calculated using simple geometry.

Obviously, the closer a star cluster is to us the easier it is to make the required measurements, as motion across the sky is easier to detect for nearby stars than for those at greater distances. The closest rich cluster is the Hyades in Taurus. A great deal of work has been devoted to determining its exact distance by the method described above. The result, 150 light years, has been confirmed by direct parallax measurements from the Hipparcos satellite.

Some moving clusters are much more spread out than the Hyades, for example the Ursa Major moving cluster. This cluster includes five of the stars in the Plough, or Dipper (β, γ, δ, ε and ζ Ursae Majoris) and also Sirius. It seems that the Sun is actually passing through the outskirts of this cluster.

TABLE 58. *The Messier objects.*

NUMBER		RA	DEC.				APPARENT	
			2000.0	CONSTELLATION	SIZE[a]	MAGNITUDE	DESCRIPTION	
M	NGC	h m	° ′		′			
1	1952	05 34.5	+22 01	Tau	6 × 4	c. 8.4	Supernova remnant	
2	7089	21 33.5	−00 49	Aqr	13	6.5	Globular cluster	
3	5272	13 42.2	+28 23	CVn	16	6.4	Globular cluster	
4	6121	16 23.6	−26 32	Sco	26	5.9	Globular cluster	
5	5904	15 18.6	+02 05	Ser	17	5.8	Globular cluster	
6	6405	17 40.1	−32 13	Sco	15	4.2	Open cluster	
7	6475	17 53.9	−34 49	Sco	80	3.3	Open cluster	
8	6523	18 03.8	−24 23	Sgr	90 × 40	c. 5.8	Diffuse nebula	
9	6333	17 19.2	−18 31	Oph	9	c. 7.9	Globular cluster	
10	6254	16 57.1	−04 06	Oph	15	6.6	Globular cluster	
11	6705	18 51.1	−06 16	Sct	14	5.8	Open cluster	
12	6218	16 47.2	−01 57	Oph	14	6.6	Globular cluster	
13	6205	16 41.7	+36 28	Her	17	5.9	Globular cluster	
14	6402	17 37.6	−03 15	Oph	12	7.6	Globular cluster	
15	7078	21 30.0	+12 10	Peg	12	6.4	Globular cluster	
16	6611	18 18.8	−13 47	Ser	7	6.0	Open cluster	
17	6618	18 20.8	−16 11	Sgr	46 × 37	7	Diffuse nebula	
18	6613	18 19.9	−17 08	Sgr	9	6.9	Open cluster	
19	6273	17 02.6	−26 16	Oph	14	7.2	Globular cluster	
20	6514	18 02.6	−23 02	Sgr	29 × 27	c. 8.5	Diffuse nebula	
21	6531	18 04.6	−22 30	Sgr	13	5.9	Open cluster	
22	6656	18 36.4	−23 54	Sgr	24	5.1	Globular cluster	
23	6494	17 56.8	−19 01	Sgr	27	5.5	Open cluster	
24	—	18 16.9	−18 29	Sgr	90	c. 4.5	See notes	
25	IC 4725	18 31.6	−19 15	Sgr	32	4.6	Open cluster	
26	6694	18 45.2	−09 24	Sct	15	8.0	Open cluster	
27	6853	19 59.6	+22 43	Vul	8 × 4	c. 8.1	Planetary nebula	
28	6626	18 24.5	−24 52	Sgr	11	c. 6.9	Globular cluster	
29	6913	20 23.9	+38 32	Cyg	7	6.6	Open cluster	
30	7099	21 40.4	−23 11	Cap	11	7.5	Globular cluster	
31	224	00 42.7	+41 16	And	178 × 63	3.4	Spiral galaxy	
32	221	00 42.7	+40 52	And	8 × 6	8.2	Elliptical galaxy	
33	598	01 33.9	+30 39	Tri	62 × 39	5.7	Spiral galaxy	
34	1039	02 42.0	+42 47	Per	35	5.2	Open cluster	
35	2168	06 08.9	+24 20	Gem	28	5.1	Open cluster	
36	1960	05 36.1	+34 08	Aur	12	6.0	Open cluster	
37	2099	05 52.4	+32 33	Aur	24	5.6	Open cluster	
38	1912	05 28.7	+35 50	Aur	21	6.4	Open cluster	
39	7092	21 32.2	+48 26	Cyg	32	4.6	Open cluster	
40	—	12 22.4	+58 05	UMa	—	8	See notes	
41	2287	06 47.0	−20 44	CMa	38	4.5	Open cluster	
42	1976	05 35.4	−05 27	Ori	66 × 60	4	Diffuse nebula	
43	1982	05 35.6	−05 16	Ori	20 × 15	9	Diffuse nebula	
44	2632	08 40.1	+19 59	Cnc	95	3.1	Open cluster	
45	—	03 47.0	+24 07	Tau	110	1.2	Open cluster	
46	2437	07 41.8	−14 49	Pup	27	6.1	Open cluster	
47	2422	07 36.6	−14 30	Pup	30	4.4	Open cluster	
48	2548	08 13.8	−05 48	Hya	54	5.8	Open cluster	
49	4472	12 29.8	+08 00	Vir	9 × 7	8.4	Elliptical galaxy	
50	2323	07 03.2	−08 20	Mon	16	5.9	Open cluster	
51	5194–5	13 29.9	+47 12	CVn	11 × 8	8.1	Spiral galaxy	
52	7654	23 24.2	+61 35	Cas	13	6.9	Open cluster	
53	5024	13 12.9	+18 10	Com	13	7.7	Globular cluster	
54	6715	18 55.1	−30 29	Sgr	9	7.7	Globular cluster	

[a] *The dimensions given are as seen on long-exposure photographs and, for galaxies in particular, are larger than the sizes that will be seen visually.*

M1 Crab Nebula	M27 Dumbbell Nebula
M8 Lagoon Nebula; contains a star cluster	M31 Andromeda Galaxy
M11 Wild Duck Cluster	M40 Faint double star Winnecke 4, mags. 9.0 and 9.6
M16 Surrounded by the Eagle Nebula	M42, M43 Orion Nebula
M17 Omega Nebula	M44 Praesepe, the Beehive Cluster
M20 Trifid Nebula	M45 The Pleiades; no NGC or IC number
M24 Star field in Sagittarius, containing the open cluster NGC 6603	M51 Whirlpool Galaxy

TABLE 58 *(continued). The Messier objects.*

NUMBER M	NGC	RA 2000.0 h m	DEC. ° ′	CONSTELLATION	SIZE[a] ′	APPARENT MAGNITUDE	DESCRIPTION
55	6809	19 40.0	−30 58	Sgr	19	7.0	Globular cluster
56	6779	19 16.6	+30 11	Lyr	7	8.2	Globular cluster
57	6720	18 53.6	+33 02	Lyr	1	c. 9.0	Planetary nebula
58	4579	12 37.7	+11 49	Vir	5 × 4	9.8	Spiral galaxy
59	4621	12 42.0	+11 39	Vir	5 × 3	9.8	Elliptical galaxy
60	4649	12 43.7	+11 33	Vir	7 × 6	8.8	Elliptical galaxy
61	4303	12 21.9	+04 28	Vir	6 × 5	9.7	Spiral galaxy
62	6266	17 01.2	−30 07	Oph	14	6.6	Globular cluster
63	5055	13 15.8	+42 02	CVn	12 × 8	8.6	Spiral galaxy
64	4826	12 56.7	+21 41	Com	9 × 5	8.5	Spiral galaxy
65	3623	11 18.9	+13 05	Leo	10 × 3	9.3	Spiral galaxy
66	3627	11 20.2	+12 59	Leo	9 × 4	9.0	Spiral galaxy
67	2682	08 50.4	+11 49	Cnc	30	6.9	Open cluster
68	4590	12 39.5	−26 45	Hya	12	8.2	Globular cluster
69	6637	18 31.4	−32 31	Sgr	7	7.7	Globular cluster
70	6681	18 43.2	−32 18	Sgr	8	8.1	Globular cluster
71	6838	19 53.8	+18 47	Sge	7	8.3	Globular cluster
72	6981	20 53.5	−12 32	Aqr	6	9.4	Globular cluster
73	6994	20 58.9	−12 38	Aqr	—	—	*See notes*
74	628	01 36.7	+15 47	Psc	10 × 9	9.2	Spiral galaxy
75	6864	20 06.1	−21 55	Sgr	6	8.6	Globular cluster
76	650–1	01 42.4	+51 34	Per	2 × 1	c. 11.5	Planetary nebula
77	1068	02 42.7	−00 01	Cet	7 × 6	8.8	Spiral galaxy
78	2068	05 46.7	+00 03	Ori	8 × 6	8	Diffuse nebula
79	1904	05 24.5	−24 33	Lep	9	8.0	Globular cluster
80	6093	16 17.0	−22 59	Sco	9	7.2	Globular cluster
81	3031	09 55.6	+69 04	UMa	26 × 14	6.8	Spiral galaxy
82	3034	09 55.8	+69 41	UMa	11 × 5	8.4	Irregular galaxy
83	5236	13 37.0	−29 52	Hya	11 × 10	c. 7.6	Spiral galaxy
84	4374	12 25.1	+12 53	Vir	5 × 4	9.3	Elliptical galaxy
85	4382	12 25.4	+18 11	Com	7 × 5	9.2	Elliptical galaxy
86	4406	12 26.2	+12 57	Vir	7 × 6	9.2	Elliptical galaxy
87	4486	12 30.8	+12 24	Vir	7	8.6	Elliptical galaxy
88	4501	12 32.0	+14 25	Com	7 × 4	9.5	Spiral galaxy
89	4552	12 35.7	+12 33	Vir	4	9.8	Elliptical galaxy
90	4569	12 36.8	+13 10	Vir	10 × 5	9.5	Spiral galaxy
91	4548	12 35.4	+14 30	Com	5 × 4	10.2	Spiral galaxy
92	6341	17 17.1	+43 08	Her	11	6.5	Globular cluster
93	2447	07 44.6	−23 52	Pup	22	c. 6.2	Open cluster
94	4736	12 50.9	+41 07	CVn	11 × 9	8.1	Spiral galaxy
95	3351	10 44.0	+11 42	Leo	7 × 5	9.7	Spiral galaxy
96	3368	10 46.8	+11 49	Leo	7 × 5	9.2	Spiral galaxy
97	3587	11 14.8	+55 01	UMa	3	c. 11.2	Planetary nebula
98	4192	12 13.8	+14 54	Com	10 × 3	10.1	Spiral galaxy
99	4254	12 18.8	+14 25	Com	5	9.8	Spiral galaxy
100	4321	12 22.9	+15 49	Com	7 × 6	9.4	Spiral galaxy
101	5457	14 03.2	+54 21	UMa	27 × 26	7.7	Spiral galaxy
102	—						*See notes*
103	581	01 33.2	+60 42	Cas	6	c. 7.4	Open cluster
104	4594	12 40.0	−11 37	Vir	9 × 4	8.3	Spiral galaxy
105	3379	10 47.8	+12 35	Leo	4 × 4	9.3	Elliptical galaxy
106	4258	12 19.0	+47 18	CVn	18 × 8	8.3	Spiral galaxy
107	6371	16 32.5	−13 03	Oph	10	8.1	Globular cluster
108	3556	11 11.5	+55 40	UMa	8 × 2	10.0	Spiral galaxy
109	3992	11 57.6	+53 23	UMa	8 × 5	9.8	Spiral galaxy
110	205	00 40.4	+41 41	And	17 × 10	8.0	Elliptical galaxy

[a] *The dimensions given are as seen on long-exposure photographs and, for galaxies in particular, are larger than the sizes that will be seen visually.*
Source: A. Hirshfeld and R. W. Sinnott (eds.), Sky Catalogue 2000.0, Vol. 2 (Sky Publishing Corp./Cambridge University Press, 1985).

M57 Ring Nebula		M97 Owl Nebula
M64 Black Eye Galaxy		M102 Duplicate of M101
M73 Small group of four stars		M104 Sombrero Galaxy

GLOBULAR CLUSTERS Unlike open clusters, which are found within the spiral arms of the Galaxy, globular clusters are situated mainly in the *galactic halo*, a spherical volume of space surrounding the Galaxy.

Globular clusters are huge, spherical concentrations of stars, tens to hundreds of light years in diameter. The stars they contain are generally much older than those in open clusters. In a globular cluster the density of stars is high, about one star per cubic light year, sufficient to resist disruption by the tidal forces exerted by the Galaxy. Globulars are among the oldest known objects in the Universe, with ages of 10 000 million years or more.

There are around 150 globular clusters known in our Galaxy. Some can be seen without optical aid, the best examples being the 4th-magnitude Omega (ω) Centauri and 47 Tucanae, both in the southern hemisphere. Omega Centauri, around 17 000 light years away, is the brightest globular in the sky, and is thought to contain more than a million stars. The brightest example in the northern sky is the 6th-magnitude M13 in Hercules. To the unaided eye, M13 appears as a fuzzy star-like object; a telescope is needed to resolve some of the million stars that comprise it.

NEBULAE

Large amounts of gas and dust are present in the Galaxy. This interstellar matter, most of which is hydrogen gas, is thought to account for around 10% of the Galaxy's total mass. Around 1% of the mass of interstellar matter is in the form of fine dust, which reveals itself through the effect it has on starlight. Light from distant stars which passes through the interstellar dust is dimmed, and also reddened because the extinction of the light is greater for shorter (bluer) wavelengths. Interstellar absorption is most conspicuous along the plane of the Milky Way, which is where most of the interstellar material is concentrated.

EMISSION NEBULAE Some interstellar matter is concentrated in clouds known as nebulae, of which there are several types. If a nebula is situated close to one or more hot, bright stars, the energy from the stars may cause it to emit light of its own. Ultraviolet radiation from the nearby stars ionizes the hydrogen in these so-called H II regions, which causes the nebula to glow. Many emission nebulae are known, the most famous being the Orion Nebula (M42), visible to the unaided eye as a misty patch of light just to the south of the Belt of Orion. The Orion Nebula lies at a distance of around 1600 light years and has a diameter of about 30 light years. Deep inside this huge gas cloud is the multiple star Theta (θ) Orionis, the four brightest components of which form a conspicuous group called the Trapezium on account of its shape. The energy from the hottest components of the Trapezium causes the gas within the Orion Nebula to shine.

REFLECTION NEBULAE contain dust that reflects light from nearby stars. They have a characteristic blue colour, in contrast to the predominantly red glow of emission nebulae. One of the best-known reflection nebulae is that surrounding the stars of the Pleiades. This nebula was once thought to be the remnants of the cloud from which the Pleiades were born, but it is now recognized to be a completely separate cloud into which the cluster has drifted by chance.

DARK NEBULAE appear as dark areas in the sky and contain no stars; in fact they blot out the light from stars behind them. They can take on a wide variety of shapes, ranging from the relatively uniform Coalsack in Crux to the long, winding Snake Nebula in Ophiuchus. Depending on their location, dark nebulae resemble either a blank region of sky or a conspicuous dark patch superimposed on a much brighter background. An example of the latter is the famous Horsehead Nebula, south of Zeta (ζ) Orionis, which is seen against the backdrop of the bright nebula IC 434. When first noticed on a photographic plate taken in 1889, the Horsehead was considered to be simply a gap in the bright nebula. The American astronomer E. E. Barnard recognized it for what it really was, and went on to compile a catalogue of dark nebulae; many of these objects are now classified by their Barnard numbers.

Bok globules are small, nearly spherical patches of dark nebulosity, thought to be stars in the very early stages of forming. They are named after Bart J. Bok, the Dutch astronomer who first drew attention to them. Bok globules can be seen against a number of nebulae, notably the Lagoon Nebula (M8) in Sagittarius and the Rosette Nebula (NGC 2237–2244) in Monoceros.

PLANETARY NEBULAE when seen through a telescope, sometimes resemble planetary disks and were so named by William Herschel in 1782. Planetary nebulae result from the ejection by old stars of their outer layers. This discarded material then takes the form of an expanding shell of gas surrounding the star. Because their hotter inner regions have been exposed, the central stars in planetary nebulae have very high surface temperatures, up to 100 000 K. Once the material has been ejected the central star begins the collapse into a white dwarf. Planetary nebulae are fairly short-lived, and within a few tens of thousands of years the expanding shell of gas dissipates into surrounding space. Not all planetary nebulae are symmetrical in shape, and there are many which have unusual appearances, including the Dumbbell Nebula (M27) in Vulpecula and the Little Dumbbell Nebula (M76), also known as the Butterfly, in Perseus. Planetary nebulae shine through the same process of ionization that takes place within emission nebulae. Over a thousand planetary nebulae are known, many within the reach of amateur telescopes.

SUPERNOVA REMNANTS Objects like the Crab Nebula (M1) in Taurus and the Veil Nebula (NGC 6960–6992) in Cygnus are the gaseous remnants of stars that have undergone supernova explosions. During these spectacular and destructive events, stars eject most of their material, creating expanding clouds of matter which eventually disperse into space. The two objects mentioned above differ considerably in appearance. The Crab Nebula is seen as a faint, diffuse patch of light and resulted from a supernova that was observed by Chinese astronomers in 1054. The Veil Nebula is part of a huge loop of material ejected around 5000 years ago. The Crab and the brightest parts of the Veil can be glimpsed in binoculars under dark-sky conditions, but larger instruments are needed to bring out the detailed structure.

OUR GALAXY

The Sun is a member of the Galaxy, a huge, spiral-shaped star system containing at least 100 000 million stars. The Galaxy has three main regions: the *central bulge*, *disk* and *halo*. The densest part of the bulge is known as the *nucleus*. The bulge itself contains old Population II stars together with small amounts of interstellar material, and has a diameter of 20 000 light years and a thickness of 10 000 light years.

In contrast to the central bulge, the disk contains much younger Population I stars, many of them in open clusters. The disk also contains much more interstellar material than the central bulge. The disk is about 100 000 light years in diameter, the Sun being roughly two-thirds of the way from the centre to the edge, and about 2000 light years thick, although this varies somewhat across its width. The stars in the disk travel around the galactic centre in orbits that are more or less circular. Stars and gas closer to the centre orbit more quickly than those farther out; the Sun, about 26 000 light years from the centre, takes around 220 million years to complete one orbit.

The halo takes the form of a spherical volume of stars surrounding the central bulge. Most of the stars of the halo are collected into globular clusters, which travel around the galactic centre in elliptical paths. The stars within the halo are very old; like the stars in the central bulge they are Population II objects, poor in heavy elements.

Interstellar extinction restricts the range at which stars can be observed along the plane of the Galaxy. The centre of the Galaxy, located in the constellation Sagittarius, is not visible optically, but observation at radio and infrared wavelengths has enabled astronomers to probe the galactic centre and to map the overall structure of the Galaxy. The galactic centre has been identified with the strong radio source Sagittarius A. The distribution of hydrogen clouds reveals that the galactic disk has a spiral structure, and that most of the Population I stars and the gas in the galactic disk are grouped into spiral arms radiating from the nucleus. In the arms, new stars are still being born from interstellar clouds.

The *Milky Way* is composed of thousands of millions of stars in the galactic disk, and may be seen as a great ring of faint light, extending right around the celestial sphere and inclined at about 63° to the plane of the ecliptic. It is brightest in Cygnus and Aquila (northern hemisphere) and in Scorpius and Sagittarius (southern hemisphere), and faintest in Monoceros. The Coalsack (see Chart 16) is the most prominent of the many 'gaps' in the Milky Way, which are dark nebulae close to the galactic plane lying between us and the star clouds beyond. One enormous dark cloud, called the Cygnus Rift or Great Rift, splits the Milky Way into two, starting in Cygnus and running southwards through Aquila into Ophiuchus (see Charts 13 and 14). It is caused by a dark nebula along the local spiral arm of the Galaxy.

Lying close to the Galaxy are two dwarf irregular galaxies – the *Large* and *Small Magellanic Clouds*. They are visible in the southern hemisphere as extensive, nebulous naked-eye objects (see Chart 15) and are, in fact, satellites of the Galaxy itself. The Large Magellanic Cloud is the closest external galaxy, lying at a distance of around 170 000 light years, on the Dorado/Mensa border. The Small Magellanic Cloud lies at a distance of about 200 000 light years in Tucana.

GALAXIES

The Galaxy is by no means unique in the Universe. Within the range of present-day telescopes lie thousands of millions of galaxies beyond our own. The best-known example of an external galaxy is the galaxy in Andromeda (M31), which is visible to the naked eye under good conditions as a faint misty patch of light. The Andromeda Galaxy is larger than our own and also has a spiral structure; it lies at a distance of around 2.5 million light years.

Although both the Milky Way Galaxy and M31 are spirals, not all galaxies have spiral structure. There are various types of galaxy, broadly classified as shown in Figure 34.

ELLIPTICAL GALAXIES are highly symmetrical systems which possess no spiral or other structure. They are denoted by E and a number from 0 to 7 to indicate the shape, which ranges from spherical (E0) to highly flattened (E7). The E0 type galaxies resemble huge globular clusters. Elliptical galaxies are deficient in interstellar matter, and the most massive ellipticals are considerably more massive than our own Galaxy. Dwarf ellipticals are denoted by the prefix d, and supergiant ellipticals by the prefix c.

NORMAL SPIRAL GALAXIES These galaxies have spiral arms containing stars, gas and dust which extend from a central nucleus. They are classified as Sa, Sb or Sc according to the relationship between the nuclear bulge and the spiral arms. An Sa system

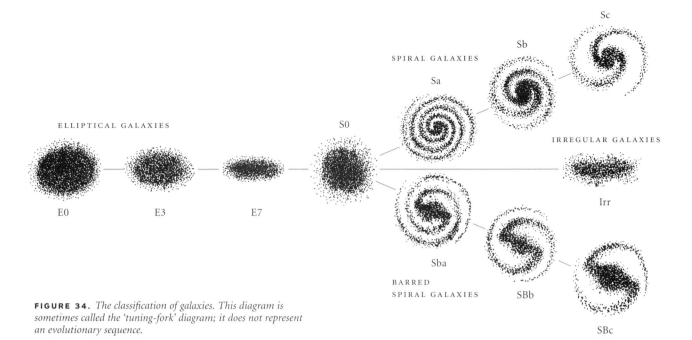

ELLIPTICAL GALAXIES

E0   E3   E7

S0

SPIRAL GALAXIES

Sa   Sb   Sc

IRREGULAR GALAXIES

Irr

BARRED
SPIRAL GALAXIES

Sba   SBb   SBc

FIGURE 34. *The classification of galaxies. This diagram is sometimes called the 'tuning-fork' diagram; it does not represent an evolutionary sequence.*

has a relatively large central bulge and tightly wound spiral arms; Sb systems have more or less equally prominent spiral arms and nuclear regions; Sc galaxies possess small nuclear masses and loose, open spiral arms. S0 galaxies, known as lenticular galaxies, are an intermediate type between elliptical and spiral galaxies.

BARRED SPIRAL GALAXIES are similar to ordinary spirals, except that the spiral arms emanate from each end of a luminous bar of material which straddles the nucleus. They are classified similarly to the spirals as SBa, SBb or SBc. Our own Galaxy shows signs of having a modest central bar.

IRREGULAR GALAXIES have no ordered structure. Small irregulars, lower in mass than our own Galaxy, in many ways resemble large star clouds. Larger irregulars probably arise from interactions and mergers between normal galaxies. They are denoted by I or Irr.

ACTIVE GALAXIES AND QUASARS A small percentage of galaxies have unusually bright centres. Apparently, these arise when infalling gas forms an intensely hot and luminous *accretion disk* around a central, massive black hole. Examples are Seyfert galaxies and BL Lac objects.

Seyfert galaxies, named after the American astronomer Carl Seyfert who discovered them in 1943, are nearly all spirals. They are thought to be active galaxies which are oriented so that we have an unobstructed view of their central luminous accretion disk. The brightest of them, at ninth magnitude, is M77 in Cetus. *BL Lac objects* are named after their prototype, BL Lacertae, which was originally classed as an unusual 14th-magnitude variable star; they are thought to be galaxies in which a jet of gas ejected along the rotation axis of the accretion disk is seen end-on.

Quasars, also known as quasi-stellar objects (QSOs), are objects of star-like appearance, each giving out as much energy as hundreds of entire galaxies from a volume of space no bigger than our Solar System. Their redshifts show that they lie far off in the Universe. There is now considerable evidence that quasars occur in galaxies which are undergoing, or have recently undergone, mergers with other galaxies; the merger feeds the central black hole with stars and gas, leading to an outburst of activity. They may be thought of as extreme forms of Seyfert galaxies. The brightest quasar is the 13th-magnitude 3C 273 in Virgo.

CLUSTERS OF GALAXIES Most galaxies belong to clusters, groups of galaxies containing anything from a few to several thousand members. Our own Galaxy is a member of the *Local Group*, a collection of around 36 galaxies including two other spirals (M31 and M33), the Magellanic Clouds, and numerous dwarf elliptical and irregular systems (Table 59).

There are many other clusters of galaxies scattered throughout space, one of the nearest being the Virgo Cluster, about 55 million light years away; the brightest members of the Virgo Cluster are visible in amateur telescopes. Both the Local Group and the Virgo Cluster are members of a much larger grouping known as the Local Supercluster, centred on the Virgo Cluster, and containing many of the other nearby galaxy clusters as well as a number of individual galaxies. The 20 brightest galaxies outside the Local Group are listed in Table 60.

THE RECESSION OF THE GALAXIES All the galaxies outside the Local Group show redshifts in their spectra, indicating that they are receding from us. Furthermore, the velocity of recession of the galaxies is proportional to their distance, a relationship established in the late 1920s by Edwin Hubble and known as *Hubble's law*. The constant relating the velocity of recession to the

TABLE 59. *The Local Group of galaxies.*

GALAXY	RA h m	DEC. 2000.0 ° ′	TYPE[a]	ABSOLUTE MAGNITUDE	APPARENT MAGNITUDE	DISTANCE FROM SUN (kpc)
Andromeda Galaxy (M31)	00 42.7	+41 16	Sb I–II	−21.2	3.4	770
Milky Way	—	—	S(B)bc I–II	−20.9	—	—
Triangulum Galaxy (M33)	01 33.9	+30 39	Sc II–III	−18.9	5.9	850
LMC[M]	05 24	−69 45	Ir III–IV	−18.5	0.2	50
SMC[M]	00 53	−72 50	Ir IV/Ir IV–V	−17.1	2.0	63
M32 (NGC 221)[A]	00 42.7	+40 52	dE2	−16.5	8.1	770
M110 (NGC 205)[A]	00 40.4	+41 41	dE5p	−16.4	8.4	830
IC 10[A]	00 20.3	+59 18	dIr IV	−16.0	10.3	660
NGC 6822 (Barnard's Galaxy)[M]	19 44.9	−14 52	dIr IV–V	−16.0	8.3	500
NGC 185[A]	00 39.0	+48 20	dE3p	−15.6	9.0	620
IC 1613[A]	01 04.8	+02 07	dIr V	−15.3	9.1	715
NGC 147[A]	00 33.2	+48 30	dE5	−15.1	9.9	755
Sagittarius I Dwarf[M]	18 55.0	−30 29	dSph	−15.0	7.7	28
Wolf–Lundmark–Melotte (DDO 221)	00 02.0	+15 28	dIr IV–V	−14.4	10.6	945
Fornax Dwarf[M]	02 40.0	−34 27	dSph	−13.1	7.7	138
Pegasus Dwarf (DDO 216, Peg dIG)[A]	23 28.6	+14 45	dIr/dSph	−12.9	12.3	760
Cassiopeia Dwarf (Andromeda VII)[A]	23 26.5	+50 42	dSph	−12.0	13.0	760
Sagittarius (Sag dIG)	19 30.0	−17 41	dIr V	−12.0	13.8	1060
Leo I (DDO 74)[M]	10 08.4	+12 18	dSph	−11.9	10.4	270
Andromeda I[A]	00 45.7	+38 00	dSph	−11.8	12.9	790
Andromeda II[A]	01 16.5	+33 25	dSph	−11.8	12.6	680
Leo A (Leo III, DDO 69)	09 59.4	+30 45	dIr V	−11.7	12.8	800
Andromeda VI (Pegasus II)[A]	23 51.8	+24 35	dSph	−11.3	13.4	775
Aquarius Dwarf (DDO 210)	20 46.9	−12 51	dIr/dSph	−10.9	14.2	950
Andromeda III[A]	00 35.6	+36 30	dSph	−10.2	14.4	760
Cetus Dwarf	00 26.2	−11 03	dSph	−10.1	14.4	775
Leo B (Leo II, DDO 93)[M]	11 13.5	+22 09	dSph	−10.1	11.5	205
Phoenix Dwarf[M]	01 51.1	−44 27	dIr/dSph	−9.8	13.3	405
Pisces Dwarf (LGS 3)[A]	01 03.9	+21 53	dIr/dSph	−9.8	14.3	620
Sculptor Dwarf[M]	01 00.1	−33 43	dSph	−9.8	10.0	88
Tucana Dwarf	22 41.8	−64 25	dSph	−9.6	15.2	870
Sextans Dwarf[M]	10 13.0	−01 37	dSph	−9.5	10.3	86
Carina Dwarf[M]	06 41.6	−50 58	dSph	−9.4	10.7	94
Draco Dwarf (DDO 208)[M]	17 20.2	+57 55	dSph	−9.4	11.0	79
Andromeda V[A]	01 10.3	+47 38	dSph	−9.1	15.9	810
Ursa Minor Dwarf (DDO 199)[M]	15 09.2	+67 13	dSph	−8.9	10.4	69

[a] dIr = dwarf irregular, dSph = dwarf spheroidal.
[M] Companion of the Milky Way; [A] Companion of the Andromeda Galaxy.
Source: Adapted from a table by Eva K. Grebel, John S. Gallagher III and Daniel Harbeck, Astronomical Journal, Vol. 125, p. 1928 (April 2003).

distance is the *Hubble constant*, H_0. The currently accepted value of H_0 is around 70 km s^{-1} per megaparsec. Hubble's law means that a galaxy's distance from us can be determined from its redshift once H_0 is accurately known.

The recession of the galaxies is usually interpreted as a general expansion of the Universe. According to the *Big Bang theory*, the entire Universe – not just matter but all of space, time and radiation – was once concentrated into an intensely hot, superdense state which was flung violently outwards by the Big Bang explosion, initiating the present expansion of the Universe. The Big Bang is currently estimated to have occurred some 14 000 million years ago.

OBSERVING DEEP-SKY OBJECTS

Star clusters, nebulae and galaxies present some of the most interesting and challenging targets for amateur astronomers. Many thousands of these objects are within the grasp of small to medium telescopes, although locating them can be a problem because they are faint and diffuse. The observer's eyes must be fully dark-adapted before serious observation can begin. A clear, moonless night is ideal.

FINDING DEEP-SKY OBJECTS One way of locating elusive deep-sky objects is to use what is sometimes called the 'drift method', whereby a star with virtually the same declination as the object being sought, but a little west of it, is brought into the field of view. Note how many minutes of arc the object is to the east of the star, then leave the telescope stationary so that the sky drifts across the field of

TABLE 60. *The 20 brightest galaxies outside the Local Group.*

GALAXY	RA 2000.0 h m	DEC. ° '	TYPE	DIMENSIONS[a] '	APPARENT MAGNITUDE
NGC 5128 = Cen A	13 25	−43 01	S0(pec)	25.7 × 20.0	6.8
NGC 3031 = M81	09 56	+69 04	Sb	26.9 × 14.1	6.9
NGC 253	00 48	−25 17	Sc	27.5 × 6.8	7.2
NGC 5236 = M83	13 37	−29 52	SBc	12.9 × 11.5	7.5
NGC 55	00 15	−39 13	Sc	32.4 × 5.6	7.9
NGC 5457 = M101	14 03	+54 21	Sc	28.8 × 26.9	7.9
NGC 4594 = M104	12 40	−11 37	Sa/b	8.7 × 3.5	8.0
NGC 300	00 54	−37 41	Sc	21.9 × 15.5	8.1
NGC 4736 = M94	12 51	+41 07	Sab	11.2 × 9.1	8.2
NGC 3034 = M82	09 56	+69 41	Amorphous	11.2 × 4.3	8.4
NGC 4258 = M106	12 19	+47 18	Sb	18.6 × 7.2	8.4
NGC 4472 = M49[V]	12 30	+08 00	E1/S0	10.2 × 8.3	8.4
NGC 5194 = M51	13 30	+47 12	Sbc	11.2 × 6.9	8.4
NGC 1291	03 17	−41 06	SBa	9.8 × 8.1	8.5
NGC 1316 = Fornax A	03 23	−37 12	Sa(pec)	12.0 × 8.5	8.5
NGC 2403	07 37	+65 36	Sc	21.9 × 12.3	8.5
NGC 4826 = M64	12 57	+21 41	Sab	10.0 × 5.4	8.5
NGC 4486 = M87[V]	12 31	+12 23	E0	8.3 × 6.6	8.6
NGC 5055 = M63	13 16	+42 02	Sbc	12.6 × 7.2	8.6
NGC 1313	03 18	−66 30	SBc	9.1 × 6.9	8.7

[a] *The dimensions are maximum sizes as measured on photographs; the sizes as seen visually will be smaller.*
[V] *Member of the Virgo Cluster.*
Sources: Magnitudes and dimensions from The Astronomical Almanac; *types from A. Sandage & G. Tammann,* Revised Shapley–Ames Catalog of Bright Galaxies *(1987).*

view for the required time until the object appears. This method, as well as calling for some patience, requires a convenient star.

A more advanced method is to use setting circles. All equatorially mounted telescopes have such circles, one on the right ascension axis and one on the declination axis. When the right ascension and declination of an object are both known, from a catalogue or star atlas, the setting circles can be used to 'dial up' the coordinates. However, for simple portable telescopes this is unsatisfactory: too much time may be spent setting up the telescope mounting and, unless the alignment is more or less precise, the object may still not appear in the field of view.

Modern computer-controlled telescopes simplify the process considerably. These instruments are fitted with sophisticated electronics which allow even portable equipment to be set up rapidly and accurately. Objects may be located using a built-in database containing thousands of positions, and tracking may be fully automatic, compensating for any drive errors.

By far the most satisfactory method, though, is to 'star-hop' to the object being sought. After locating a fairly bright nearby star, the observer literally 'hops' from one star to another with the aid of the finderscope, switching to the main telescope once the target region is reached. When an object has been observed several times, it becomes much easier to locate again.

A good finderscope is essential for locating deep-sky objects, particularly when star-hopping; at least 7 × 50 is recommended.

EQUIPMENT FOR DEEP-SKY OBSERVING Deep-sky observing requires a range of eyepieces with different powers and fields of view. Low-power eyepieces with wide fields are essential for large open clusters or large areas of nebulosity. Certain deep-sky objects can appear very small or even stellar, so for these a high power is necessary. A good selection for deep-sky observing is a 32-mm wide-field eyepiece, a 10-mm and a 6-mm plus a ×2 Barlow, giving six different magnifications in all.

One of the biggest problems facing the deep-sky observer is light-pollution, which swamps fainter objects. To combat this, special filters are available, often referred to as *nebula filters.* As their name implies, these filters are primarily intended for observing nebulae; apart from the broadband variety, they have little if any effect on galaxies and star clusters.

A nebula filter darkens the background field of view, thereby enhancing the contrast between object and sky, but it also cuts out light from stars, making it difficult on occasion to identify the area of interest. Most nebula filters work best with low-power, wide-field eyepieces as these give brighter, higher-contrast images of deep-sky objects. Nebula filters are expensive, but using one is like doubling the size of your telescope, or being transported to a dark-sky site. Such filters can also help with imaging of deep-sky objects on CCD and film (see p. 46).

There are four kinds of nebula filter on the market. The most general is the *broadband* or *light-pollution-reduction* (LPR) filter,

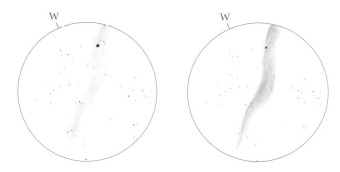

FIGURE 35. *Nebula filters in action. The Veil Nebula in Cygnus, seen at left as it appears through a 220-mm reflector without a filter and at right with an O III filter. Through the filter the nebula appears brighter, with wispy structure and well-defined edges, but surrounding stars are dimmed.*

which boosts the contrast of all objects. It has a relatively wide bandpass of around 90 nm, and consequently still lets some light-pollution through.

A second type, the *narrow-bandpass* filter, has on average about a 25-nm bandpass centred on the two O III lines emitted by oxygen. This type of filter significantly improves the appearance of emission and planetary nebulae, allowing intricate detail to be seen through even a modest-sized telescope (see Figure 35). It is a good choice of filter to start with.

The two other types of filters, called *line filters*, are centred on specific emission lines. The O III filter has an 11-nm bandpass that allows through only the O III emission lines, giving detailed views of nebulosity. The Hβ filter has a 9-nm bandpass that transmits only the single hydrogen-beta emission line; it is used for the faintest nebulae.

METHODS OF OBSERVING DEEP-SKY OBJECTS Open clusters make good targets for small telescopes, and binoculars can reveal many objects of this type. Because the stars in open clusters are spread out, a wide field of view is essential. This is even more important when it comes to trying to pick out an open cluster against a bright starry background. Try counting the number of stars in the cluster and note any doubles or coloured stars. When sketching them, position the brightest stars accurately before adding in the fainter ones.

Globular clusters, although containing more stars, are much more compact and generally appear as faint, diffuse objects. A higher magnification can help to bring out the brightest individual members; experiment with different magnifications and fields of view. When sketching a globular, start with a circular glow to represent the halo, then add individual stars, noting any features such as chains, clumps or dark lanes. In well-resolved clusters it will be necessary to give a general impression of the distribution of stars near the centre rather than trying to plot every one.

Some diffuse nebulae, such as the Orion Nebula (M42), can be observed with either high or low magnification, although many are large and faint, contrasting little with the surrounding sky. Averted vision may be useful here. The nebula may reveal itself more readily if the telescope is swept back and forth across the area of sky. Both these techniques can help to enhance the contrast between the nebula and the sky. Higher magnifications may be needed to exclude the glare from nearby bright stars when seeking out certain nebulae. The same general rules apply to galaxies.

Low to medium magnifications can be used with planetary nebulae, which are generally small and relatively distinct. With very low powers, certain planetary nebulae appear star-like and can be difficult to identify straight away. Emission nebulae and planetary nebulae display such complex detail, especially when seen through a large aperture or with a nebula filter, that they can be very difficult to draw. The best method is the same as that for galaxies.

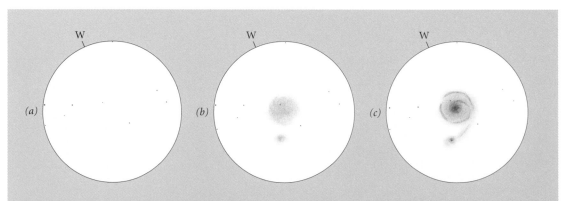

Sketching Deep-sky Objects

When sketching a deep-sky object (the galaxy M51 in this case), begin by drawing a circle about 100 mm across to represent the telescope's field of view. At the eyepiece, mark in the brighter stars followed by the fainter ones (*a*). Next, draw in the shape of the object with the side of a soft pencil and spread the graphite with a finger to give a nebulous effect (*b*). Continue laying graphite and spreading it to build up a negative image of the object, i.e. the brightest areas will be darkest. Add any interesting details such as a starlike nucleus or bright gas clouds (H II regions) in the spiral arms with the pencil, and use an eraser to mark any dark dust lanes (*c*). To orient the sketch, let the object drift through the field of view; the side on which it exits is west. Add the time of observation, size of instrument, magnification and the observing conditions.

CHAPTER FIVE

Star Charts

Index to the Constellations

CONSTELLATION	CHART	VISIBILITY RANGE[a] FULL	VISIBILITY RANGE[a] PARTIAL	ON THE MERIDIAN AT 10 P.M.[b]
Andromeda	3, 5	90°N–37°S	37°S–68°S	Oct., Nov.
Antlia	8, 10	49°N–90°S	65°N–49°N	Mar., Apr.
Apus	16	7°N–90°S	22°N–7°N	May, June, July
Aquarius	4, 14	65°N–86°S	90°N–65°N	Sept., Oct.
Aquila	13, 14	78°N–71°S	90°N–78°N, 71°S–90°S	Aug.
Aries	5	90°N–58°S	58°S–79°S	Nov., Dec.
Auriga	5, 7	90°N–34°S	34°S–62°S	Dec., Jan.
Boötes	11	90°N–35°S	35°S–82°S	May, June
Caelum	6	41°N–90°S	62°N–41°N	Dec., Jan.
Camelopardalis	1, 2	90°N–3°S	3°S–37°S	Dec., Jan., Feb.
Cancer	7	90°N–57°S	57°S–83°S	Feb., Mar.
Canes Venatici	9	90°N–37°S	37°S–62°S	Apr., May
Canis Major	8	56°N–90°S	78°N–56°N	Jan., Feb.
Canis Minor	7	89°N–77°S	77°S–90°S	Feb.
Capricornus	14	62°N–90°S	78°N–62°N	Aug., Sept.
Carina	8, 16	14°N–90°S	39°N–14°N	Jan., Feb., Mar., Apr.
Cassiopeia	2, 3	90°N–12°S	12°S–43°S	Oct., Nov., Dec.
Centaurus	10, 12, 16	35°N–90°S	59°N–35°N	Apr., May, June
Cepheus	2	90°N–1°S	1°S–36°S	Sept., Oct.
Cetus	4, 6	65°N–79°S	90°N–65°N, 79°S–90°S	Oct., Nov., Dec.
Chamaeleon	16	7°N–90°S	14°N–7°N	Feb., Mar., Apr., May
Circinus	16	19°N–90°S	34°N–19°N	May, June
Columba	6, 8	46°N–90°S	62°N–46°N	Jan.
Coma Berenices	9	90°N–57°S	57°S–77°S	Apr., May
Corona Australis	14	44°N–90°S	53°N–44°N	July, Aug.
Corona Borealis	11	90°N–50°S	50°S–64°S	June
Corvus	10	65°N–90°S	78°N–65°N	Apr., May
Crater	10	65°N–90°S	83°N–65°N	Apr.
Crux	16	25°N–90°S	34°N–25°N	Apr., May
Cygnus	13	90°N–28°S	28°S–62°S	Aug., Sept.
Delphinus	13	90°N–69°S	69°S–87°S	Aug., Sept.
Dorado	15	20°N–90°S	41°N–20°N	Dec., Jan.
Draco	1, 2, 13	90°N–4°S	4°S–42°S	Apr., May, June, July, Aug.
Equuleus	13	90°N–77°S	77°S–87°S	Sept.
Eridanus	6	32°N–89°S	90°N–32°N	Nov., Dec., Jan.
Fornax	6	50°N–90°S	66°N–50°N	Nov., Dec.
Gemini	7	90°N–55°S	55°S–80°S	Jan., Feb.
Grus	4	33°N–90°S	53°N–33°N	Sept., Oct.
Hercules	11, 13	90°N–38°S	38°S–86°S	June, July
Horologium	6, 15	23°N–90°S	50°N–23°N	Nov., Dec.
Hydra	8, 10, 12	54°N–83°S	90°N–54°N	Mar., Apr., May
Hydrus	15	8°N–90°S	32°N–8°N	Oct., Nov., Dec.
Indus	14, 15	15°N–90°S	43°N–15°N	Sept., Oct.
Lacerta	3	90°N–33°S	33°S–54°S	Sept., Oct.
Leo	7, 9	82°N–57°S	57°S–90°S	Mar., Apr.
Leo Minor	9	90°N–48°S	48°S–67°S	Mar., Apr.
Lepus	6	62°N–90°S	79°N–62°N	Jan.
Libra	12	60°N–90°S	89°N–60°N	May, June
Lupus	12	34°N–90°S	60°N–34°N	May, June
Lynx	1, 7	90°N–28°S	28°S–57°S	Feb., Mar.
Lyra	13	90°N–42°S	42°S–64°S	July, Aug.
Mensa	15, 16	5°N–90°S	20°N–5°N	Dec., Jan., Feb.
Microscopium	14	45°N–90°S	62°N–45°N	Aug., Sept.
Monoceros	7, 8	78°N–78°S	90°N–78°N, 78°S–90°S	Jan., Feb.
Musca	16	14°N–90°S	25°N–14°N	Apr., May
Norma	12	29°N–90°S	48°N–29°N	June
Octans	15, 16	0°–90°S	25°N–0°	Aug., Sept., Oct.
Ophiuchus	11, 12	59°N–75°S	90°N–59°N, 75°S–90°S	June, July
Orion	5, 6	79°N–67°S	90°N–79°N, 67°S–90°S	Jan.
Pavo	15	15°N–90°S	33°N–15°N	July, Aug., Sept.
Pegasus	3, 13	90°N–53°S	53°S–87°S	Sept., Oct.
Perseus	5	90°N–31°S	31°S–59°S	Nov., Dec.
Phoenix	4	32°N–90°S	50°N–32°N	Oct., Nov.
Pictor	6, 15, 16	26°N–90°S	47°N–26°N	Jan.
Pisces	3	83°N–56°S	56°S–90°S	Oct., Nov.
Piscis Austrinus	4, 14	53°N–90°S	65°N–53°N	Sept., Oct.
Puppis	8	39°N–90°S	78°N–39°N	Jan., Feb.
Pyxis	8	52°N–90°S	72°N–52°N	Feb., Mar.
Reticulum	15	23°N–90°S	37°N–23°N	Dec.
Sagitta	13	90°N–69°S	69°S–73°S	Aug.
Sagittarius	14	46°N–90°S	78°N–46°N	July, Aug.
Scorpius	12	44°N–90°S	81°N–44°N	June, July
Sculptor	4	50°N–90°S	65°N–50°N	Oct., Nov.
Scutum	14	74°N–90°S	86°N–74°N	July, Aug.
Serpens (Caput)	11, 12	86°N–64°S	90°N–86°N, 64°S–90°S	June
Serpens (Cauda)	13, 14	73°N–83°S	90°N–73°N, 83°S–90°S	July, Aug.
Sextans	9, 10	78°N–83°S	90°N–78°N, 83°S–90°S	Mar., Apr.
Taurus	5	88°N–58°S	58°S–90°S	Dec., Jan.
Telescopium	14	33°N–90°S	44°N–33°N	July, Aug.
Triangulum	5	90°N–52°S	52°S–64°S	Nov., Dec.
Triangulum Australe	16	19°N–90°S	29°N–19°N	June, July
Tucana	15	14°N–90°S	33°N–14°N	Sept., Oct., Nov.
Ursa Major	1, 7, 9	90°N–16°S	16°S–62°S	Feb., Mar., Apr., May
Ursa Minor	1	90°N–0°	0°–24°S	May, June, July
Vela	8, 10	32°N–90°S	52°N–32°N	Feb., Mar., Apr.
Virgo	9, 10, 11, 12	67°N–75°S	90°N–67°N, 75°S–90°S	Apr., May, June
Volans	16	14°N–90°S	25°N–14°N	Feb., Mar.
Vulpecula	13	90°N–61°S	61°S–71°S	Aug., Sept.

[a] This column gives the range of latitudes from which each constellation is fully or partially visible. Objects will be considerably dimmed by atmospheric extinction when the constellation is close to the horizon.

[b] This column gives the month(s) in which each constellation is on or near the meridian at around 10 p.m., and hence is best placed for observation from the latitudes at which it is visible.

Northern Index Chart

EPOCH 2000.0

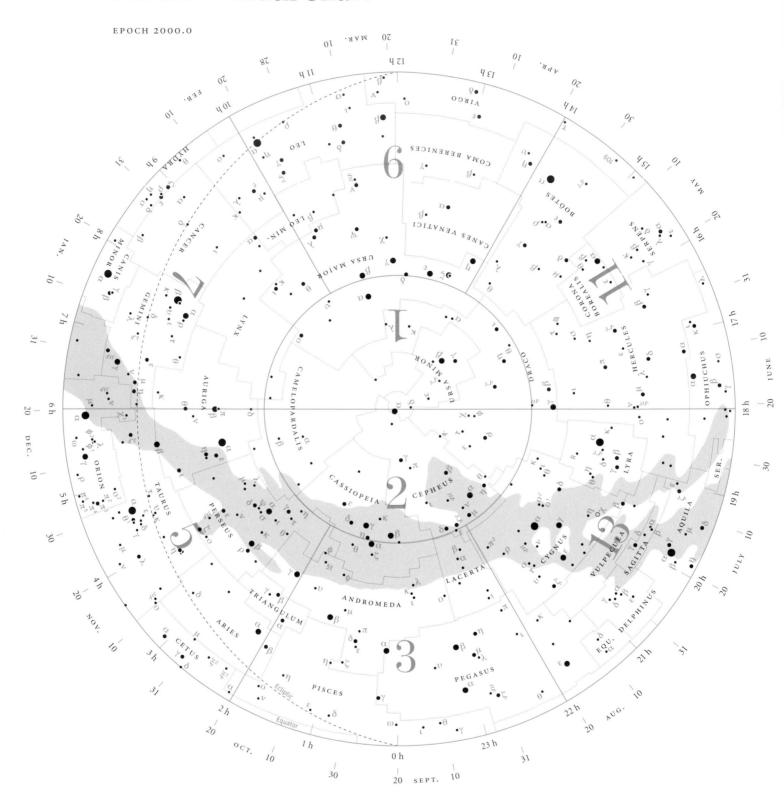

*The dates around the margin of the chart indicate
when that particular meridian lies due south at midnight,
as seen from the northern hemisphere*

Southern Index Chart

EPOCH 2000.0

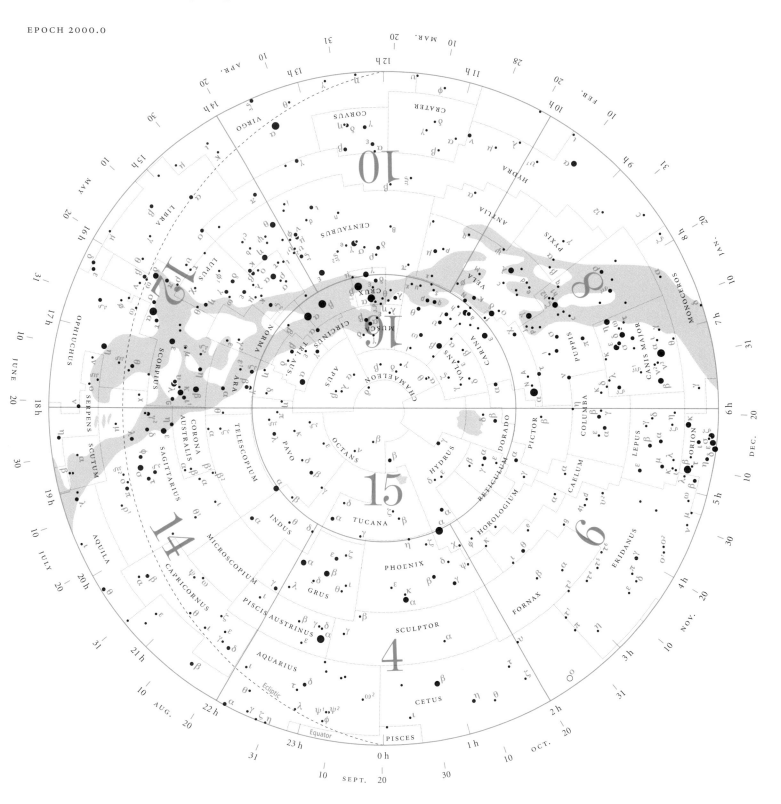

The dates around the margin of the chart indicate
when that particular meridian lies due north at midnight,
as seen from the southern hemisphere

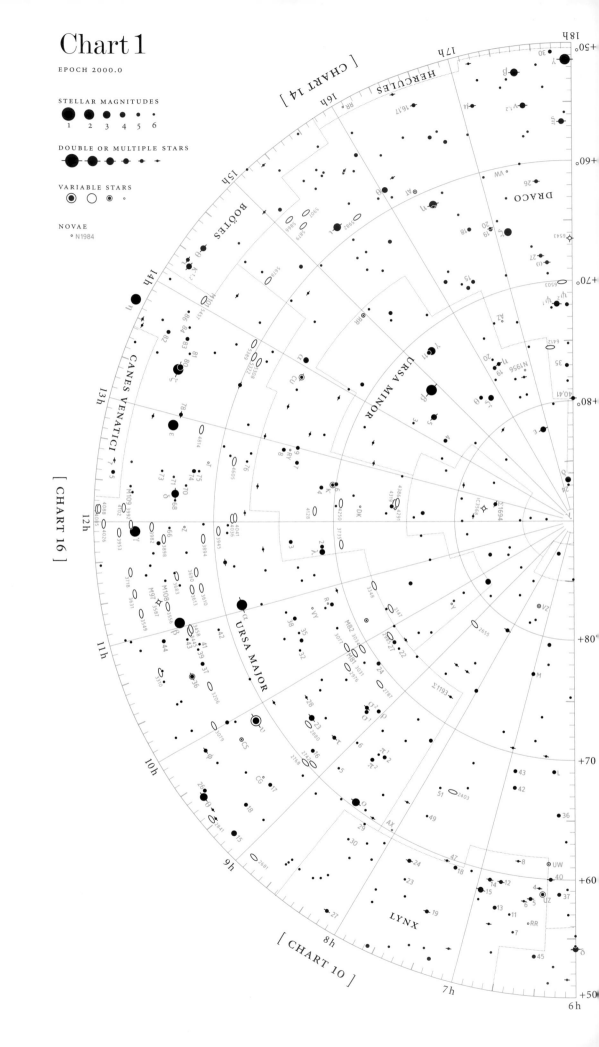

Chart 1

EPOCH 2000.0

STELLAR MAGNITUDES

1 2 3 4 5 6

DOUBLE OR MULTIPLE STARS

VARIABLE STARS

NOVAE

° N1984

[CHART 14]

[CHART 16]

[CHART 10]

HERCULES

DRACO

URSA MINOR

BOÖTES

CANES VENATICI

URSA MAJOR

LYNX

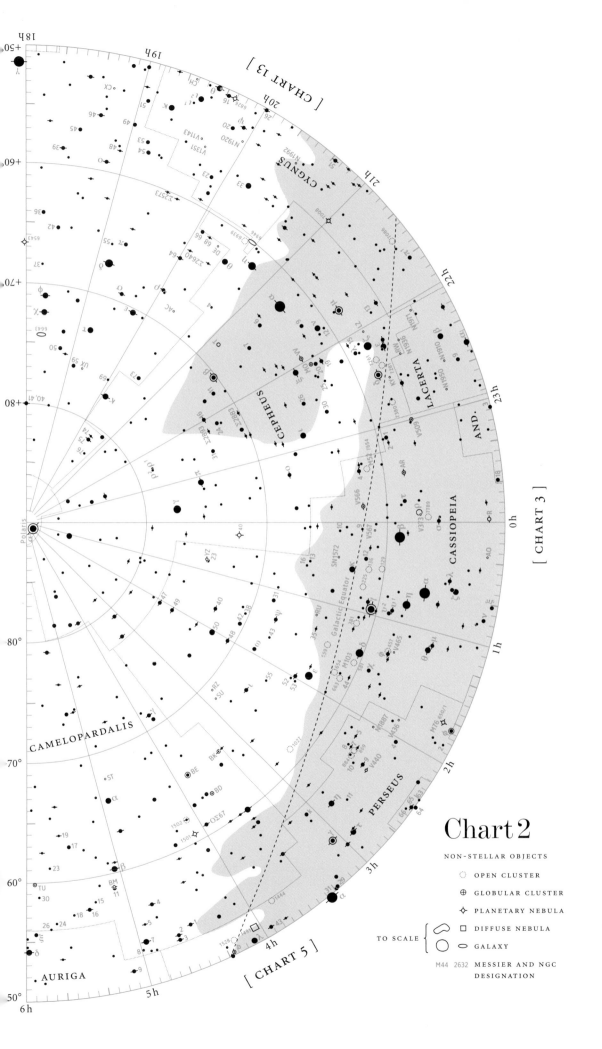

[CHART 13]

18ʰ

19ʰ

20ʰ

21ʰ

22ʰ

23ʰ

0ʰ

1ʰ

2ʰ

3ʰ

4ʰ

5ʰ

6ʰ

+50°

+60°

+70°

+80°

80°

70°

60°

50°

CYGNUS

CEPHEUS

LACERTA

AND.

CASSIOPEIA

PERSEUS

CAMELOPARDALIS

AURIGA

Polaris

Galactic Equator

SN1572

SN1604

M103

M52

M76

[CHART 3]

[CHART 5]

Chart 2

NON-STELLAR OBJECTS

⊙ OPEN CLUSTER

⊕ GLOBULAR CLUSTER

✧ PLANETARY NEBULA

TO SCALE { ▢ DIFFUSE NEBULA

◯ GALAXY

M44 2632 MESSIER AND NGC
DESIGNATION

Interesting Objects, Charts 1 and 2

Dec. +60° to +90°

DOUBLE STARS

ADS	STAR	RA 2000.0 h m	DEC. ° '	MAGNITUDES		PA °	DIST. "	NOTES
624	HN 122	00 45.7	+74 59	var.	10.4	161	35.8	Optical; fixed. A is YZ Cas
782	γ Cas	00 56.7	+60 43	var.	10.9	252	2.3	Little change
1129	ψ Cas	01 25.9	+68 08	4.7	9.2	123	22.1	Closing, PA increasing. B is double: 9.4, 10.0; 252°, 2".6; closing slowly
1598	48 Cas	02 02.0	+70 54	4.7	6.7	267	0.8	Binary, 60 years[a]
1860	ι Cas	02 29.1	+67 24	4.6	6.9	231	2.8	AB binary, 600 years? C slowly closing, PA increasing. Fine object in 100 mm
					9.1	116	7.3	
1477	α UMi	02 31.8	+89 16	2.0	9.0	216	17.8	Polaris. Physical pair.
2867	OΣ67 Cam	03 57.1	+61 07	5.3	8.1	49	1.7	Fixed; yellowish, greenish (by contrast)
3615	β Cam	05 03.4	+60 27	4.1	7.4	209	82.4	Fixed
4177	19 Cam	05 37.3	+64 09	6.2	9.8	56	1.5	Slowly widening
6724	Σ1193 UMa	08 20.7	+72 24	6.2	9.7	90	42.7	Little change
7203	σ² UMa	09 10.4	+67 08	4.9	8.9	352	3.9	Binary, 1100 years?
7402	23 UMa	09 31.5	+63 04	3.7	9.2	268	22.9	Fixed
8197	OΣ235 UMa	11 32.3	+61 05	5.7	7.6	349	0.6	Binary, 73 years[a]
8682	Σ1694 Cam	12 49.2	+83 25	5.3	5.7	329	21.5	Fixed
10058	η Dra	16 24.0	+61 31	2.8	8.2	139	4.8	PA decreasing, separation increasing
10279	20 Dra	16 56.4	+65 02	7.1	7.3	68	1.1	Binary, 420 years[a]
10660	26 Dra	17 35.0	+61 52	5.3	8.5	331	1.6	Binary, 76 years[a]
10759	ψ¹ Dra	17 41.9	+72 09	4.6	5.6	17	30.2	Fixed
11061	40/41 Dra	18 00.2	+80 00	5.7	6.0	232	18.6	Little change; orange pair
12789	Σ2573 Dra	19 40.2	+60 30	6.5	8.9	26	18.4	Fixed
13007	ε Dra	19 48.2	+70 16	4.0	6.9	20	3.2	PA increasing
13371	Σ2640 Dra	20 04.7	+63 53	6.3	9.5	15	5.5	Opening slowly with decrease of PA
13524	κ Cep	20 08.9	+77 43	4.4	8.3	121	7.3	Little change
15032	β Cep	21 28.7	+70 34	3.2	8.6	250	13.3	Fixed
15600	ξ Cep	22 03.8	+64 38	4.5	6.4	275	7.9	Binary, 3800 years?
15719	Σ2883 Cep	22 10.6	+70 08	5.6	8.6	252	14.6	Fixed
15764	Σ2893 Cep	22 12.9	+73 18	6.2	7.9	347	28.8	Fixed
16538	π Cep	23 07.9	+75 23	4.6	6.8	350	1.1	Binary, 160 years[a]
16666	o Cep	23 18.6	+68 07	5.0	7.3	219	3.3	Binary, 1500 years? Test for 50 mm
17022	6 Cas	23 48.8	+62 13	5.7	8.0	197	1.4	Fixed

[a] Orbital elements for these binaries are given in Table 53. PA and Dist. are from recent observations.

VARIABLE STARS

STAR	RA 2000.0 h m	DEC. 2000.0 ° ′	TYPE	RANGE (mags)	PERIOD (d)	SPECTRAL TYPE	NOTES
YZ Cas	00 45.7	+74 59	EA/DM	5.7–6.1	4.47	A2 + F2	See Double stars
γ Cas	00 56.7	+60 43	GCAS	1.6–3.0	—	B0	X-ray source. See Double stars
RZ Cas	02 48.9	+69 38	EA/SD	6.2–7.7	1.20	A3	
SU Cas	02 52.0	+68 53	DCEPS	5.7–6.2	1.95	F	
ST Cam	04 51.2	+68 10	SRB	6–8	300?	C	
Y Dra	09 42.4	+77 51	M	6.2–15.0	325.79	M	
R UMa	10 44.6	+68 47	M	6.5–13.7	301.62	M	Mean range 7.5–13.0
VY UMa	10 45.1	+67 25	LB	5.9–7.0	—	C	
RY Dra	12 56.4	+66 00	SRB?	6.0–8.0	200?	C	
AZ Dra	16 40.7	+72 40	LB	6.4–7.2	—	M	
VW Dra	17 16.5	+60 40	SRD?	6.0–7.0	170?	K	
UX Dra	19 21.6	+76 34	SRA?	5.9–7.1	168	C	Eclipsing?
T Cep	21 09.5	+68 29	M	5.2–11.3	388.14	M	Mean range 6.0–10.3
VV Cep	21 56.7	+63 38	EA/GS + SRC	4.8–5.4	7430	M + B8	Main oscillations with period of 118 d, also 25, 58 and 150 d. Eclipses too shallow for visual detection

CLUSTERS, NEBULAE AND GALAXIES

NGC	M	RA 2000.0 h m	DEC. 2000.0 ° ′	NOTES
225	—	00 43	+61 47	7th-mag. open cluster in Cassiopeia
581	103	01 33	+60 42	Open cluster in Cassiopeia; 7th mag., consists of faint stars
663	—	01 46	+61 15	Open cluster in Cassiopeia; good binocular object
1502	—	04 08	+62 20	6th-mag. open cluster in Camelopardalis
2403	—	07 37	+65 36	Spiral galaxy in Camelopardalis; 8th mag.
3031	81	09 56	+69 04	Spiral galaxy in Ursa Major; 7th mag.
3034	82	09 56	+69 41	Peculiar galaxy in Ursa Major seen edge-on; 8th mag.; forms a pair with M81
6543	—	17 59	+66 38	Planetary nebula in Draco; 9th mag., one of the brightest of its kind
7654	52	23 24	+61 35	Open cluster in Cassiopeia; 7th mag.

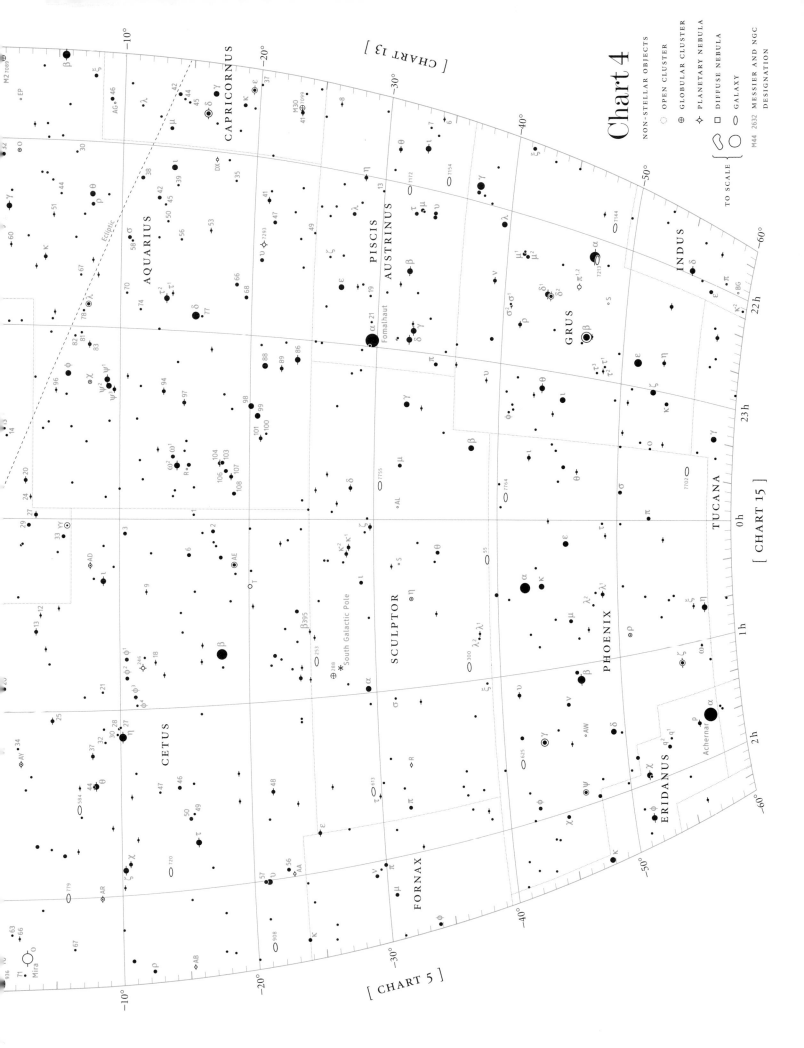

Interesting Objects, Charts 3 and 4

RA 22h to 02h, Dec. +60° to −60°

DOUBLE STARS

ADS	STAR	RA 2000.0 h m	DEC. ° ′	MAGNITUDES		PA °	DIST. ″	NOTES
15536	η PsA	22 00.8	−28 27	5.7	6.8	113	1.8	Fixed
15753	41 Aqr	22 14.3	−21 04	5.6	6.7	112	5.1	Little change
15828	Σ2894 Lac	22 18.9	+37 46	6.2	8.9	194	15.8	Fixed
15934	53 Aqr	22 26.6	−16 45	6.3	6.4	14	1.6	Closing, slow increase of PA. Binary, 3500 years?
15971	ζ Aqr	22 28.8	−00 01	4.3	4.5	183	1.9	Binary, 760 years[a]. Opening with decrease of PA. Test for 60 mm
15987	δ Cep	22 29.2	+58 25	var.	6.1	191	40.9	Fixed; yellow, bluish
	β PsA	22 31.5	−32 21	4.3	7.1	172	30.3	Fixed; optical
16095	8 Lac	22 35.9	+39 38	5.7	6.3	185	22.2	Fixed. C is 9.1 at 168°, 48″.9; D is 10.5 at 144°, 81″.6
16268	τ¹ Aqr	22 47.7	−14 03	5.7	9.0	126	21.5	Optical; closing, PA increasing
	γ PsA	22 52.5	−32 53	4.5	5.0	258	4.0	Gradually opening; PA slowly decreasing
	θ Gru	23 06.9	−43 31	4.5	6.6	107	1.4	Closing; PA increasing
16633	ψ¹ Aqr	23 15.9	−09 05	4.4	9.9	312	49.6	C.p.m.; B is very close double, separation 0″.4
16672	94 Aqr	23 19.1	−13 28	5.3	7.0	351	12.3	Little change; yellowish, bluish
16836	72 Peg	23 34.0	+31 20	5.7	6.1	97	0.6	Binary, 246 years[a]
	θ Phe	23 39.5	−46 38	6.5	7.3	276	3.9	Little change
16957	78 Peg	23 44.0	+29 22	5.1	8.1	268	0.8	Binary, 600 years?
16979	107 Aqr	23 46.0	−18 41	5.7	6.5	136	6.7	Slowly opening
17140	σ Cas	23 59.0	+55 45	5.0	7.2	328	2.4	Fixed; greenish, bluish. Fine field in low power
17175	85 Peg	00 02.2	+27 05	5.8	8.9	194	0.7	Binary, 26 years[a]
111	κ¹ Scl	00 09.3	−27 59	6.1	6.2	260	1.4	Separation increasing, PA decreasing
191	35 Psc	00 15.0	+08 49	6.1	7.5	147	11.6	Fixed
434	λ Cas	00 31.8	+54 31	5.3	5.6	199	0.4	Binary, 500 years[a]
513	π And	00 36.9	+33 43	4.3	7.1	173	36.0	Fixed
520	β395 Cet	00 37.3	−24 46	6.2	6.6	288	0.5	Binary, 25 years[a]
558	55 Psc	00 39.9	+21 26	5.6	8.5	195	6.6	Fixed; orange, bluish
671	η Cas	00 49.1	+57 49	3.5	7.4	319	12.8	Binary, 480 years[a]; yellow and red
683	65 Psc	00 49.9	+27 43	6.3	6.3	296	4.3	PA increasing
755	36 And	00 55.0	+23 38	6.1	6.5	313	0.9	Binary, 168 years[a]
899	ψ¹ Psc	01 05.6	+21 28	5.3	5.5	159	29.6	Fixed
	β Phe	01 06.1	−46 43	4.1	4.2	258	0.3	Binary, 200 years? Closest in 2003
	ζ Phe	01 08.4	−55 15	var.	8.2	242	6.8	C.p.m. Star A is a close double, separation 0″.6
996	ζ Psc	01 13.7	+07 35	5.2	6.2	62	22.7	C.p.m.
1003	37 Cet	01 14.4	−07 55	5.2	7.9	331	49.4	Fixed
1081	42 Cet	01 19.8	−00 31	6.5	7.0	18	1.6	C.p.m. PA increasing. B is a very close binary
	p Eri	01 39.8	−56 12	5.8	5.9	192	11.3	Binary, 500 years? Both orange
1394	ε Scl	01 45.6	−25 03	5.4	8.5	25	5.1	Binary, 1200 years?
1457	1 Ari	01 50.1	+22 17	6.3	7.2	165	2.8	Little change. Test for 50 mm
1507	γ Ari	01 53.5	+19 18	4.5	4.6	1	7.5	Slowly closing. Beautiful, very easy pair
1538	Σ186 Cet	01 55.9	+01 51	6.8	6.8	64	1.0	Binary, 160 years[a]
1563	λ Ari	01 57.9	+23 36	4.8	6.7	47	37.1	Fixed

[a] Orbital elements for these binaries are given in Table 53. PA and Dist. are from recent observations.

VARIABLE STARS

STAR	RA 2000.0 h m	DEC. ° '	TYPE	RANGE (mags)	PERIOD (d)	SPECTRAL TYPE	NOTES
DX Aqr	22 02.4	−16 58	EA/KE?	6.4–6.8	0.95	A2	Secondary minimum 6.7
AR Lac	22 08.7	+45 44	EA/AR/RS	6.1–6.8	1.98	G2 + K0	Secondary minimum 6.4
π¹ Gru	22 22.7	−45 57	SRB	5.4–6.7	150?	S	
RW Cep	22 23.1	+55 58	SRD	6.2–7.6	346?	K	
S Gru	22 26.1	−48 26	M	6.0–15.0	401.51	M	Mean range 7.7–14.4
δ Cep	22 29.2	+58 25	DCEP	3.5–4.4	5.37	G	See Double stars
KY Cep	22 32.3	+57 40	★	4?–13?	—	Pec	Flare of 65 seconds
V509 Cas	23 00.1	+56 57	SRD	4.8–5.5	—	G + B1	Pulsations with period 3 years; shell ejected 1975
β Peg	23 03.8	+28 05	LB	2.3–2.7	—	M	
R Aqr	23 43.8	−15 17	M	5.8–12.4	386.96	M + Pec	Range varies; possible cycle of 24 years
TX Psc	23 46.4	+03 29	LB	4.8–5.2	—	C	
ρ Cas	23 54.4	+57 30	SRD	4.1–6.2	—	G	Usually 4.4–5.2, but peculiar fade to deep minimum in 1945–47
V373 Cas	23 55.6	+57 25	E?/GS	5.9–6.3	13.42	B0 + B0	Unique binary with possible eclipses and physical variation of components. Normally range is only 0.1 mag.
R Cas	23 58.4	+51 24	M	4.7–13.5	430.46	M	Mean range 7.0–12.6
S Scl	00 15.4	−32 03	M	5.5–13.6	362.57	M	Mean range 6.7–12.9
T Cet	00 21.8	−20 03	SRC	5.0–6.9	158.9	M	
R And	00 24.0	+38 35	M	5.8–14.9	409.33	S	Mean range 6.9–14.3
TV Psc	00 28.0	+17 54	SR	4.7–5.4	49.1	M	
ζ Phe	01 08.4	−55 15	EA/DM	3.9–4.4	1.67	B6 + B9	Secondary minimum 4.2
V465 Cas	01 18.2	+57 48	SRB	6.2–7.2	60	M	
AA Cet	01 59.0	−22 55	EW/KE	6–6.5	0.54	F2	Secondary minimum 6.5

CLUSTERS, NEBULAE AND GALAXIES

NGC	M	RA 2000.0 h m	DEC. ° '	NOTES
7293	—	22 30	−20 48	Planetary nebula, the Helix, in Aquarius; the largest planetary, 0°.2 across, best seen with binoculars in a dark sky
7662	—	23 26	+42 33	Planetary nebula in Andromeda, 9th mag.; one of the easiest planetaries for small telescopes, appears star-like at low powers
55	—	00 15	−39 11	Spiral galaxy in Sculptor, edge-on; 8th mag.
205	110	00 40	+41 41	Elliptical galaxy, larger of the two companions of the Andromeda Galaxy but less easy to see; 8th mag.
221	32	00 43	+40 52	8th-mag. elliptical companion to the Andromeda Galaxy
224	31	00 43	+41 16	The Andromeda Galaxy, naked-eye spiral 2.5 million l.y. away; ideal for binoculars and telescopes with low powers
253	—	00 48	−25 17	7th-mag. spiral galaxy in Sculptor, seen edge-on
457	—	01 19	+58 20	Open cluster in Cassiopeia, including 5th-mag. φ Cas
598	33	01 34	+30 39	Spiral galaxy in Triangulum, large but with low surface brightness; best in binoculars
628	74	01 37	+15 47	Spiral galaxy in Pisces; 9th mag., but one of the most difficult Messier objects to observe
650–1	76	01 42	+51 34	Planetary nebula in Perseus, the Little Dumbbell; 12th mag., the faintest Messier object, given as a double nebula in the NGC
752	—	01 58	+37 41	Large open cluster in Andromeda

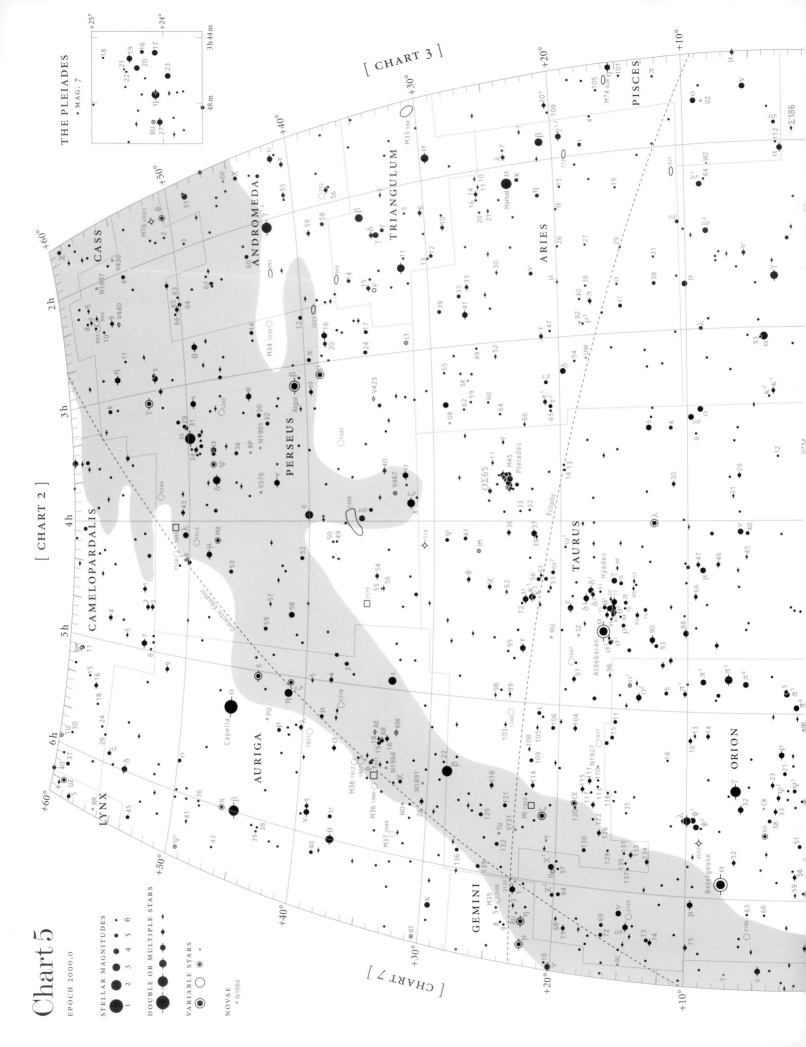

Chart 5

EPOCH 2000.0

STELLAR MAGNITUDES

DOUBLE OR MULTIPLE STARS

VARIABLE STARS

NOVAE
● N1984

THE PLEIADES
• MAG. 7

[CHART 3]

[CHART 2]

[CHART 7]

CASS

ANDROMEDA

TRIANGULUM

PISCES

ARIES

PERSEUS

TAURUS

CAMELOPARDALIS

AURIGA

LYNX

GEMINI

ORION

Pleiades

Hyades

Capella

Aldebaran

Betelgeuse

Hamal

Algol

Ecliptic

Galactic Equator

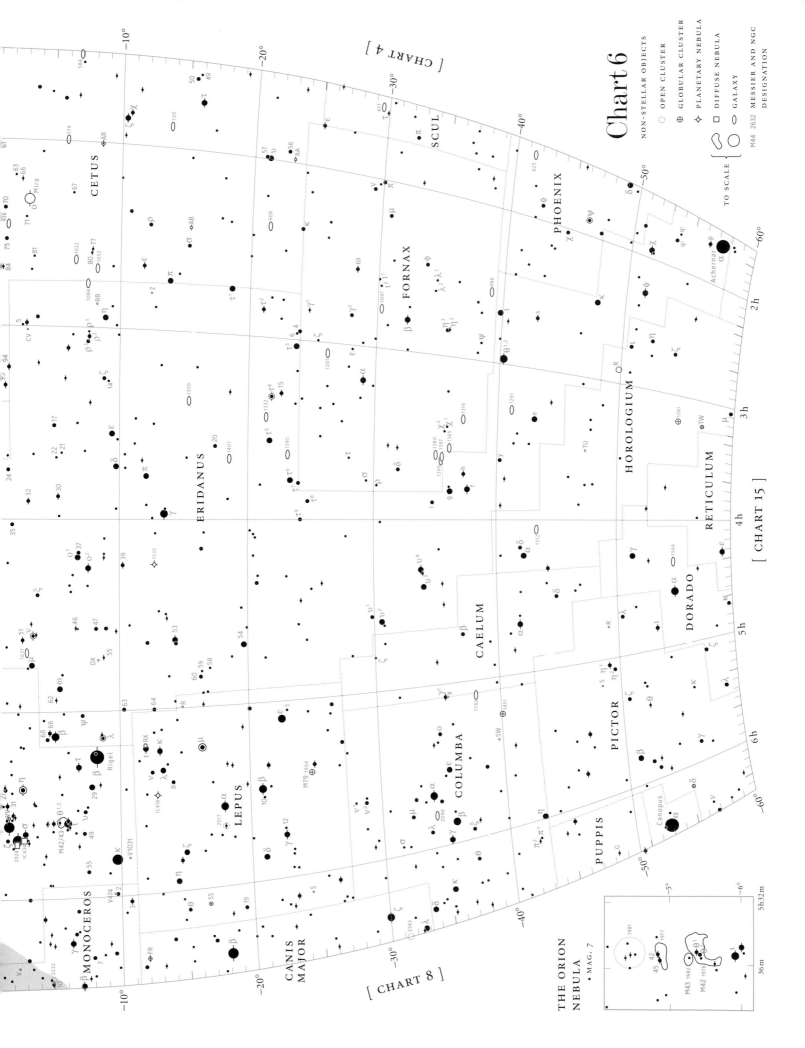

Chart 6

[CHART 4]

CETUS

SCUL

PHOENIX

FORNAX

HOROLOGIUM

RETICULUM

[CHART 15]

ERIDANUS

DORADO

CAELUM

PICTOR

COLUMBA

MONOCEROS

LEPUS

CANIS MAJOR

PUPPIS

[CHART 8]

NON-STELLAR OBJECTS

⊙ OPEN CLUSTER
⊕ GLOBULAR CLUSTER
✧ PLANETARY NEBULA
□ DIFFUSE NEBULA
◯ GALAXY

TO SCALE

M44   2632   MESSIER AND NGC
              DESIGNATION

THE ORION NEBULA
• MAG. 7

M43 1982
M42 1976

45   1981
42   1977

M43 1982   θ¹
M42 1976   θ²

5h32m

36m

Mira

Rigel

Canopus

Achernar

Interesting Objects, Charts 5 and 6

RA 02h to 06h, Dec. +60° to −60°

DOUBLE STARS

ADS	STAR	RA 2000.0 h m	DEC. ° ′	MAGNITUDES	PA °	DIST. ″	NOTES
1615	α Psc	02 02.0	+02 46	4.1 5.2	271	1.8	Binary, 900 years[a]
1631	10 Ari	02 03.7	+25 56	5.8 7.9	235	1.2	Binary, 325 years[a]
1630	γ And	02 03.9	+42 20	2.3 5.0	63	9.6	Little change. Superb pair; orange, bluish. B is a close binary, 64 years
1683	59 And	02 10.9	+39 02	6.1 6.7	36	16.5	Fixed
1697	6 Tri	02 12.4	+30 18	5.3 6.7	69	3.9	Yellowish, bluish
1703	66 Cet	02 12.8	−02 24	5.7 7.7	234	16.7	Fixed. Yellow, blue; fine pair
1778	o Cet	02 19.3	−02 59	var. var.	110	0.6	Mira. Binary, 500 years? B is VZ Cet, 9.5–12
1954	ω For	02 33.8	−28 14	5.0 7.7	245	10.6	Widening
2080	γ Cet	02 43.3	+03 14	3.6 6.2	299	2.3	Little change
2157	η Per	02 50.7	+55 54	3.8 8.5	301	28.5	Fixed; orange, bluish, in an attractive field
2200	20 Per	02 53.7	+38 20	5.0 9.7	237	14.1	Fixed. A is close binary, 31.5 years
	θ Eri	02 58.3	−40 18	3.2 4.1	90	8.3	Little change; fine blue–white pair
2257	ε Ari	02 59.2	+21 20	5.2 5.6	209	1.4	Increase in PA and separation. Test for 100 mm
2312	ρ² Eri	03 02.7	−07 41	5.4 8.9	66	1.5	PA and distance decreasing
2362	β Per	03 08.2	+40 57	var. 10.5	193	82.2	Algol. Optical; fixed
2402	α For	03 12.1	−28 59	4.0 7.2	300	4.6	Binary, 270 years[a]
2616	7 Tau	03 34.4	+24 28	6.6 6.9	356	0.7	Binary, 520 years[a]. C is physical
				9.9	54	22.1	
2799	OΣ65 Tau	03 50.3	+25 35	5.7 6.5	30	0.2	Binary, 61 years[a]. Widest around 2032 (0″.7)
2843	ζ Per	03 54.1	+31 53	2.9 9.2	208	12.2	Fixed
2850	32 Eri	03 54.3	−02 57	4.8 5.9	349	6.9	Fixed; yellow, greenish (by contrast)
2888	ε Per	03 57.9	+40 01	2.9 8.9	10	9.0	Optical; fixed
3079	39 Eri	04 14.4	−10 15	5.0 8.5	144	6.4	Slowly closing
3093	o² Eri	04 15.2	−07 39	4.4 9.7	104	83.0	Fixed. B is binary, 250 years, consisting of a white dwarf, mag. 9.5, and a red dwarf, mag. 11.2, separation 8″.9
3137	φ Tau	04 20.4	+27 21	5.1 7.5	256	49.2	Optical; closing, PA increasing; orange, white
3161	χ Tau	04 22.6	+25 38	5.4 8.5	25	19.6	Fixed; blue, yellow
3321	α Tau	04 35.9	+16 31	0.9 11.3	32	131.3	Aldebaran. Opening
	ι Pic	04 50.9	−53 28	5.6 6.2	58	12.2	Fixed; easy
3572	4 Aur	04 59.3	+37 53	5.0 8.2	3	4.7	Slowly closing
	γ¹ Cae	05 04.4	−35 29	4.7 8.2	306	3.2	Little change
3800	κ Lep	05 13.2	−12 56	4.4 6.8	0	2.3	Gradually closing
3797	ρ Ori	05 13.3	+02 52	4.6 8.5	65	7.0	Fixed; other stars in the field
3823	β Ori	05 14.5	−08 12	0.2 6.8	203	9.2	Rigel. Fixed. Test for 50 mm
4002	η Ori	05 24.5	−02 24	3.6 4.9	78	1.7	Separation increasing. Test for 100 mm
	θ Pic	05 24.8	−52 19	6.2 6.7	288	38.1	C.p.m. θ¹ is close binary, 190 years
4066	β Lep	05 28.2	−20 46	3.0 7.5	346	2.3	Closing; PA increasing
4134	δ Ori	05 32.0	−00 18	2.4 6.8	0	52.8	A is close double, 0″.3
4179	λ Ori	05 35.1	+09 56	3.5 5.5	44	4.3	Fixed. Very fine region
4186	θ¹ Ori	05 35.3	−05 23	5.1 6.7			The Trapezium; two components are eclipsing binaries. Two other faint stars (mags. 11.1, 11.5) are test for 100 mm. Fine fixed multiple group
				6.7 (var.) 7.9 (var.)			
4241	σ Ori	05 38.7	−02 36	3.7 6.3			A is close binary, 155 years[a]. Rich telescopic object, with triple Σ761 (mags. 7.9, 8.4, 8.6) in same field
				6.6 8.8			
4263	ζ Ori	05 40.8	−01 57	1.9 3.7	165	2.4	Binary, 1500 years? Test for 75 mm
4334	γ Lep	05 44.5	−22 27	3.6 6.3	350	97.0	Little change; yellow, orange
4566	θ Aur	05 59.7	+37 13	2.7 7.2	309	3.8	Test for 100 mm

[a] Orbital elements for these binaries are given in Table 53. PA and Dist. are from recent observations.

VARIABLE STARS

STAR	RA 2000.0 h m	DEC. ° '	TYPE	RANGE (mags)	PERIOD (d)	SPECTRAL TYPE	NOTES
o Cet	02 19.3	−02 59	M	2.0–10.1	331.96	M	Mira. Mean range 3.5–9.1. See Double stars
R Tri	02 37.0	+34 16	M	5.4–12.6	266.9	M	Mean range 6.2–11.7
Z Eri	02 47.9	−12 28	SRB	5.6–7.2	80	M	Secondary period 746.4 d
RR Eri	02 52.2	−08 16	SRB	6.3–8.1	97	M	
R Hor	02 53.9	−49 53	M	4.7–14.3	407.6	M	Mean range 6.0–13.0
γ Per	03 04.8	+53 30	EA/GS	2.9–3.2	5330	G5 + A2	Longest-period eclipsing binary known, 14.6 years
ρ Per	03 05.2	+38 50	SRB	3.3–4.0	50?	M	Mean mag. varies?
β Per	03 08.2	+40 57	EA/SD	2.1–3.4	2.87	B8	Algol. Weak X-ray source. See Double stars
TW Hor	03 12.6	−57 19	SRB	5.5–6.0	158?	C	
BU Tau	03 49.2	+24 08	GCAS	4.8–5.5	—	B8	Pleione, in the Pleiades
X Per	03 55.4	+31 03	GCAS + XP	6.0–7.0	—	O	
λ Tau	04 00.7	+12 29	EA/DM	3.4–3.9	3.95	B3 + A4	
SZ Tau	04 37.2	+18 33	DCEPS	6.3–6.8	3.15	F	In halo of open cluster NGC 1647
HU Tau	04 38.3	+20 41	EA/SD?	5.9–6.7	2.06	B8	
R Pic	04 46.2	−49 15	SR	6.4–10.1	170.9	M	
R Lep	04 59.6	−14 48	M	5.5–11.7	427.07	C	Amplitude varies, period over 40 years? Maxima can be as faint as 9.5
ε Aur	05 02.0	+43 49	EA/GS	2.9–3.8	9892	F + B	Fluctuations of 0.2 mag. in cycle of about 110 d
ζ Aur	05 02.5	+41 05	EA/GS	3.7–4.0	972.16	K5 + B7	
W Ori	05 05.4	+01 11	SRB	5.9–7.7	212	C	Secondary period of 2450 d
S Pic	05 11.0	−48 30	M	6.5–14.0	428.0	M	Mean range 8.1–13.8
RX Lep	05 11.4	−11 51	SRB	5.0–7.4	60?	M	
μ Lep	05 12.9	−16 12	ACV	3.0–3.4	2?	B9	
AR Aur	05 18.3	+33 46	EA/DM	6.2–6.8	4.13	Ap + B9	Secondary minimum 6.7
CK Ori	05 30.3	+04 12	SR?	5.9–7.1	120?	K	
TU Tau	05 45.2	+24 25	SRB	5.9–9.2	190?	C + A2	
Y Tau	05 45.7	+20 42	SRB	6.5–9.2	241.5	C	
V1031 Ori	05 47.4	−10 32	EA/DM	6.0–6.4	3.41	A4	
α Ori	05 55.2	+07 24	SRC	0.0–1.3	2335	M	Betelgeuse. Also waves of 200–400 d
U Ori	05 55.8	+20 10	M	4.8–13.0	368.3	M	Mean range 6.3–12.0
V474 Mon	05 59.0	−09 23	DSCT	5.9–6.4	0.14	F2	

CLUSTERS, NEBULAE AND GALAXIES

NGC	M	RA 2000.0 h m	DEC. ° '	NOTES
869	—	02 19	+57 09	Double Cluster in Perseus, also known as h and χ Persei, each cluster covering 0°.5;
884	—	02 22	+57 07	NGC 869 is the richer. Naked-eye and binocular object
1039	34	02 42	+42 47	5th-mag. open cluster in Perseus, 0°.5 wide
1068	77	02 43	−00 01	Spiral galaxy of Seyfert variety (bright nucleus) in Cetus; 9th mag.
—	45	03 47	+24 07	Pleiades open cluster in Taurus; at least five stars visible to naked eye; covers nearly 2°; ideal binocular object
1904	79	05 25	−24 33	Globular cluster in Lepus; 8th mag. In same field as multiple star h 3752
1912	38	05 29	+35 50	6th-mag. open cluster in Auriga; one of a chain with M36 and M37
1952	1	05 35	+22 01	The Crab Nebula in Taurus, mag. 8.4; remnant of a supernova; covers 6' × 4'
1976	42	05 35	−05 37	Orion Nebula, covering over 1°; superb in all apertures. At its centre is the multiple star θ¹ Orionis
1981	—	05 35	−04 26	Open cluster north of the Orion Nebula
1977	—	05 36	−04 52	Nebula surrounding 42 Orionis
1982	43	05 36	−05 16	Part of the Orion Nebula, just to the north of the main cloud
1960	36	05 36	+34 08	Smallest of the three open clusters in Auriga; the most prominent in binoculars
2068	78	05 47	+00 03	Nebulosity in Orion
2099	37	05 52	+32 33	Largest and richest of the three open clusters in Auriga; 0°.4 across

Chart 7

EPOCH 2000.0

STELLAR MAGNITUDES

DOUBLE OR MULTIPLE STARS

VARIABLE STARS

NOVAE
N1984

[CHART 1]

[CHART 5]

CAM

AURIGA

LYNX

URSA MAJOR

LEO MINOR

CANCER

GEMINI

TAURUS

ORION

Betelgeuse

CANIS MINOR

Procyon

LEO

Regulus

Castor

Pollux

[CHART 9]

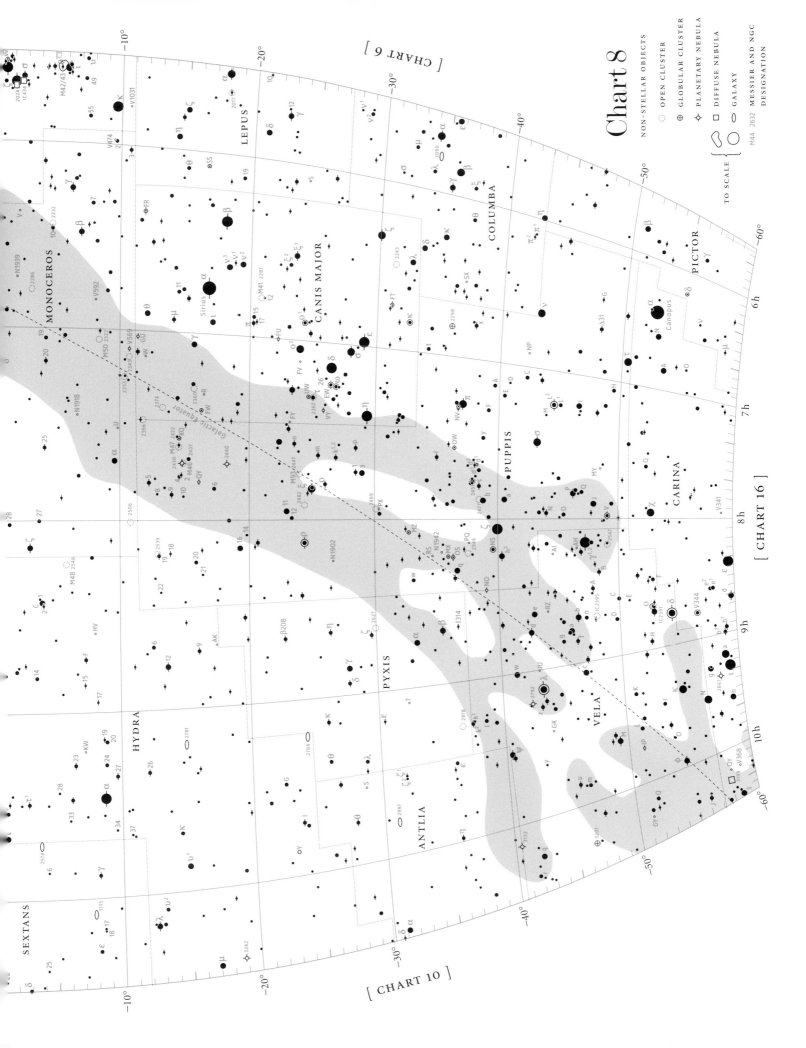

Interesting Objects, Charts 7 and 8

RA 06h to 10h, Dec. +60° to −60°

DOUBLE STARS

ADS	STAR	RA 2000.0 h m	DEC. 2000.0 ° ′	MAGNITUDES		PA °	DIST. ″	NOTES
4773	41 Aur	06 11.6	+48 43	6.2	6.9	357	7.6	Little change
4841	η Gem	06 14.9	+22 30	var.	6.2	258	1.7	Binary, 500 years?
4990	μ Gem	06 22.9	+22 31	2.9	9.4	141	121.7	Wide optical pair, fixed. B is double: 9.8, 10.7; 265°, 0″.6
5012	ε Mon	06 23.8	+04 36	4.4	6.6	29	12.3	Optical, little change. Fine field with low power
5107	β Mon	06 28.8	−07 02	4.6	5.0	133	7.1	BC is 108°, 2″.9. Fine fixed triple
					5.4	125	9.8	
	μ Pic	06 32.0	−58 45	5.6	9.3	230	2.5	Fixed
5166	20 Gem	06 32.3	+17 47	6.3	6.9	211	19.7	Fixed; yellowish, bluish
5253	ν¹ CMa	06 36.4	−18 40	5.8	7.4	264	17.8	Fixed
	Δ31 Pup	06 38.6	−48 13	5.1	7.4	321	12.9	Fixed
5423	α CMa	06 45.1	−16 43	−1.4	8.5	190	3.7	Sirius. Binary, 50 years[a]
5400	12 Lyn	06 46.2	+59 27	5.4	6.0	72	1.8	AB is binary, 700 years[a]; test for 75 mm. C is physical
					7.1	309	8.7	
5514	14 Lyn	06 53.1	+59 27	6.0	6.5	287	0.3	Binary, 290 years[a]. Closest (0″.2) around 2002
5559	38 Gem	06 54.6	+13 11	4.8	7.8	145	7.2	Binary, 2000–3000 years? Opening, PA decreasing
5605	μ CMa	06 56.1	−14 03	5.3	7.1	343	2.8	Gradually closing; yellowish, bluish
5654	ε CMa	06 58.6	−28 58	1.5	7.5	161	7.0	Fixed
5961	λ Gem	07 18.1	+16 32	3.6	10.7	33	9.8	Fixed. Easy test for 75 mm
5983	δ Gem	07 20.1	+21 59	3.6	8.2	225	5.4	Binary, 1200 years?
	σ Pup	07 29.2	−43 18	3.3	8.8	74	22.2	Separation gradually increasing, PA decreasing
6190	n Pup	07 34.3	−23 28	5.8	5.9	117	9.8	C.p.m.
6175	α Gem	07 34.6	+31 53	1.9	3.0	64	4.1	Castor. Binary, 450 years[a]. Castor C (YY Gem), mag. 8.9–9.6, lies at 164°, 71″; fixed
6255	k Pup	07 38.8	−26 48	4.4	4.6	318	9.9	Fixed. Fine pair
6321	κ Gem	07 44.4	+24 24	3.7	8.2	241	7.2	Slow increase in PA and separation
6420	9 Pup	07 51.8	−13 54	5.6	6.5	328	0.2	Binary, 23 years[a]. Widest (0″.5) around 2013
	γ Vel	08 09.5	−47 20	1.8	4.1	221	41.4	Optical pair. A is the brightest Wolf–Rayet star. Two wider companions: C mag. 7.3 at 62″, D mag. 9.4 at 94″
6650	ζ Cnc	08 12.2	+17 39	5.1	6.2	73	5.9	Binary, 1100 years[a]. Brighter component is itself binary: 5.3, 6.3; 92°, 0″.9; 60 years[a]
	h² Pup	08 14.0	−40 21	4.4	9.5	341	51.1	Fixed
6815	φ² Cnc	08 26.8	+26 56	6.2	6.2	218	5.2	Slowly widening
6914	β208 Pyx	08 39.1	−22 40	5.4	6.8	216	1.3	Binary, 123 years[a]
	I314 Pyx	08 39.4	−36 36	6.4	7.9	223	0.5	Binary, 67 years[a]
	δ Vel	08 44.7	−54 43	var.	5.1	344	1.1	Binary, 140 years[a]. A is closer double, 5°, 0″.7.
6988	ι Cnc	08 46.7	+28 46	4.1	6.0	307	30.7	Fixed; easy
6993	ε Hya	08 46.8	+06 25	3.5	6.7	299	2.9	Binary, 1000 years[a]; yellow and blue
	H Vel	08 56.3	−52 43	4.7	7.7	335	2.6	Slowly closing
7114	ι UMa	08 59.2	+48 03	3.1	9.2	24	3.1	Binary, 800 years? B is itself binary, 40 years
7292	38 Lyn	09 18.8	+36 48	3.9	6.1	225	2.7	Little change
7307	Σ1338 Lyn	09 21.0	+38 11	6.7	7.1	291	1.1	Binary, 300 years[a]
7351	κ Leo	09 24.7	+26 11	4.5	9.7	211	2.4	Little change
7390	ω Leo	09 28.5	+09 03	5.7	7.3	91	0.6	Binary, 118 years[a]
	ψ Vel	09 30.7	−40 28	3.9	5.1	226	0.8	Binary, 34 years[a]; 1″.1 at widest
	ζ¹ Ant	09 30.8	−31 53	6.2	6.8	212	7.8	Fixed
7545	φ UMa	09 52.1	+54 04	5.3	5.4	263	0.3	Binary, 105 years[a]; 0″.5 at widest
7555	γ Sex	09 52.5	−08 06	5.4	6.4	59	0.5	Binary, 78 years[a]

[a] Orbital elements for these binaries are given in Table 53. PA and Dist. are from recent observations.

VARIABLE STARS

STAR	RA 2000.0 h m	DEC. 2000.0 ° ′	TYPE	RANGE (mags)	PERIOD (d)	SPECTRAL TYPE	NOTES
S Lep	06 05.8	−24 11	SRB	6.0–7.6	89.0	M	Secondary period 890 d
BU Gem	06 12.3	+22 54	LC	5.7–8.1	—	M	
η Gem	06 14.9	+22 30	SRA + EA	3.2–3.9	232.9	M	Deep minima (eclipses?) every 8.2 years. See Double stars
V Mon	06 22.7	−02 12	M	6.0–13.9	340.5	M	Mean range 7.0–13.1
ψ¹ Aur	06 24.9	+49 17	LC	4.8–5.7	—	M	
T Mon	06 25.2	+07 05	DCEP	5.6–6.6	27.02	G	
BL Ori	06 25.5	+14 43	LB	6.3–6.9	—	C	
RR Lyn	06 26.4	+56 17	EA/DM	5.5–6.0	9.95	A7 + F3	Secondary minimum 5.9

VARIABLE STARS (continued)

STAR	RA 2000.0 h m	DEC. 2000.0 ° ′	TYPE	RANGE (mags)	PERIOD (d)	SPECTRAL TYPE	NOTES
RT Aur	06 28.6	+30 30	DCEP	5.0–5.8	3.73	G	
WW Aur	06 32.5	+32 27	EA/DM	5.8–6.5	2.53	A3 + A3	Secondary minimum 6.4
W Gem	06 35.0	+15 20	DCEP	6.5–7.4	7.91	G	
UU Aur	06 36.5	+38 27	SRB	5.1–6.8	234	C	
IS Gem	06 49.7	+32 36	SRC	5.3–6.0	47?	K	
ζ Gem	07 04.1	+20 34	DCEP	3.6–4.2	10.15	G	
R Gem	07 07.4	+22 42	M	6.0–14.0	369.91	S	Mean range 7.1–13.5
W CMa	07 08.1	−11 55	LB	6.4–7.9	—	C	
BQ Gem	07 13.4	+16 10	SRB	5.1–5.5	50?	M	
L² Pup	07 13.5	−44 39	SRB	2.6–6.2	140.6	M	
EW CMa	07 14.3	−26 21	GCAS	4.4–4.8	—	B3	
ω CMa	07 14.8	−26 46	GCAS	3.6–4.2	—	B2	
UW CMa	07 18.7	−24 34	EB/KE?	4.8–5.3	4.39	O7 + OB	Secondary minimum 5.3
R CMa	07 19.5	−16 24	EA/SD	5.7–6.3	1.14	F1	
VY CMa	07 23.0	−25 46	★	6.5–9.6	—	M	Unique variable in reflection nebula near young open cluster NGC 2362. Cyclic variations and slow fade since 1801
FW CMa	07 24.7	−16 12	GCAS	5.0–5.5	—	B3	
U Mon	07 30.8	−09 47	RVB	5.9–7.8	91.32	G	Secondary period 2320 d
QY Pup	07 47.6	−15 59	SRD	6.2–6.7	—	K	
PX Pup	07 56.4	−30 17	LB?	6.0–6.5	—	M	
V341 Car	07 56.8	−59 08	L	6.2–7.1	—	M	In variable reflection nebula IC 2220, and near open cluster NGC 2516
V Pup	07 58.2	−49 15	EB/SD	4.4–4.9	1.45	B1 + B3	Secondary minimum 4.8
RS Pup	08 13.1	−34 35	DCEP	6.5–7.7	41.39	G	
AI Vel	08 14.1	−44 34	DSCT	6.2–6.8	0.11	F	
R Cnc	08 16.6	+11 44	M	6.1–11.8	361.60	M	Mean range 6.8–11.2
NO Pup	08 26.3	−39 04	EA/KE?	6.5–7.0	1.26	B8	
RZ Vel	08 37.0	−44 07	DCEP	6.4–7.6	20.40	G	
AK Hya	08 39.9	−17 18	SRB	6.3–6.9	75?	M	
δ Vel	08 44.7	−54 43	EA	1.9–2.3	45	A1 + ?	
BO Cnc	08 52.5	+28 16	LB?	5.9–6.4	—	M	
X Cnc	08 55.4	+17 14	SRB	5.6–7.5	195?	C	
T Pyx	09 04.7	−32 23	NR	6.5–15.3	(7000)	Pec	Outbursts in 1890, 1902, 1920, 1944, 1966
RS Cnc	09 10.6	+30 58	SRC?	5.1–7.0	120?	M	Secondary period 700 d
KW Hya	09 12.4	−07 07	EA/DM	6.1–6.6	7.75	A3 + A0	Secondary minimum 6.4
IN Hya	09 20.6	+00 11	SRB	6.3–6.9	65?	M	
CG UMa	09 21.7	+56 42	LB	5.5–6.0	—	M	
S Ant	09 32.3	−28 38	EW/KE?	6.4–6.9	0.65	A9	Secondary minimum 6.9
R LMi	09 45.6	+34 31	M	6.3–13.2	372.19	M	Mean range 7.1–12.6
R Leo	09 47.6	+11 26	M	4.4–11.3	309.95	M	Mean range 5.8–10.0
Y Hya	09 51.1	−23 01	SRB	5–8	302.8	C	Mean mag. varies

CLUSTERS, NEBULAE AND GALAXIES

NGC	M	RA 2000.0 h m	DEC. 2000.0 ° ′	NOTES
2168	35	06 09	+24 20	Open cluster in Gemini; large and rich, good binocular object
2232	—	06 27	−04 45	Large, scattered cluster in Monoceros including 5th-mag. star 10 Mon
2244	—	06 32	+04 52	Large, elongated open cluster in Monoceros; surrounded by the faint Rosette Nebula, NGC 2237, 1° across, visible in binoculars under dark skies
2264	—	06 41	+09 53	Arrow-shaped open cluster in Monoceros including the 5th-mag. star 15 Mon. The surrounding nebulosity, including the dark Cone Nebula, shows up well only on photographs
2287	41	06 47	−20 44	Naked-eye open cluster in Canis Major, 0°.6 wide, good in binoculars
2323	50	07 03	−08 20	Open cluster in Monoceros; 6th mag.
2392	—	07 29	+20 55	Planetary nebula in Gemini, called the Eskimo; reported mags. vary from 8 to 10
2422	47	07 37	−14 30	Large naked-eye cluster in Puppis, brightest stars of 6th mag.
2437	46	07 42	−14 49	Open cluster in Puppis, makes a contrasting binocular pair with M47
2447	93	07 45	−23 52	Open cluster in Puppis
2451	—	07 45	−37 58	Large, scattered open cluster in Puppis containing the 4th-mag. star c Pup
2477	—	07 52	−38 33	Open cluster in Puppis, like a large globular through binoculars
2547	—	08 11	−49 16	Open cluster in Vela
2548	48	08 14	−05 48	Large open cluster in Hydra
2632	44	08 40	+19 59	Praesepe, the Beehive Cluster, in Cancer; covers 1°.5; visible to the naked eye as a misty patch; best seen in binoculars
IC 2391	—	08 40	−53 04	Large, scattered cluster in Vela containing the 4th-mag. star o Vel
IC 2395	—	08 41	−48 12	Open cluster in Vela; 5th mag.
2682	67	08 50	+11 49	Rich open cluster in Cancer

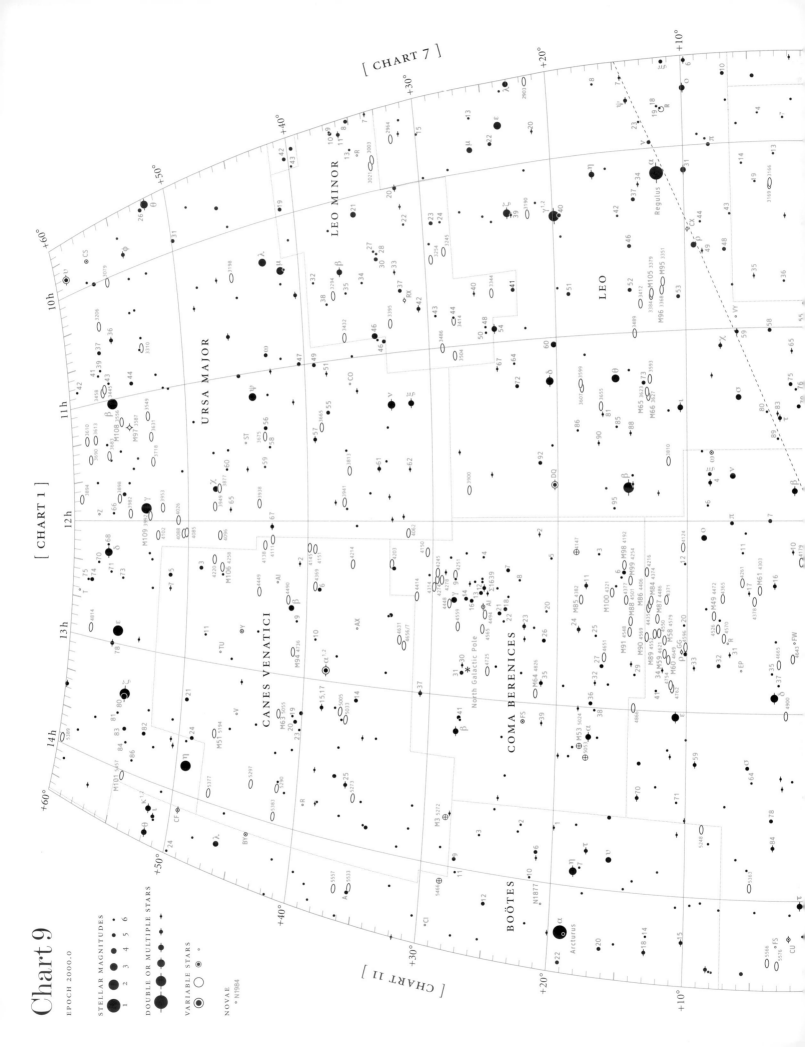

Interesting Objects, Charts 9 and 10

RA 10h to 14h, Dec. +60° to −60°

DOUBLE STARS

ADS	STAR	RA 2000.0 h m	DEC. ° ′	MAGNITUDES		PA °	DIST. ″	NOTES
7654	α Leo	10 08.4	+11 58	1.4	8.2	307	175.3	Regulus. Fixed
7724	γ Leo	10 20.0	+19 51	2.4	3.6	125	4.5	Binary, 600 years[a]. Fine yellow pair
	J Vel	10 20.9	−56 03	4.5	7.2	102	7.2	Little change
	δ Ant	10 29.6	−30 36	5.6	9.8	226	11.1	Little change
7846	β411 Hya	10 36.1	−26 41	6.7	7.8	311	1.3	Binary, 170 years[a]
	μ Vel	10 46.8	−49 25	2.8	5.7	51	2.1	Binary, 140 years?
8119	ξ UMa	11 18.2	+31 32	4.3	4.8	261	1.8	Binary, 60 years[a]. Fine yellow pair; opening
8123	ν UMa	11 18.5	+33 06	3.5	10.1	147	7.1	Fixed
8148	ι Leo	11 23.9	+10 32	4.1	6.7	107	1.6	Binary, 186 years[a]. Opening
8153	γ Crt	11 24.9	−17 41	4.1	7.9	93	5.3	Little change
8175	57 UMa	11 29.1	+39 20	5.4	10.7	357	5.5	Little change
8196	88 Leo	11 31.7	+14 22	6.3	9.1	330	15.9	Little change
	β Hya	11 52.9	−33 54	4.7	5.5	36	0.7	Closing, PA increasing
8406	2 Com	12 04.3	+21 28	6.2	7.5	236	3.7	Fixed
	D Cen	12 14.0	−45 43	5.8	7.0	243	2.9	Slowly closing
8539	Σ1639 Com	12 24.4	+25 35	6.7	7.8	325	1.7	Binary, 600 years[a]
8572	δ Crv	12 29.9	−16 31	3.0	8.5	216	24.7	C.p.m. Companion described as 'purplish'
8573	β28 Crv	12 30.1	−13 24	6.5	9.6	334	2.0	Binary, 150 years[a]
	γ Cru	12 31.2	−57 07	1.8	6.5	26	125.4	Optical; widening, PA decreasing. There is also a mag. 9.5 star at 82°, 155″
8600	24 Com	12 35.1	+18 23	5.1	6.3	271	20.7	Fixed; orange, bluish
	γ Cen	12 41.5	−48 58	2.8	2.9	347	0.9	Binary, 84 years[a]. Closest between 2010 and 2020
8630	γ Vir	12 41.7	−01 27	3.5	3.5	244	1.0	Binary, 169 years[a]. Closest around 2005
8695	35 Com	12 53.3	+21 15	5.2	7.1	187	1.0	Binary, 360 years[a]. C is physical
					9.8	127	28.7	
8706	α CVn	12 56.0	+38 19	2.9	5.5	229	19.3	Cor Caroli. Fixed
8801	θ Vir	13 09.9	−05 32	4.4	9.4	341	7.1	Fixed. Test for 75 mm. A is a close double, 338°, 0″.5. Mag. 10.4 star at 300°, 71″.1; fixed.
8891	ζ UMa	13 23.9	+54 56	2.2	3.9	152	14.6	Mizar. Physical. Naked-eye pair with Alcor (80 UMa), mag. 4.0: 71°, 708″; c.p.m.
8974	25 CVn	13 37.5	+36 18	5.0	7.0	99	1.7	Binary, 230 years[a]
	Q Cen	13 41.7	−54 34	5.2	6.5	163	5.4	Separation increasing
9000	84 Vir	13 43.1	+03 32	5.6	8.3	228	2.8	Little change. Test for 75 mm
	3 Cen	13 51.8	−33 00	4.5	6.0	106	7.9	C.p.m.
	4 Cen	13 53.2	−31 56	4.7	8.5	185	14.8	Fixed

[a] *Orbital elements for these binaries are given in Table 53. PA and Dist. are from recent observations.*

VARIABLE STARS

STAR	RA 2000.0 h m	DEC. ° ′	TYPE	RANGE (mags)	PERIOD (d)	SPECTRAL TYPE	NOTES
U Ant	10 35.2	−39 34	LB	5–6	—	C	
U Hya	10 37.6	−13 23	SRB	4.3–6.5	114.8	C	
η Car	10 45.1	−59 41	SDOR	−0.8–7.9	—	Pec	Very massive young star in emission nebula NGC 3372 and cluster Tr 16. Maximum brightness 1843. Since 1880 range has been 5.9–7.9, with brightening to around 5.2 in 1999
U Car	10 57.8	−59 44	DCEP	5.7–7.0	38.77	G	
ST UMa	11 27.8	+45 11	SRB	6.0–7.6	110?	M	
o¹ Cen	11 31.8	−59 27	SRD	4.7–5.5	200?	G	
Z UMa	11 56.5	+57 52	SRB	6.2–9.4	195.5	M	Usually double maxima and minima
SS Vir	12 25.3	+00 48	SRA	6.0–9.6	364.14	C	
R Vir	12 38.5	+06 59	M	6.1–12.1	145.63	M	Mean range 6.9–11.5
Y CVn	12 45.1	+45 26	SRB	5.2–6.6	157?	C	La Superba. Secondary period 2000 d
S Cru	12 54.4	−58 26	DCEP	6.2–6.9	4.69	G	
TU CVn	12 54.9	+47 12	SRB	5.6–6.6	50?	M	

VARIABLE STARS *(continued)*

STAR	RA 2000.0 h	m	DEC. °	′	TYPE	RANGE (mags)	PERIOD (d)	SPECTRAL TYPE	NOTES
FS Com	13	06.4	+22	37	SRB	5.3–6.1	58?	M	
SW Vir	13	14.1	−02	48	SRB	6.4–7.9	150?	M	
V CVn	13	19.5	+45	32	SRA	6.5–8.6	191.89	M	
R Hya	13	29.7	−23	17	M	3.5–10.9	388.87	M	Mean range 4.5–9.5; period shortening from about 500 d in 17th century
S Vir	13	33.0	−07	12	M	6.3–13.2	375.10	M	Mean range 7.0–12.7
V744 Cen	13	40.0	−49	57	SRB	5.1–6.6	90?	M	
T Cen	13	41.8	−33	36	SRA	5.5–9.0	90.44	K	
R CVn	13	49.0	+39	33	M	6.5–12.9	328.53	M	Mean range 7.7–11.9
W Hya	13	49.0	−28	22	SRA	6–9	361?	M	Amplitude and shape of light curve vary strongly
μ Cen	13	49.6	−42	28	GCAS	2.9–3.5	—	B2	
V767 Cen	13	53.9	−47	08	GCAS	5.9–6.3	—	B2	
V412 Cen	13	57.5	−57	43	LB	6.5–8.5	—	M	

CLUSTERS, NEBULAE AND GALAXIES

NGC	M	RA 2000.0 h m	DEC. ° ′	NOTES
3132	—	10 08	−40 26	Planetary nebula on Vela–Antlia border; 8th mag.
3242	—	10 25	−18 38	Planetary nebula in Hydra called the 'Ghost of Jupiter'; 9th mag.
3351	95	10 44	+11 42	} Pair of spiral galaxies in Leo, 10th and 9th mags. respectively
3368	96	10 47	+11 49	
3379	105	10 48	+12 35	9th-mag. elliptical galaxy in Leo
3532	—	11 06	−58 40	Naked-eye cluster in Carina covering 0°.9
3587	97	11 15	+55 01	Planetary nebula in Ursa Major, the Owl; large (3′) and faint (11th mag.), needs at least 75 mm aperture
3623	65	11 19	+13 05	} Pair of 9th-mag. spiral galaxies in Leo tilted at an angle to us
3627	66	11 20	+12 59	
3918	—	11 50	−57 11	Planetary nebula in Centaurus, known as the Blue Planetary; 8th mag.
4258	106	12 19	+47 18	Spiral galaxy in Canes Venatici; 8th mag., 17′ × 10′
4374	84	12 25	+12 53	Elliptical galaxy in Virgo; 9th mag.
4382	85	12 25	+18 11	9th-mag. elliptical galaxy in Coma
4406	86	12 26	+12 57	9th-mag. elliptical galaxy in Virgo
4472	49	12 30	+08 00	8th-mag. elliptical galaxy in Virgo
4486	87	12 31	+12 24	9th-mag. elliptical galaxy, centre of the Virgo Cluster of galaxies
4501	88	12 32	+14 25	10th-mag. spiral galaxy in Coma
4552	89	12 36	+12 33	10th-mag. elliptical galaxy in Virgo
4565	—	12 36	+25 59	10th-mag. spiral galaxy in Coma, seen edge-on
4569	90	12 37	+13 10	9th-mag. spiral galaxy in Virgo
4579	58	12 38	+11 49	10th-mag. spiral galaxy in Virgo
4590	68	12 40	−26 45	Globular cluster in Hydra; 8th mag.
4594	104	12 40	−11 37	Spiral galaxy in Virgo, the Sombrero, seen edge-on; 8th mag.
4621	59	12 42	+11 39	10th-mag. elliptical galaxy in Virgo
4649	60	12 44	+11 33	9th-mag. elliptical galaxy in Virgo
4736	94	12 51	+41 07	8th-mag. spiral galaxy in Canes Venatici
4826	64	12 57	+21 41	The Black Eye spiral galaxy in Coma; 9th mag.; the 'black eye' feature probably needs 150 mm aperture to be seen
5024	53	13 13	+18 10	Globular cluster in Coma; 8th mag.
5055	63	13 16	+42 02	9th-mag. spiral galaxy in Canes Venatici
5128	—	13 26	−43 01	Centaurus A, a large 7th-mag. elliptical galaxy; a strong radio source
5139	—	13 27	−47 29	ω Centauri, the largest and brightest globular cluster in the sky, mag. 3.7, diameter 0°.6
5194	51	13 30	+47 12	The Whirlpool Galaxy in Canes Venatici, an 8th-mag. spiral with a small satellite galaxy, NGC 5195, at the end of one arm
5236	83	13 37	−29 52	Face-on spiral galaxy in Hydra; 8th mag.
5272	3	13 42	+28 23	Globular cluster in Canes Venatici; 6th mag.

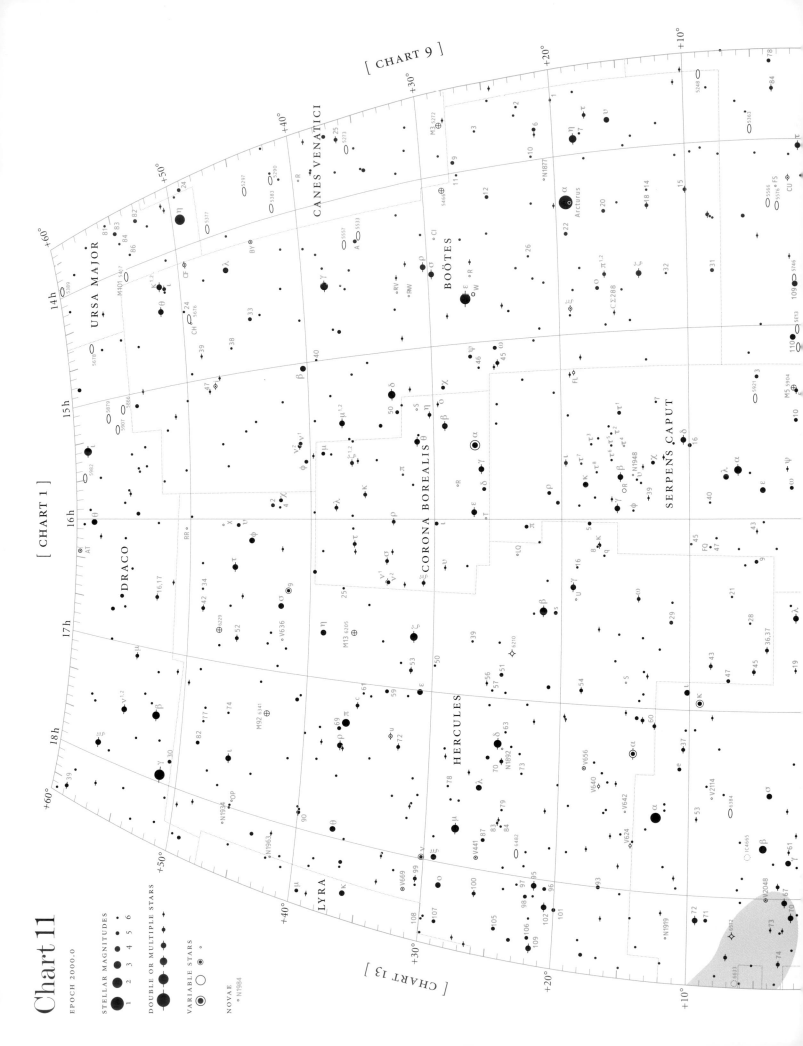

Chart 11

EPOCH 2000.0

STELLAR MAGNITUDES

DOUBLE OR MULTIPLE STARS

VARIABLE STARS

NOVAE

[CHART 1]

[CHART 9]

[CHART 13]

URSA MAJOR

CANES VENATICI

BOÖTES

DRACO

CORONA BOREALIS

SERPENS CAPUT

HERCULES

LYRA

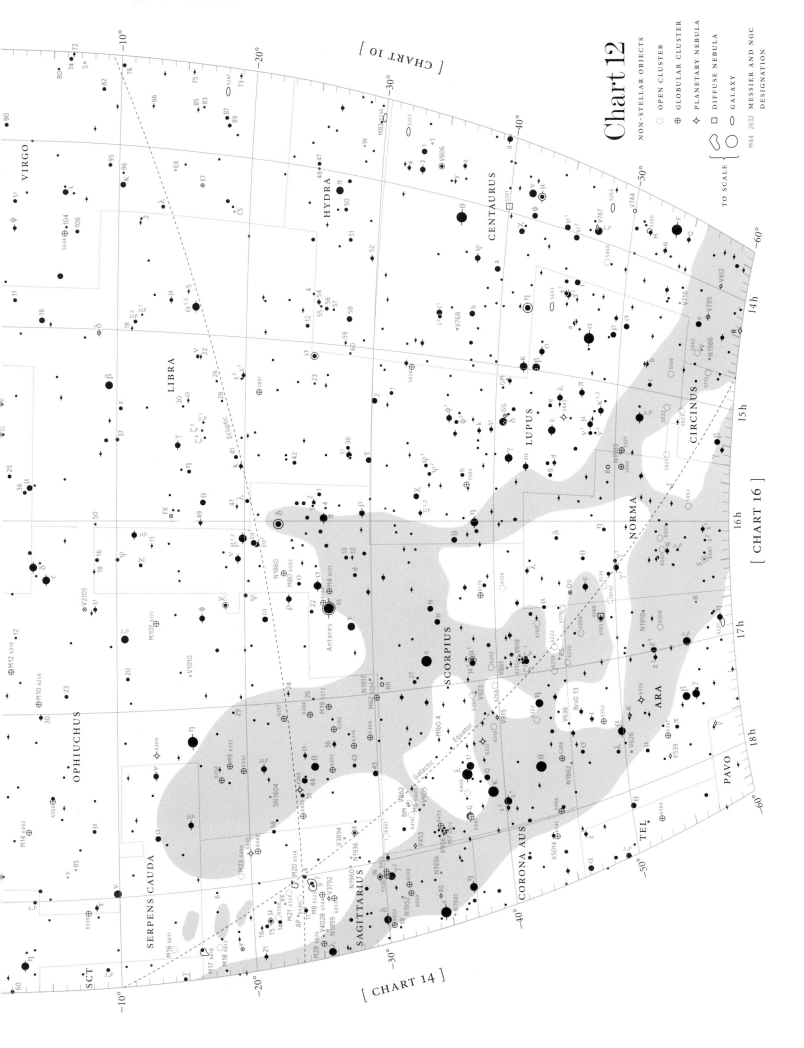

[CHART 10]

Chart 12

NON-STELLAR OBJECTS

☉ OPEN CLUSTER
⊕ GLOBULAR CLUSTER
✦ PLANETARY NEBULA
□ DIFFUSE NEBULA
◯ GALAXY

TO SCALE

M44 2632 MESSIER AND NGC
DESIGNATION

VIRGO

HYDRA

CENTAURUS

LIBRA

LUPUS

CIRCINUS

OPHIUCHUS

SCORPIUS

NORMA

SERPENS CAUDA

SCT

SAGITTARIUS

ARA

CORONA AUS

PAVO

TEL

Antares

Ecliptic

Equator

Galactic

[CHART 16]

[CHART 14]

DOUBLE STARS

ADS	STAR	RA 2000.0 h m	DEC. ° ′	MAGNITUDES		PA °	DIST. ″	NOTES
9085	τ Vir	14 01.6	+01 33	4.3	9.5	290	81.0	Optical; fixed
9173	κ Boo	14 13.5	+51 47	4.5	6.6	235	13.5	Optical; fixed
9198	ι Boo	14 16.2	+51 22	4.8	7.4	33	39.9	Fixed
9273	φ Vir	14 28.2	−02 14	4.9	10.0	112	5.3	Slowly widening. Test for 75 mm
9338	π Boo	14 40.7	+16 25	4.9	5.8	110	5.6	Little change
9343	ζ Boo	14 41.1	+13 44	4.5	4.6	296	0.7	Binary, 123 years[a]. Closing
9372	ε Boo	14 45.0	+27 04	2.6	4.8	344	2.8	PA increasing; yellowish, bluish. Test for 75 mm
9396	μ Lib	14 49.3	−14 09	5.6	6.6	3	1.9	Easy test for 75 mm
9406	39 Boo	14 49.7	+48 43	6.3	6.7	47	2.9	Slowly closing
9413	ξ Boo	14 51.4	+19 06	4.8	7.0	316	6.5	Binary, 152 years[a]. Yellow and orange
9425	OΣ288 Boo	14 53.4	+15 42	6.9	7.6	165	1.3	Binary, 300 years[a]
9494	44,i Boo	15 03.8	+47 39	5.2	var.	56	2.1	Binary, 206 years[a]. Closest in 2019 (0″.2)
	π Lup	15 05.1	−47 03	4.6	4.6	71	1.6	Opening, PA decreasing
	κ Lup	15 11.9	−48 44	3.8	5.5	143	26.5	Fixed
9532	ι¹ Lib	15 12.2	−19 47	4.5	9.9	109	57.3	Fixed. B is double, 10.4, 10.9; 14°, 2″.0
	μ Lup	15 18.5	−47 53	4.9	5.0	323	1.0	PA and distance of AB decreasing. C is physical
					6.3	129	23.9	
9584	5 Ser	15 19.3	+01 46	5.1	10.1	36	11.4	Little change. Near cluster M5. Mag. 9.2 star at 40°, 127″
9617	η CrB	15 23.2	+30 17	5.6	6.0	82	0.6	Binary, 42 years[a]
	γ Cir	15 23.4	−59 19	4.9	5.7	196	0.8	Binary, 270 years? Widening
9626	μ Boo	15 24.5	+37 23	4.3	6.6	170	107.1	μ² is binary, 257 years[a]; 7.1, 7.6; 9°, 2″.2
9701	δ Ser	15 34.8	+10 32	4.2	5.2	173	4.0	Binary, 3000 years?
	γ Lup	15 35.1	−41 10	3.5	3.6	276	0.8	Binary, 190 years[a]
9737	ζ CrB	15 39.4	+36 38	5.0	5.9	306	6.4	C.p.m., little change
	η Lup	16 00.1	−38 24	3.4	7.5	20	14.9	Fixed
9909	ξ Sco	16 04.4	−11 22	4.3	7.3	46	7.5	A is a close binary, 46 years[a]. In the same field is Σ1999, mags. 7.5, 8.1; 99°, 12″.0
9913	β Sco	16 05.4	−19 48	2.6	4.5	20	13.8	C.p.m. β¹ is binary, 600 years? Companion is mag. 10.6, at 171°, 0″.3
9951	ν Sco	16 12.0	−19 28	4.2	6.1	336	41.5	Both components are close doubles. ν¹: 4.4, 5.3; 2°, 1″.2; fixed. ν²: 6.6, 7.2; 53°, 2″.3; widening, PA increasing
9979	σ CrB	16 14.7	+33 52	5.6	6.5	238	7.1	Binary, 900 years[a]. Both yellow
10049	ρ Oph	16 25.6	−23 27	5.1	5.7	340	2.9	Closing, PA decreasing. Physical
10074	α Sco	16 29.4	−26 26	var.	5.4	277	2.8	Antares. Binary, 900 years? Red, green (by contrast)
10087	λ Oph	16 30.9	+01 59	4.2	5.2	31	1.6	Binary, 129 years[a]
10157	ζ Her	16 41.3	+31 36	3.0	5.4	0	0.8	Binary, 34.5 years[a]
10345	μ Dra	17 05.3	+54 28	5.7	5.7	16	2.2	Binary, 670 years[a]. Opening
10418	α Her	17 14.6	+14 23	var.	5.4	104	4.8	Binary, 3600 years? Reddish, greenish (by contrast)
10424	δ Her	17 15.0	+24 50	3.1	8.3	282	11.0	Optical; closing, PA increasing
	MlbO 4 Sco	17 19.0	−34 59	6.4	7.4	272	2.1	Binary, 42 years[a]. C is physical, mag. 10.8 at 138°, 32″.5
	BrsO 13 Ara	17 19.1	−46 38	5.6	8.9	250	8.7	Binary, 2000 years?
10526	ρ Her	17 23.7	+37 09	4.5	5.4	321	3.9	PA slowly increasing
10628	ν Dra	17 32.2	+55 11	4.9	4.9	312	62.1	Binocular pair, very wide and easy
10786	μ Her	17 46.5	+27 43	3.4	9.8	248	35.5	C.p.m. B is close binary, 43 years
10875	90 Her	17 53.3	+40 00	5.3	8.8	116	1.5	Little change; yellowish, bluish

[a] Orbital elements for these binaries are given in Table 53. PA and Dist. are from recent observations.

VARIABLE STARS

STAR	RA 2000.0 h m	DEC. ° ′	TYPE	RANGE (mags)	PERIOD (d)	SPECTRAL TYPE	NOTES
V716 Cen	14 13.7	−54 38	EB/KE	6.0–6.5	1.49	B5	
R Cen	14 16.6	−59 55	M	5.3–11.8	546.2	M	Double maxima (mean 5.8 and 6.0) and minima (11.1 and 8.3)
V Cen	14 32.5	−56 53	DCEP	6.4–7.2	5.49	G	Near open cluster NGC 5662
R Boo	14 37.2	+26 44	M	6.2–13.1	223.40	M	Mean range 7.2–12.3
RV Boo	14 39.3	+32 32	SRB	6.3–8.0	137?	M	
RW Boo	14 41.2	+31 34	SRB	6.4–7.9	209?	M	
W Boo	14 43.4	+26 32	SRB?	4.7–5.4	—	M	Periods of 30 and 450 d have been reported
δ Lib	15 01.0	−08 31	EA/SD	4.9–5.9	2.33	A0	
44,i Boo B	15 03.8	+47 39	EW/KW	5.8–6.4	0.27	G2 + G2	Secondary minimum 6.3. See Double stars

STAR	RA 2000.0 h m	DEC. ° ′	TYPE	RANGE (mags)	PERIOD (d)	SPECTRAL TYPE	NOTES
GG Lup	15 18.9	−40 47	EA/DM	5.6–6.1	1.85	B7	Secondary minimum 5.8
S CrB	15 21.4	+31 22	M	5.8–14.1	360.26	M	Mean range 7.3–12.9
R Nor	15 36.0	−49 30	M	5–12	507.50	M	Double maxima and minima
τ⁴ Ser	15 36.5	+15 06	SRB	5.9–7.1	100?	M	
T Nor	15 44.1	−54 59	M	6.2–13.6	240.7	M	Mean range 7.4–13.2
R CrB	15 48.6	+28 09	RCB	5.7–14.8	—	C	Major fades in 1962, 1972 and 1977
R Ser	15 50.7	+15 08	M	5.2–14.4	356.41	M	Mean range 6.9–13.4
T CrB	15 59.5	+25 55	NR	2.0–10.8	—	M3 + Pec	The Blaze Star. Outbursts in 1866, 1946
δ Sco	16 00.3	−22 37	GCAS	1.6–2.3	—	B0	First observed outburst in 2000
X Her	16 02.7	+47 14	SRB	6.3–7.4	95.0	M	Secondary period 746 d
RR Her	16 04.2	+50 30	SRB	6–10	239.7	C	
AT Dra	16 17.3	+59 45	LB	5.3–6.0	—	M	
S Nor	16 18.9	−57 54	DCEP	6.1–6.8	9.75	G	In centre of open cluster NGC 6087
U Her	16 25.8	+18 54	M	6.4–13.4	406.1	M	Mean range 7.5–12.5
χ Oph	16 27.0	−18 27	GCAS	4.2–5.0	—	B2	
30,g Her	16 28.6	+41 53	SRB	4.3–6.3	89.2	M	Also slow pulsations (period about 2.4 years) and irregular variations
α Sco	16 29.4	−26 26	LC	0.9–1.2	—	M	Antares. See Double stars
R Ara	16 39.7	−57 00	EA/DM?	6.0–6.9	4.43	B9	
V1010 Oph	16 49.5	−15 40	EB/KE	6.1–7.0	0.66	A5	Secondary minimum 6.5
S Her	16 51.9	+14 56	M	6.4–13.8	307.28	M	Mean range 7.6–12.6
RS Sco	16 55.6	−45 06	M	6.2–13.0	319.91	M	Mean range 7.0–12.2
RR Sco	16 56.6	−30 35	M	5.0–12.4	281.45	M	Mean range 5.9–11.8
κ Oph	16 57.7	+09 22	LB?	2.8–3.6	—	K	
V915 Sco	17 14.5	−36 03	?	6.2–6.6	—	G5	Faded between 1978 and 1979
α Her	17 14.6	+14 23	SRC	2.7–4.0	—	M	Slow variations (period about 6 years) and quicker changes (over about 100 d). See Double stars
U Oph	17 16.5	+01 13	EA/DM	5.8–6.6	1.68	B5 + B5	Secondary minimum 6.5
68,u Her	17 17.3	+33 06	EB/SD	4.7–5.4	2.05	B2 + B5	
V636 Sco	17 22.8	−45 37	DCEP	6.4–6.9	6.80	G	
V862 Sco	17 40.0	−32 12	GCAS?	2–8.5	—	B	In open cluster M6. Flare of 40 min on 1965 July 3
BM Sco	17 41.0	−32 13	SRD	5.0–6.9	815?	K	Brightest star in open cluster M6
V Pav	17 43.3	−57 43	SRB	6.3–8.2	225.4	C	Secondary period 3735 d
X Sgr	17 47.6	−27 50	DCEP	4.2–4.9	7.01	G	
RS Oph	17 50.2	−06 43	NR	4.3–12.5	—	OB + M	Outbursts in 1898, 1933, 1958, 1967, 1985
V539 Ara	17 50.5	−53 37	EA/DM	5.7–6.2	3.17	B2 + B3	
Y Oph	17 52.6	−06 90	DCEPS	5.9–6.5	17.12	G	
OP Her	17 56.8	+45 21	SRB	5.9–6.7	120.5	M	

CLUSTERS, NEBULAE AND GALAXIES

NGC	M	RA 2000.0 h m	DEC. ° ′	NOTES
5457	101	14 03	+54 21	Spiral galaxy in Ursa Major, seen face-on; 8th mag.
5460	—	14 08	- 48 19	Open cluster in Centaurus; 6th mag.
5822	—	15 05	−54 21	Open cluster in Lupus; rich, with faint stars; 0°.6 across
5904	5	15 19	+02 05	Globular cluster in Serpens; 6th mag.
6067	—	16 13	−54 13	Open cluster in Norma; 6th mag.
6093	80	16 17	−22 59	Globular cluster in Scorpius; 7th mag.
6121	4	16 24	−26 32	Globular cluster in Scorpius; large (0°.4) but low surface brightness
6193	—	16 41	−48 46	Open cluster in Ara; 5th mag.
6205	13	16 42	+36 28	Globular cluster in Hercules, the finest in northern skies, easy in binoculars; 6th mag., 0°.25 across
6210	—	16 45	+23 49	Planetary nebula in Hercules; 9th mag.
6218	12	16 47	−01 57	} Pair of 7th-mag. globular clusters in Ophiuchus, each covering 0°.25
6254	10	16 57	−04 06	
6231	—	16 54	−41 48	Outstanding cluster in Scorpius for small telescopes, brightest stars of 5th mag.
6266	62	17 01	−30 07	} Pair of 7th-mag. globular clusters in Ophiuchus
6273	19	17 03	−26 16	
6341	92	17 17	+43 08	Globular cluster in Hercules, smaller and fainter than M13
6333	9	17 19	−18 31	Globular cluster in Ophiuchus; 8th mag.
6402	14	17 38	−03 15	Globular cluster in Ophiuchus; 8th mag.
6405	6	17 40	−32 13	Open cluster in Scorpius, 4th mag., 0°.25, good binocular object
6397	—	17 41	−53 40	6th-mag. globular cluster in Ara, 0°.4 diameter, scattered stars
IC 4665	—	17 46	+05 43	Loose binocular cluster in Ophiuchus
6475	7	17 54	−34 49	Outstanding naked-eye cluster in Scorpius, over 1° across, an excellent binocular pairing with M6
6494	23	17 57	−19 01	6th-mag. cluster of faint stars in Sagittarius covering nearly 0°.5

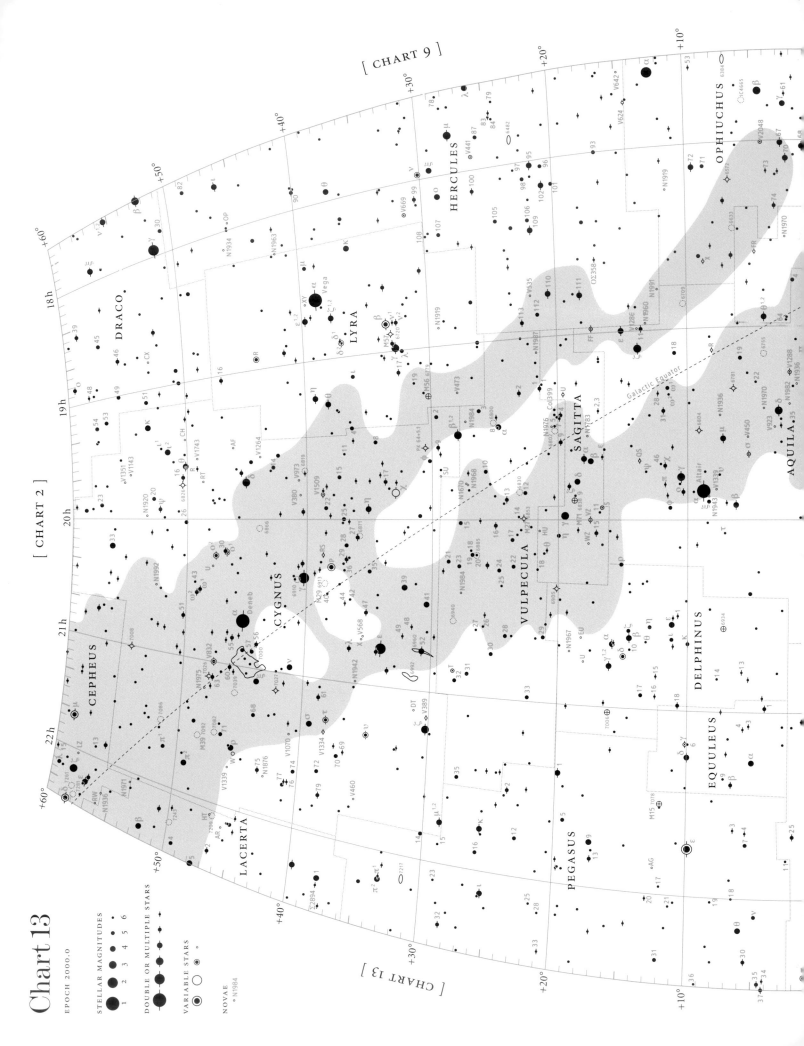

Interesting Objects, Charts 13 and 14

RA 18h to 22h, Dec. +60° to −60°

DOUBLE STARS

ADS	STAR	RA 2000.0 h m	DEC. 2000.0 ° ′	MAGNITUDES		PA °	DIST. ″	NOTES
11005	τ Oph	18 03.1	−08 11	5.3	5.9	282	1.6	Binary, 260 years[a]. Closing
11046	70 Oph	18 05.5	+02 30	4.2	6.2	143	4.4	Binary, 88 years[a]. Easy yellow and orange pair
	h5014 CrA	18 06.8	−43 25	5.7	5.7	8	1.7	Binary, 450 years[a]
	κ CrA	18 33.4	−38 44	5.6	6.2	358	21.4	Optical; little change
11483	OΣ358 Her	18 35.9	+16 59	6.9	7.1	154	1.6	Binary, 380 years[a]
11635	ε Lyr	18 44.3	+39 40	4.7	4.6	174	210.5	Fixed naked-eye pair. Both binary: ε¹ 5.0, 6.1; 350°, 2″.4; 1200 years[a]. ε² 5.3, 5.4; 82°, 2″.3; 700 years[a]
11639	ζ Lyr	18 44.8	+37 36	4.3	5.6	150	44.0	Fixed. Very easy pair
11745	β Lyr	18 50.1	+33 22	var.	6.7	149	46.0	Fixed; cream and blue
11853	θ Ser	18 56.2	+04 12	4.6	4.9	104	22.5	Fixed. Fine, easy pair. Mag. 7.9 star at 58°, 26″.0
	γ CrA	19 06.4	−37 04	4.5	6.4	62	1.3	Binary, 120 years[a]
12197	η Lyr	19 13.8	+39 09	4.4	8.6	80	28.4	Fixed. Fine low-power field
12540	β Cyg	19 30.7	+27 58	3.4	4.7	54	34.6	Albireo. Fixed. Glorious pair: amber, greenish (by contrast)
12880	δ Cyg	19 45.0	+45 08	2.9	6.3	224	2.6	Binary, 800 years[a]. Test for 100 mm
12962	π Aql	19 48.7	+11 49	6.3	6.8	107	1.4	Slow decrease of PA. Test for 75 mm
13148	ψ Cyg	19 55.6	+52 26	5.0	7.5	176	2.8	Slowly closing with decrease of PA
13442	θ Sge	20 09.9	+20 55	6.6	8.9	331	11.5	Slowly closing. Mag. 7.5 star at 222°, 89″.2
13632	α¹ Cap	20 17.6	−12 30	4.2	9.6	221	46.0	Optical; widening
13645	α² Cap	20 18.1	−12 33	3.8	10.6	172	6.6	Fainter component is itself double: 11.2, 11.5; 243°, 1″.3; fixed. Naked-eye pair with α¹ at 292°, 381″; optical
13765	γ Cyg	20 22.2	+40 15	2.2	9.6	196	41.2	Optical; fixed. B is itself double: 10.0, 11.0; 302°, 1″.9; fixed
	κ² Sgr	20 23.9	−42 25	5.9	7.3	268	0.4	Binary, 700 years?
13887	ρ Cap	20 28.9	−17 49	5.0	6.9	194	1.3	Binary, 300 years?
14158	49 Cyg	20 41.0	+32 18	5.8	8.1	47	3.1	Fixed; yellowish, bluish
14259	52 Cyg	20 45.7	+30 43	4.2	8.7	69	6.4	Little change. In nebula NGC 6960
14279	γ Del	20 46.7	+16 07	4.4	5.0	266	9.2	Binary, 3200 years; yellow, greenish (by contrast). Σ2725, mags. 7.5, 8.2; 11°, 6″.0, in the same field of view
14296	λ Cyg	20 47.4	+36 29	4.7	6.3	9	0.9	Binary, 400 years[a]
14360	4 Aqr	20 51.4	−05 38	6.4	7.4	21	0.9	Binary, 194 years[a]
14499	1 Equ	20 59.1	+04 18	5.4	7.1	67	10.4	Physical. A is close binary, 101 years[a]
14636	61 Cyg	21 06.9	+38 45	5.4	6.1	150	31.1	Binary, 660 years[a]. Both orange. Large p.m.
14787	τ Cyg	21 14.8	+38 03	3.8	6.6	287	0.7	Binary, 50 years[a]
15270	μ Cyg	21 44.1	+28 45	4.8	6.2	312	2.0	Binary, 800 years[a]
15281	κ Peg	21 44.6	+25 39	4.1	10.8	291	14.2	Optical. A is close binary, 11.6 years

[a] Orbital elements for these binaries are given in Table 53. PA and Dist. are from recent observations.

VARIABLE STARS

STAR	RA 2000.0 h m	DEC. 2000.0 ° ′	TYPE	RANGE (mags)	PERIOD (d)	SPECTRAL TYPE	NOTES
W Sgr	18 05.0	−29 35	DCEP	4.3–5.1	7.60	G	
VX Sgr	18 08.1	−22 13	SRC	6.5–14.0	732?	M	
V3792 Sgr	18 08.9	−25 28	EB/DM	6.4–6.9	2.25	B5	
AP Sgr	18 13.0	−23 07	DCEP	6.5–7.4	5.06	G	
RS Sgr	18 17.6	−34 06	EA/SD	6.0–7.0	2.42	B3 + A	Secondary minimum 6.3
Y Sgr	18 21.4	−18 52	DCEP	5.3–6.2	5.77	G	
U Sgr	18 31.9	−19 07	DCEP	6.3–7.2	6.75	G	In open cluster M25
XY Lyr	18 38.1	+39 40	LC	5.8–6.4	—	M	
X Oph	18 38.3	+08 50	M	5.9–9.2	328.85	M	Mean range 6.8–8.8
V3879 Sgr	18 42.9	−19 17	SRB	6.1–6.6	50?	M	
R Sct	18 47.5	−05 42	RVA	4.2–8.6	146.5	K	
β Lyr	18 50.1	+33 22	EB	3.3–4.4	12.94	B8	Secondary minimum 3.9. See Double stars
R Lyr	18 55.3	+43 57	SRB	3.9–5.0	46?	M	
FF Aql	18 58.2	+17 22	DCEPS	5.2–5.7	4.47	F	
R Aql	19 06.4	+08 14	M	5.5–12.0	284.2	M	Mean range 6.1–11.5; period shortening
TT Aql	19 08.2	+01 18	DCEP	6.5–7.7	13.75	G	
RY Sgr	19 16.5	−33 31	RCB	5.8–14.0	—	G	Pulsations up to 1.5 mag. in period 38 d
U Sge	19 18.8	+19 37	EA/SD	6.5–9.3	3.38	B8 + G2	
CH Cyg	19 24.5	+50 14	ZAND+SR	5.6–8.5	—	M + B	Pulsations with period 97 d, secondary period 4700 d; also cycle of 725 d, flares and eclipses
U Aql	19 29.4	−07 03	DCEP	6.1–6.9	7.02	F	
AF Cyg	19 30.2	+46 09	SRB	6.4–8.4	92.5	M	Secondary periods 175.8 and 941.2 d

VARIABLE STARS *(continued)*

STAR	RA 2000.0 h m	DEC. ° ′	TYPE	RANGE (mags)	PERIOD (d)	SPECTRAL TYPE	NOTES
AQ Sgr	19 34.3	−16 22	SRB	6–8	199.6	C	
R Cyg	19 36.8	+50 12	M	6.1–14.4	426.45	S	Mean range 7.5–13.9
V1143 Cyg	19 38.7	+54 58	EA/DM	5.9–6.4	7.64	F5 + F5	
RT Cyg	19 43.6	+48 47	M	6.0–13.1	190.28	M	Mean range 7.3–11.8
V973 Cyg	19 44.8	+40 43	SRB	6.2–7.0	40?	M	
SU Cyg	19 44.8	+29 16	DCEP	6.4–7.2	3.85	G + B7	
χ Cyg	19 50.6	+32 55	M	3.3–14.2	408.05	S	Mean range 5.2–13.4
η Aql	19 52.5	+01 00	DCEP	3.5–4.4	7.18	G	
V505 Sgr	19 53.1	−14 36	EA/SD	6.5–7.5	1.18	A2 + F6	
RR Sgr	19 55.9	−29 11	M	5.4–14.0	336.33	M	Mean range 6.8–13.2
S Sge	19 56.0	+16 38	DCEP	5.2–6.0	8.38	G	
RU Sgr	19 58.7	−41 51	M	6.0–13.8	240.49	M	Mean range 7.2–12.8
HU Sge	20 03.7	+21 30	LB	6.3–7.3	—	M	
RS Cyg	20 13.4	+38 44	SRA	6.5–9.5	417.39	C	Shape of light curve varies strongly, maxima sometimes double
RT Cap	20 17.2	−21 20	SRB	6–9	393?	C	
RT Sgr	20 17.7	−39 07	M	6.0–14.1	306.46	M	Mean range 7.0–13.3
P Cyg	20 17.8	+38 02	SDOR	3–6	—	B1	'Nova' of 1600. Since 18th century range has been 4.6–5.6
U Cyg	20 19.6	+47 54	M	5.9–12.1	463.24	C	Mean range 7.2–10.7
EU Del	20 37.9	+18 16	SRB	5.8–6.9	59.7	M	
X Cyg	20 43.4	+35 35	DCEP	5.9–6.9	16.39	G	
U Del	20 45.5	+18 05	SRB	5.6–7.5	110?	M	Mean brightness varies with a period of *c.*1100 d
T Vul	20 51.5	+28 15	DCEP	5.4–6.1	4.44	G	
T Ind	21 20.2	−45 01	SRB	5–6.5	320?	C	
V1070 Cyg	21 22.8	+40 56	SRB	6.5–8.5	73.5	M	
W Cyg	21 36.0	+45 22	SRB	5.0–7.6	131.1	M	Secondary period 235.3 d
V460 Cyg	21 42.0	+35 31	SRB	5.6–7.0	180?	C	
V1339 Cyg	21 42.1	+45 46	SRB	5.9–7.1	35?	M	
μ Cep	21 43.5	+58 47	SRC	3.4–5.1	730?	M	The Garnet Star. Secondary period 4400 d
ε Peg	21 44.2	+09 52	LC	0.7–3.5	—	K	Normally 2.3–2.4; unconfirmed flare 1972 Sept. 26/27
EP Aqr	21 46.5	−02 13	SRB	6.4–6.8	55?	M	
AG Peg	21 51.0	+12 38	NC	6.0–9.4	—	W + M3	Maximum in 1870. In recent years range has been 8.0–8.6 in period of about 800 d

CLUSTERS, NEBULAE AND GALAXIES

NGC	M	RA 2000.0 h m	DEC. ° ′	NOTES
6514	20	18 03	−23 02	Trifid Nebula in Sagittarius, with embedded stars; 9th mag.
6523	8	18 04	−24 23	Lagoon Nebula in Sagittarius, 1°.5 × 0°.6, containing star cluster NGC 6530
6531	21	18 05	−22 30	6th-mag. open cluster in Sagittarius, in same field as M20
6572	—	18 12	+06 51	Planetary nebula in Ophiuchus; 9th mag.
6603	24	18 18	−18 30	Small, faint star cluster in rich Milky Way star field in Sagittarius
6611	16	18 19	−13 47	Open cluster in Serpens; appears hazy since it is embedded in the Eagle Nebula, which shows up well in photographs
6613	18	18 20	−17 08	7th-mag. cluster in Sagittarius, in same binocular field as M17
6618	17	18 21	−16 11	Omega Nebula in Sagittarius, noticeably elongated, containing a star cluster
6633	—	18 28	+06 34	5th-mag. star cluster in Ophiuchus
IC 4725	25	18 32	−19 15	5th-mag. open cluster in Sagittarius, containing U Sgr
6656	22	18 36	−23 54	Globular cluster in Sagittarius, large (0°.4) and bright (5th mag.); excellent binocular object
6705	11	18 51	−06 16	Wild Duck Cluster in Scutum, 5th mag., fan-shaped; superb in all apertures
6720	57	18 54	+33 02	Ring Nebula in Lyra, elliptical 9th-mag. planetary nebula, 1′ across
6715	54	18 55	−30 29	8th-mag. globular cluster in Sagittarius
6779	56	19 17	+30 11	8th-mag. globular cluster in Lyra
—	—	19 25	+20 11	Collinder 399 in Vulpecula, also known as Brocchi's Cluster, consisting of ten stars in the shape of a coathanger
6809	55	19 40	−30 58	7th-mag. globular cluster in Sagittarius
6826	—	19 45	+50 31	Planetary nebula in Cygnus, called the 'blinking planetary' because it seems to wink in and out of view; 9th mag.
6838	71	19 54	+18 47	8th-mag. globular cluster in Sagitta
6853	27	20 00	+22 43	The Dumbbell Nebula, planetary nebula in Vulpecula; visible in binoculars as a rounded haze; telescopes show its twin lobes; 8th mag., 6′ long
6992	—	20 56	+31 43	Brightest part of the Veil Nebula supernova remnant in Cygnus, visible in binoculars under dark skies
7000	—	21 00	+44 20	North America Nebula in Cygnus; 2° long, but with low surface brightness; visible in binoculars under dark skies
7009	—	21 04	−11 22	Saturn Nebula, 8th-mag. planetary nebula in Aquarius; large telescopes are needed to show the Saturn-like shape
7078	15	21 30	+12 10	Globular cluster in Pegasus; 6th mag.
7092	39	21 32	+48 26	Open cluster in Cygnus; 5th mag.; stars thinly scattered over 0°.5
7089	2	21 34	−00 49	Globular cluster in Aquarius; stars densely packed; mag. 6.5
7099	30	21 40	−23 11	8th-mag. globular cluster in Capricornus

Chart 15

EPOCH 2000.0

STELLAR MAGNITUDES

1 2 3 4 5 6

DOUBLE OR MULTIPLE STARS

VARIABLE STARS

NOVAE
° N1984

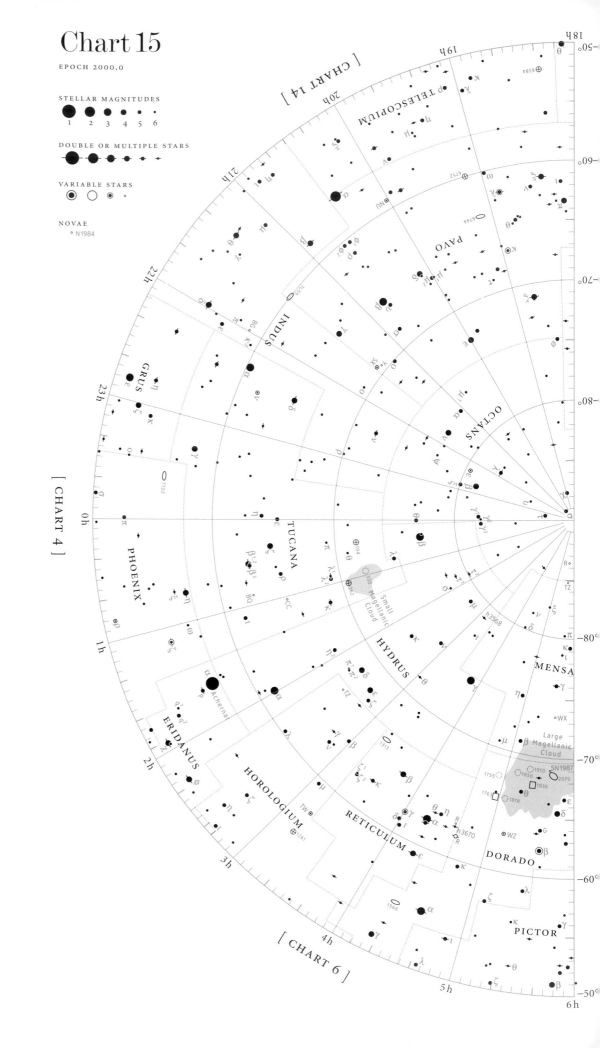

[CHART 14]

[CHART 4]

[CHART 6]

TELESCOPIUM

PAVO

INDUS

GRUS

OCTANS

TUCANA

PHOENIX

Small Magellanic Cloud

HYDRUS

MENSA

Large Magellanic Cloud

SN1987

ERIDANUS

Achernar

HOROLOGIUM

RETICULUM

DORADO

PICTOR

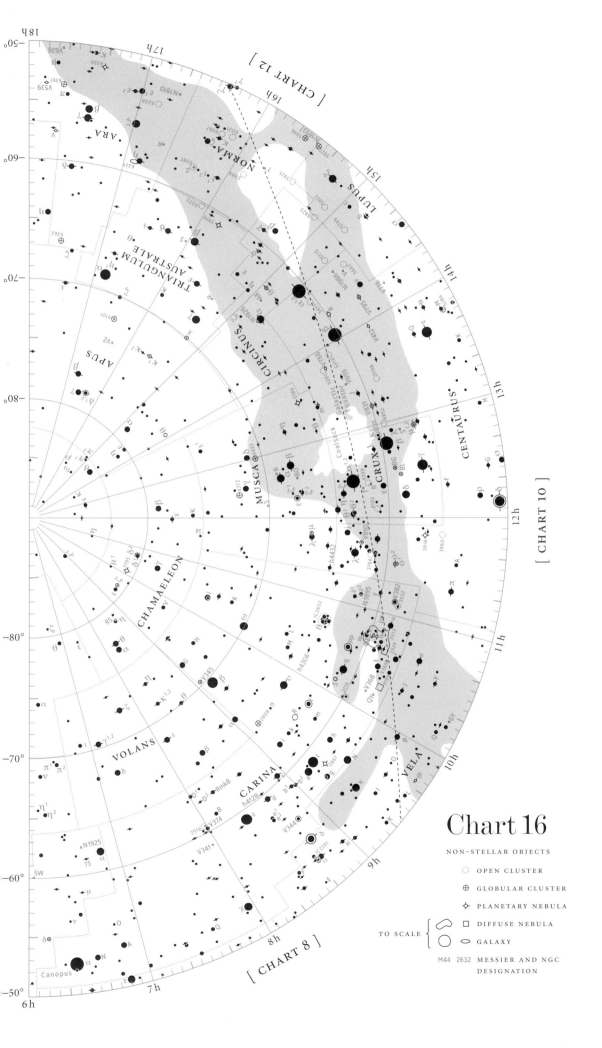

Chart 16

NON-STELLAR OBJECTS

⊙ OPEN CLUSTER

⊕ GLOBULAR CLUSTER

✧ PLANETARY NEBULA

☐ DIFFUSE NEBULA

◯ GALAXY

TO SCALE

M44 2632 MESSIER AND NGC
DESIGNATION

Interesting Objects, Charts 15 and 16

Dec. −60° to −90°

DOUBLE STARS

STAR	RA 2000.0 h m	DEC. ° ′	MAGNITUDES		PA °	DIST. ″	NOTES
β¹,² Tuc	00 31.5	−62 58	4.3	4.5	169	27.0	C.p.m. Both double. β¹: 4.4, 13.5; 153°, 2″.6; fixed. β²: 4.6, 6.5; 277°, 0″.6; binary, 45 years, closing
λ¹ Tuc	00 52.4	−69 30	6.7	7.4	81	20.5	Optical; little change
κ Tuc	01 15.8	−68 53	5.0	7.7	322	5.0	Binary, 1200 years?
h3568 Hyi	03 07.5	−78 59	5.7	7.7	224	15.3	Fixed
θ Ret	04 17.7	−63 15	6.0	7.7	2	4.3	Slowly closing
h3670 Ret	04 33.6	−62 49	5.9	9.3	100	31.8	Optical; fixed
I 5 Pic	06 38.0	−61 32	6.3	8.8	255	0.3	Closing
γ Vol	07 08.8	−70 30	3.9	5.4	298	14.1	Physical pair, but fixed; gold and cream
ε Vol	08 07.9	−68 37	4.4	7.3	24	6.1	PA decreasing
Rmk 8	08 15.3	−62 55	5.3	7.6	69	4.0	Little change
θ Vol	08 39.1	−70 23	5.2	10.3	105	42.0	Optical; little change
h4128 Car	08 39.2	−60 19	6.8	7.5	202	1.2	Slowly closing; PA decreasing
υ Car	09 47.1	−65 04	3.0	6.0	129	5.0	Fixed
h4306 Car	10 19.1	−64 41	6.3	6.5	133	2.4	Slowly widening
h4432 Mus	11 23.4	−64 57	5.4	6.6	308	2.4	PA increasing slowly
ε Cha	11 59.6	−78 13	5.3	6.0	211	0.4	Closing, PA increasing
α Cru	12 26.6	−63 06	1.3	1.6	114	3.9	Slowly closing. Very easy. C is mag. 4.8 at 202°, 90″. All three stars are c.p.m.
ι Cru	12 45.6	−60 59	4.7	10.2	8	28.1	Opening with decrease of PA
β Mus	12 46.3	−68 06	3.5	4.0	40	1.1	Binary, 400 years?
θ Mus	13 08.1	−65 18	5.7	7.6	187	5.3	Slowly widening. B is a Wolf–Rayet star
J Cen	13 22.6	−60 59	4.5	6.2	345	60.8	Fixed. Wide, easy pair.
β Cen	14 03.8	−60 22	0.6	4.0	234	0.9	PA slowly decreasing
α Cen	14 39.6	−60 50	0.0	1.3	224	13.3	Superb binary, 80 years[a]. Both yellow
α Cir	14 42.5	−64 59	3.2	8.5	227	15.6	Separation increasing, PA decreasing
ι TrA	16 28.0	−64 03	5.3	9.4	11	18.5	Optical pair; PA slowly decreasing
ξ Pav	18 23.2	−61 30	4.4	8.1	156	3.4	Little change
λ Oct	21 50.9	−82 43	5.6	7.3	62	3.3	Little change
δ Tuc	22 27.3	−64 58	4.5	8.7	281	7.0	Slowly closing

[a] *Orbital elements for these binaries are given in Table 53. PA and Dist. are from recent observations.*

VARIABLE STARS

STAR	RA 2000.0 h m	DEC. ° ′	TYPE	RANGE (mags)	PERIOD (d)	SPECTRAL TYPE	NOTES
R Ret	04 33.5	−63 02	M	6.5–14.2	278.46	M	Mean range 7.6–13.3
R Dor	04 36.8	−62 05	SRB	4.8–6.6	338?	M	
R Oct	05 26.1	−86 23	M	6.4–13.2	405.39	M	Mean range 7.9–12.4
TZ Men	05 30.2	−84 47	EA/D	6.2–6.9	8.57	A1 + B9	
β Dor	05 33.6	−62 29	DCEP	3.5–4.1	9.84	G	
RS Cha	08 43.2	−79 04	EA + DSCT	6.0–6.7	1.67	A5 + A7	Secondary minimum 6.5
R Car	09 32.2	−62 47	M	3.9–10.5	308.71	M	Mean range 4.6–9.6
l Car	09 45.2	−62 30	DCEP	3.3–4.2	35.54	G	
S Car	10 09.4	−61 33	M	4.5–9.9	149.49	M	Mean range 5.7–8.5
S Mus	12 12.8	−70 09	DCEP	5.9–6.5	9.66	G	
T Cru	12 21.4	−62 17	DCEP	6.3–6.8	6.73	G	Near open cluster NGC 4349
R Cru	12 23.6	−61 38	DCEP	6.4–7.2	5.83	G	Near open cluster NGC 4349
BO Mus	12 34.9	−67 45	LB	5.9–6.6	—	M	
R Mus	12 42.1	−69 24	DCEP	5.9–6.7	7.51	G	
V766 Cen	13 47.2	−62 35	SDOR?	6.2–7.5	—	G8	
θ Aps	14 05.3	−76 48	SRB	5–7	119?	M	
AX Cir	14 52.6	−63 49	DCEP	5.7–6.1	5.27	G + B4	
θ Cir	14 56.7	−62 47	GCAS	5.0–5.4	—	B3	
X TrA	15 14.3	−70 05	LB	5.0–6.4	—	C	
R TrA	15 19.8	−66 30	DCEP	6.3–7.0	3.39	G	
S TrA	16 01.2	−63 47	DCEP	6.0–6.8	6.32	G	
VZ Aps	16 16.3	−74 02	M	6–15	385?	M	
κ Pav	18 56.9	−67 14	CEP	3.9–4.8	9.09	G	
Y Pav	21 24.3	−69 44	SRB	5.6–7.3	233.3	C	
SX Pav	21 28.7	−69 30	SRB	5.3–6.0	50?	M	
ε Oct	22 20.0	−80 26	SRB	4.6–5.3	55?	M	

CLUSTERS, NEBULAE AND GALAXIES

NGC	RA 2000.0 h m	DEC. ° ′	NOTES
104	00 24	−72 05	Globular cluster 47 Tucanae; 4th mag., diameter 0°.5, the second most prominent globular, after ω Centauri
362	01 03	−70 51	7th-mag. globular cluster in Tucana, near the Small Magellanic Cloud but not part of it
2070	05 39	−69 06	Tarantula Nebula, also called 30 Doradus, in the Large Magellanic Cloud, with embedded stars; 0°.5 across, visible to the naked eye
2516	07 58	−60 52	Rich naked-eye cluster in Carina, 0°.5 across; brightest star 5th mag.
2808	09 12	−64 52	6th-mag. globular cluster in Carina; large, with bright centre
3114	10 03	−60 07	4th-mag. open cluster in Carina
IC 2602	10 43	−64 24	Naked-eye cluster nearly 1° across centred on 3rd-mag. θ Carinae
3372	10 45	−59 50	Naked-eye nebula around the star η Carinae; 2° wide with embedded stars; contains dark nebula called the Keyhole
3766	11 36	−61 37	5th-mag. open cluster in Centaurus
—	12 50	−63 00	The Coalsack, a dark nebula in front of the Milky Way in Crux, 6°.5 × 5°
4755	12 54	−60 20	The Jewel Box cluster, a collection of glittering coloured stars including the 6th-mag. blue supergiant κ Crucis
4833	13 00	−70 53	7th-mag. globular cluster in Musca
6025	16 04	−60 30	5th-mag. open cluster in Triangulum Australe
6752	19 11	−59 59	Large 5th-mag. globular cluster in Pavo

Chart 17

GALACTIC CHART

LONGITUDE: 0° TO 180°; LATITUDE: 50°N TO 50°S

MAGNITUDES:

1 2 3 4

The shaded area represents the general position of the Milky Way.

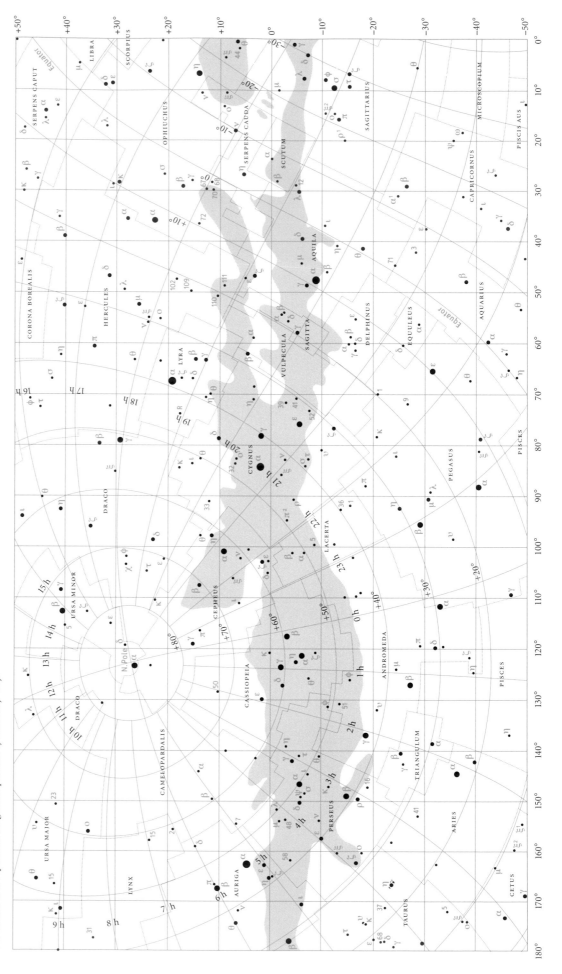

Chart 18

GALACTIC CHART

LONGITUDE: 180° TO 360°; LATITUDE: 50°N TO 50°S

MAGNITUDES:

1 2 3 4

The shaded area represents the general position of the Milky Way.

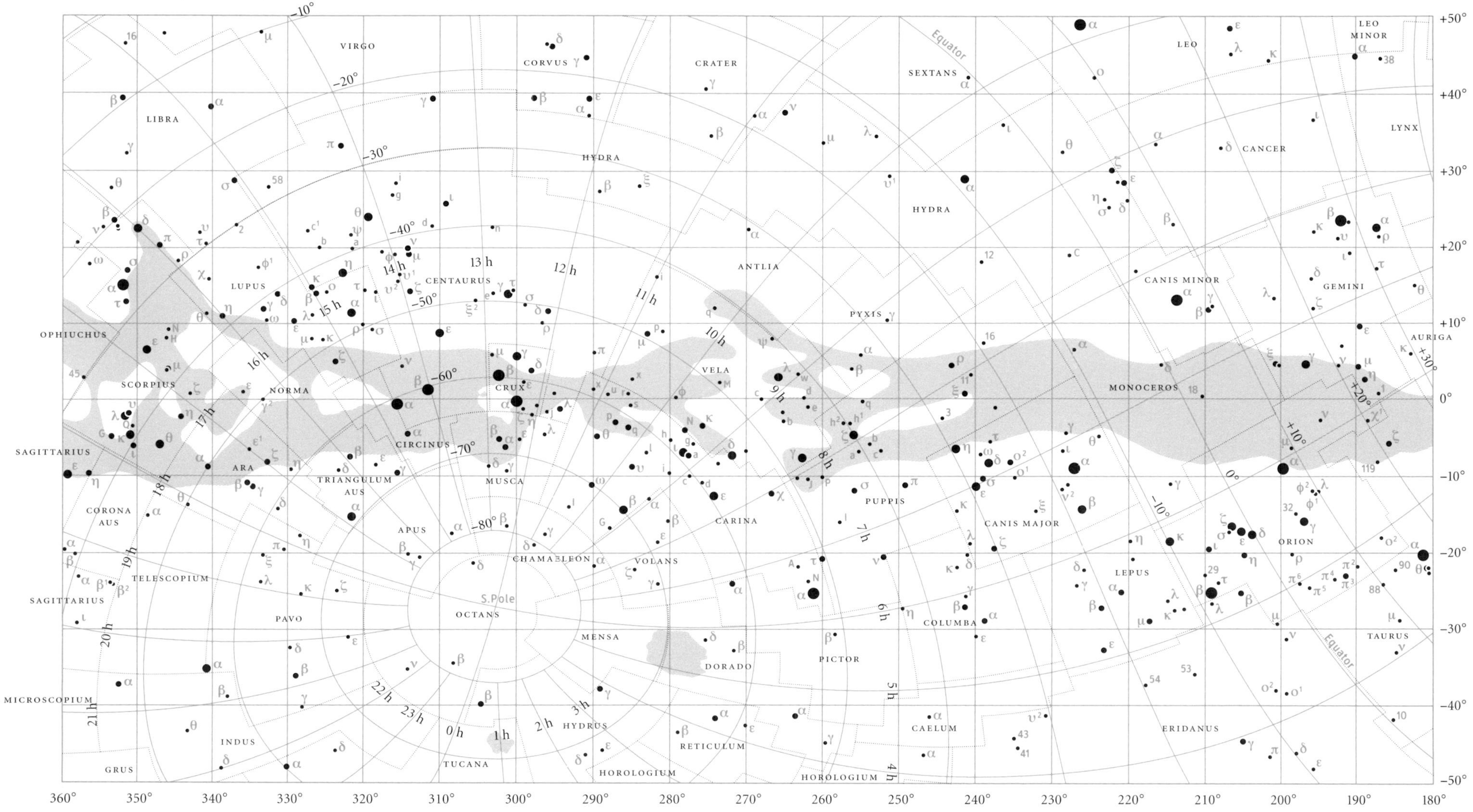

Appendix

Units and Notation

The International System of Units (SI) has been established by worldwide agreement as a common standard for all scientific disciplines. Astronomers, however, still work with a mixture of incompatible units, many of them unique to astronomy.

The International Astronomical Union (IAU) is responsible for setting standards within astronomy. At its General Assembly in 1988, the IAU strongly urged that only SI units should be used, together with a few other units recognized for use in astronomy. The units and notation described in this section follow the recommendations of the IAU and the Royal Society, tempered by the conventions of current usage.

THE INTERNATIONAL SYSTEM OF UNITS

In the SI system there are seven *base units* and two dimensionless *supplementary units*, listed in Table 61.

For every physical quantity there is an SI *derived unit* that can be formed from simple combinations of the base and supplementary units. For example, the SI unit of area is the square metre, written m^2, and the unit of velocity is the metre per second, written either m/s or $m\,s^{-1}$ (the latter is preferred). A small number of derived units are given special names and symbols. The SI unit of force, for example, is the newton (N), defined as $1\,N = 1\,kg\,m\,s^{-2}$.

Any SI unit can be modified by a prefix indicating a decimal multiple of the unit (Table 62). The prefixes may also be attached to certain units which are not part of SI. Examples are $10^{-9}\,m = 1$ nanometre = $1\,nm$; $10^3\,m = 1$ kilometre = $1\,km$; $10^6\,pc = 1$ megaparsec = $1\,Mpc$. Note that although the kilogram is the unit of mass, prefixes are attached to the symbol g (gram) and not to kg (e.g. mg, not μkg).

TABLE 61. *Names and symbols for the SI base and supplementary units.*

PHYSICAL QUANTITY	NAME OF UNIT	SYMBOL
Base units		
Length	metre	m
Mass	kilogram	kg
Time	second	s
Electric current	ampere	A
Thermodynamic temperature	kelvin	K
Luminous intensity	candela	cd
Amount of substance	mole	mol
Supplementary units		
Plane angle	radian	rad
Solid angle	steradian	sr

TABLE 62. *Prefixes for use with SI units. The multiples enclosed in brackets are no longer recommended for scientific usage.*

MULTIPLE	PREFIX	SYMBOL	MULTIPLE	PREFIX	SYMBOL
[10^{-1}	deci-	d]	[10	deca-	da]
[10^{-2}	centi-	c]	[10^2	hecto-	h]
10^{-3}	milli-	m	10^3	kilo-	k
10^{-6}	micro-	μ	10^6	mega-	M
10^{-9}	nano-	n	10^9	giga-	G
10^{-12}	pico-	p	10^{12}	tera-	T
10^{-15}	femto-	f	10^{15}	peta-	P
10^{-18}	atto-	a	10^{18}	exa-	E

PRACTICAL UNITS USED IN ASTRONOMY

Few of the SI units are commonly used in astronomy. It is still quite usual to work with a mixture of units traditional to astronomy and units derived from the obsolescent c.g.s. system (centimetre, gram, second) that preceded SI. Table 63 lists some of the more common named units used in astronomy, and shows how they are related to the corresponding SI units. The magnitude system, for measuring the brightness of stars, is dealt with on pages 107 to 110; the various systems of time used in astronomy are discussed on pages 7 to 21. In complex calculations it is often helpful to reduce all quantities to SI units to lessen the risk of conversion errors. Some conversion factors are given in Table 64.

The units for other astronomical quantities, which may not have special names, can be constructed by appropriately combining

TABLE 63. *Named units commonly used in astronomy.*

PHYSICAL QUANTITY	NAME OF UNIT (SYMBOL)[a]	RELATION TO SI UNIT[b]	NOTES AND ASTRONOMICAL USAGE
Length	metre (m)	(SI unit)	
	astronomical unit (AU, au)[c]	$1\,\mathrm{AU} = 1.495\,978\,70 \times 10^{11}$ m	Astronomical unit of length;[d] approximately the semi-major axis of the Earth's orbit. Mainly Solar System work
	parsec (pc)	$1\,\mathrm{pc} = 3.0857 \times 10^{16}$ m	*Parallax second* (distance at which 1 AU subtends an angle of 1″). Mainly stellar and galactic distances. Also kpc, Mpc, Gpc
	light year (l.y.)[c]	$1\,\mathrm{l.y.} = 9.4605 \times 10^{15}$ m	Distance traversed in one year by electromagnetic waves in free space. Mainly in popular writing
	solar radius ($R_\odot$)	$1\,R_\odot = 6.960 \times 10^8$ m	
	angstrom (Å)	$1\,\mathrm{Å} \equiv 10^{-10}$ m	Formerly used for optical wavelengths, atomic and molecular dimensions. Has largely given way to the nanometre (10^{-9} m). Non-IAU
	micron (μm, μ)	$1\,\mathrm{\mu m} \equiv 10^{-6}$ m	Common name for the micrometre (the non-IAU symbol μ is obsolete)
Time[e]	second (s)	(SI unit)	Also ms, μs, ns
	minute (min, m)	$1\,\mathrm{min} \equiv 60$ s	The symbol m can be used where there is no risk of confusion with metre
	hour (h, hr)	$1\,\mathrm{h} \equiv 60\,\mathrm{min} \equiv 3600$ s	
	day (d)	$1\,\mathrm{d} \equiv 24\,\mathrm{h} \equiv 86\,400$ s	Astronomical unit of time[d]
	year (yr, y, a)	$1\,\mathrm{y} \equiv 365.25\,\mathrm{d} = 3.1558 \times 10^7$ s	Julian year, unless otherwise specified. The symbol 'a' is recommended but rarely used
Mass	kilogram (kg)	(SI unit)	Prefixes must be attached to g, not kg
	solar mass unit ($M_\odot$)	$1\,M_\odot = 1.9891 \times 10^{30}$ kg	Astronomical unit of mass[d]
Thermodynamic temperature	kelvin (K)	(SI unit)	Origin of scale is absolute zero, i.e. 0 K = −273.15°C. As a unit of temperature difference, it is identical to the degree Celsius
Angle (or 'distance' on the celestial sphere)	radian (rad)	See Table 55 (SI unit)	$\frac{1}{2}\pi$ of a circle. Also mrad
	second of arc (″, arcsec)	$1'' \equiv 4.8481 \times 10^{-6}$ rad	
	minute of arc (′, arcmin)	$1' \equiv 60'' = 2.9089 \times 10^{-4}$ rad	
	degree (°, deg)	$1° \equiv 60' = 1.7453 \times 10^{-2}$ rad	
Solid angle (or 'area' on the celestial sphere)	steradian (sr)	(SI unit)	$\frac{1}{4}\pi$ of a sphere
	square degree (deg^2)	$1\,\mathrm{deg}^2 = 3.0462 \times 10^{-4}$ sr	
Frequency	hertz (Hz)	$1\,\mathrm{Hz} \equiv 1\,\mathrm{s}^{-1}$ (SI unit)	Formerly known as 'cycle per second', c/s. Also kHz, MHz, GHz
Force	newton (N)	$1\,\mathrm{N} \equiv 1\,\mathrm{kg\,m\,s^{-2}}$ (SI unit)	About the weight of an apple
	dyne (dyn)	$1\,\mathrm{dyn} \equiv 10^{-5}$ N	C.g.s. unit, in decline. Non-IAU
Energy	joule (J)	$1\,\mathrm{J} \equiv 1\,\mathrm{N\,m}$ (SI unit)	
	erg (erg)	$1\,\mathrm{erg} \equiv 10^{-7}$ J	C.g.s. unit. Non-IAU
	electron-volt (eV)	$1\,\mathrm{eV} = 1.6022 \times 10^{-19}$ J	Energies of photons and particles. Also keV, MeV, GeV
Power	watt (W)	$1\,\mathrm{W} \equiv 1\,\mathrm{J\,s^{-1}}$ (SI unit)	Also kW, MW, GW
	erg per second ($\mathrm{erg\,s^{-1}}$)	$1\,\mathrm{erg\,s^{-1}} \equiv 10^{-7}$ W	C.g.s. unit. Non-IAU
	solar luminosity ($L_\odot$)	$1\,L_\odot = 3.90 \times 10^{26}$ W	Bolometric luminosity of the Sun (radiated power over all wavelengths).
Pressure	pascal (Pa)	$1\,\mathrm{Pa} \equiv 1\,\mathrm{N\,m^{-2}}$ (SI unit)	
	bar (bar)	$1\,\mathrm{bar} \equiv 10^5$ Pa	Also mbar (or mb). Non-IAU
	atmosphere (atm)	$1\,\mathrm{atm} \equiv 101\,325$ Pa	International standard atmosphere. Non-IAU
	torr (Torr)	$1\,\mathrm{Torr} \equiv \frac{1}{760}\,\mathrm{atm} = 133.32$ Pa	Formerly millimetre of mercury (mmHg)
Spectral flux density	jansky (Jy)	$1\,\mathrm{Jy} \equiv 10^{-26}\,\mathrm{W\,m^{-2}\,Hz^{-1}}$	Radio astronomy. Also mJy
Magnetic flux density	tesla (T)	$1\,\mathrm{T} \equiv 1\,\mathrm{V\,s\,m^{-2}}$ (SI unit)	
	gauss (G)	$1\,\mathrm{G} \equiv 10^{-4}$ T	C.g.s. unit, giving way to the tesla. Non-IAU

Units marked 'non-IAU' the IAU recommends should no longer be used in astronomy.

[a] Where more than one symbol is given, they are generally in order of preference.

[b] The identity sign $\equiv$ means 'exactly equal to, by definition'; the equals sign $=$ means 'equal to' (to the accuracy given).

[c] Abbreviation; there is no standard international symbol.

[d] This unit is defined in the IAU (1976) System of Astronomical Constants which was adopted for ephemerides in 1984; see The Astronomical Almanac and its Explanatory Supplement for a full list.

[e] See pages 7–21 for a full discussion of the systems of time used in astronomy.

TABLE 64. *Some conversion factors.*

1 inch (in.) ≡ 25.4 mm
1 foot (ft) ≡ 12 in. ≡ 0.3048 m
1 yard (yd) ≡ 3 ft ≡ 0.9144 m
1 mile (mi) = 1.6093 km
1 UK nautical mile ≡ 6080 ft = 1.8532 km
1 international nautical mile ≡ 1.852 km
1 mi h^{-1} = 0.447 04 ms^{-1}
1 litre (1) ≡ 10^{-3} m³
1 ounce (oz) = 28.350 g
1 pound (lb) ≡ 16 oz = 0.453 592 kg
1 ton (ton) ≡ 2240 lb = 1016.0 kg
1 tonne (t) ≡ 1000 kg = 2204.6 lb = 0.9842 ton
1 parsec (pc) = 3.2616 l.y. = 206 265 AU
1 light year (l.y.) = 0.3066 pc = 63 240 AU
1 radian (rad) = 57°.2958
1 sphere = 41 252.961 deg² = 12.5664 sr

Temperature conversions

$T(°C) \equiv T(K) - 273.15 \equiv (T(°F) - 32)/1.8$

$T(°F) \equiv 1.8T(°C) + 32 = 1.8T(K) - 459.67$

$T(K) \equiv T(°C) + 273.15 = T(°F)/1.8 + 255.37$

The sign ≡ means 'exactly equal to, by definition'.

TABLE 65. *Some astronomical constants.*

CONSTANT	SYMBOL AND VALUE
Speed of light	$c = 299\,792\,458$ m s^{-1}
Gaussian gravitational constant	$k = 0.017\,202\,098\,95$
Constant of gravitation	$G = 6.672 \times 10^{-11}$ N m² kg^{-2}
Astronomical unit	$A = 1.495\,978\,70 \times 10^{11}$ m
Light time for unit distance	$\tau_A = 499.004\,782$ s
Solar parallax	$\pi_\odot = 8''.794\,148$
Mass of the Sun	$M_\odot = 1.9891 \times 10^{30}$ kg
Heliocentric gravitational constant	$GM_\odot = 1.327\,124\,38 \times 10^{20}$ N m² kg^{-1}
Mass of the Earth	$M_\oplus = 3.003\,490 \times 10^{-6}\,M_\odot = 5.9742 \times 10^{24}$ kg
Geocentric gravitational constant	$GM_\oplus = 3.986\,005 \times 10^{14}$ N m² kg^{-1}
Equatorial radius of the Earth	$a_e = 6\,378\,140$ m
Flattening factor of the Earth	$f = 0.003\,352\,81 = 1/298.257$
Mass of the Moon	$M_\mathbb{C} = 0.012\,300\,02\,M_\oplus = 7.3483 \times 10^{22}$ kg

At standard epoch 2000.0

Obliquity of the ecliptic	$\varepsilon = 23° 26' 21''.448$
General precession in longitude	$\rho = 50''.290\,966$ y^{-1}
Constant of nutation	$N = 9''.2025$
Constant of aberration	$\kappa = 20''.495\,52$

the units in Table 63. For example, density can be measured in kg m^{-3} (SI), g cm^{-3} (c.g.s.) or $M_\odot$ pc^{-3}; Hubble's constant is conventionally measured in units of km s^{-1} Mpc^{-1}.

Note that the unit symbols are not abbreviations and should not be given full stops: so 5 km, not 5 km. or 5 k.m., and 100 Mpc, not 100 M.p.c. Similarly, the letter s should not be added to the symbol to form a plural: three parsecs is written 3 pc, not 3 pcs.

WRITING NUMBERS

Very large or very small numbers are best written in *exponential notation*, i.e. in the form $a \times 10^b$, where a is a number between 1 and 10, and b is an integer (whole number). For example, the mass of the electron is 9.11×10^{-31} kg, the velocity of light 2.998×10^8 m s^{-1}.

Where there are more than four significant figures, digits may be grouped in threes from the decimal point for clarity. Commas should not be used for this purpose as the comma is the symbol for a decimal point in most European countries. For example: 12345.678901 km can be written 12 345.678 901 km, but not 12,345.678901 km.

A leading decimal point should be preceded by a zero for clarity: 0.1234, not .1234.

ERRORS AND UNCERTAINTIES

Every measurement has an uncertainty, or error, which is often denoted by a ± (plus or minus) sign; for example, a distance of 5.34 ± 0.25 kpc. Care should be taken in the interpretation of the error. Unless stated otherwise, the quantity following the ± is assumed to be a *standard error* (s.e.), such that the probability of the true value being in that range is about 68%. (An earlier convention was to quote a *probable error* (p.e.) such that the probability was 50%.) But it is often not possible to estimate an accurate standard error, and the figure may be just a rough indication of the uncertainty.

Astronomical Constants

In 1976 the IAU adopted a consistent set of constants to be used in astronomical computations. The IAU (1976) System of Astronomical Constants came into effect in 1984. Although values of individual constants will become better known as techniques of measurement improve, astronomers should keep to the IAU (1976) System in the interests of consistency. Table 65 lists a selection of constants derived from the IAU (1976) System. A full list is given every year in *The Astronomical Almanac*.

The constant of gravitation, G, is the least well-determined of the fundamental

physical constants, with an uncertainty of 1 part in 6000. It follows that the masses of the Sun and all other astronomical bodies have a similar uncertainty. Fortunately, the product GM can be determined with a much greater precision than either G or M alone, and for this reason the constants $GM_\odot$ and $GM_\oplus$ should be used for computing heliocentric and geocentric orbits. Similarly, the ratio of two masses, e.g. $M_\oplus/M_\odot$, can be measured more accurately than can the individual masses.

Symbols and Abbreviations

Abbreviations and symbols used in this book or commonly encountered in practical astronomy (see also Tables 41, 54, 61, 62 and 63).

a	semi-major axis; altitude
A	azimuth; extinction
b	galactic or heliocentric latitude
B_0	heliographic latitude of the centre of the Sun's disk
BC	bolometric correction
c	speed of light
CM	central meridian
c.p.m.	common proper motion

D	aperture (of a telescope)	T	time of perihelion passage (in an orbit); temperature
dec.	declination	T_c	colour temperature
e	eccentricity	T_{eff}	effective temperature
ET	Ephemeris Time	TAI	International Atomic Time
f	following	TT	Terrestrial Time
F	focal length	UT	Universal Time
g	acceleration of free fall (acceleration due to gravity)	UTC	Coordinated Universal Time
GHA	Greenwich hour angle	z	zenith distance; redshift
GMT	Greenwich Mean Time	ZHR	zenithal hourly rate
GST	Greenwich sidereal time		
h	altitude	α	right ascension
H_0	Hubble constant	β	celestial latitude
HR	Hertzsprung–Russell (diagram)	δ	declination
i	inclination	Δ	geocentric distance, in AU. When used with a suffix it means a correction, as in ΔT
IC	Index Catalogue		
JD	Julian date		
l	galactic or heliocentric longitude	ε	obliquity of the ecliptic
L_0	heliographic longitude of the centre of the Sun's disk	λ	wavelength; longitude
L	luminosity; longitude at the epoch	μ	proper motion
		ν	frequency
LHA	local hour angle	π	parallax
LMC	Large Magellanic Cloud	τ	light travel time
LST	local sidereal time	ϕ	latitude
m	apparent magnitude	ω	argument of perihelion
m_{bol}	apparent bolometric magnitude	ϖ	longitude of perihelion
m_{pg}	apparent photographic magnitude	Ω	longitude of the ascending node
m_{pv}	apparent photovisual magnitude	♈	first point of Aries (vernal equinox)
m_v	apparent visual magnitude	♎	first point of Libra (autumnal equinox)
m_V	photometric visual magnitude	☊	ascending node
M	absolute magnitude; magnification; mass	☋	descending node
M	Messier catalogue	○	full moon
MJD	modified Julian date	●	new moon
NGC	New General Catalogue	◗ or ◖	gibbous moon
NPD	north polar distance	◑ or ☽	first quarter
p	preceding	◐ or ☾	last quarter
P	period; position angle	☉	Sun
PA	position angle	⊕ or ♁	Earth
q	perihelion distance	☿	Mercury
Q	aphelion distance	♀	Venus
r	radius vector (i.e. distance from Sun in AU)	♂	Mars
		♃	Jupiter
RA	right ascension	♄	Saturn
SMC	Small Magellanic Cloud	♅ or ⛢	Uranus
t	time	♆ or ♆	Neptune
		♇	Pluto

Useful Addresses

U.K. AND COMMONWEALTH

Royal Astronomical Society,
Burlington house, Piccadilly,
London W1J 0BQ, U.K.
EMAIL: info@ras.org.uk
WWW: http://www.ras.org.uk/

British Astronomical Association,
Burlington House, Piccadilly,
London W1J 0DU, U.K.
EMAIL: office@britastro.com
WWW: http://www.britastro.org/

Society for Popular Astronomy,
36 Fairway, Keyworth,
Nottingham NG12 5DU, U.K.
EMAIL: secretary@popastro.com
WWW: http://www.popastro.com/

Federation of Astronomical Societies,
10 Glan y Llyn, North Cornelly,
Bridgend, CF33 4EF, U.K.
EMAIL: secretary@fedastro.org.uk
WWW: http://www.fedastro.org.uk/

Royal Astronomical Society of Canada,
136 Dupont Street, Toronto,
Ontario M5R 1V2, Canada.
EMAIL: rasc@vela.astro.utoronto.ca
WWW: http://www.rasc.ca/

British Astronomical Association
New South Wales branch,
Sydney Observatory, Watson Road,
Sydney, NSW 2000, Australia.
EMAIL: honsecretary@baansw.asn.au
WWW: http://www.baansw.asn.au/

Royal Astronomical Society
of New Zealand, P.O. Box 3181,
Wellington, New Zealand.
EMAIL: rasnz@rasnz.org.nz
WWW: http://www.rasnz.org.nz/

Astronomical Society of Southern Africa,
P.O. Box 9, Observatory, 7935,
South Africa.
WWW: http://www.saao.ac.za/assa/
index.htm

REPUBLIC OF IRELAND

Irish Astronomical Society,
P.O. Box 2547,
Dublin 14, Ireland.
EMAIL: ias@esatclear.ie
WWW: http://www.esatclear.ie/~ias/

Astronomy Ireland,
P.O. Box 2888, Dublin 5, Ireland.
EMAIL: info@astronomy.ie
WWW: http://www.astronomy.ie/

Irish Federation of Astronomical Societies.
WWW: http://www.irishastronomy.org/

U.S.A.

The American Association
of Amateur Astronomers,
P.O. Box 7981, Dallas,
TX 75209-0981, U.S.A.
EMAIL: aaaa@corvus.com
WWW: http://www.corvus.com/

Astronomical League,
9201 Ward Parkway, Suite #100,
Kansas City, MO 64114, U.S.A.
EMAIL: execsec@astroleague.org
WWW: http://www.astroleague.org/

Astronomical Society of the Pacific,
390 Ashton Avenue,
San Francisco, CA 94112, U.S.A.
EMAIL: publicinfo@astrosociety.org
WWW: http://www.astrosociety.org/

SPECIALIST ORGANIZATIONS

American Association
of Variable Star Observers,
25 Birch Street,
Cambridge, MA 02138, U.S.A.
EMAIL: aavso@aavso.org
WWW: http://www.aavso.org/

Association of Lunar
and Planetary Observers,
P.O. Box 13456,
Springfield, IL 62791-3456, U.S.A.
WWW: http://www.lpl.arizona.edu/alpo/

The Planetary Society,
65 North Catalina Avenue,
Pasadena, CA 91106-2301, U.S.A.
EMAIL: tps@planetary.org
WWW: http://www.planetary.org/

International Lunar Occultation Center,
5-3-1 Tsukiji, Chuo-ku,
Tokyo 104-0045, Japan.
EMAIL: iloc@jodc.go.jp
WWW: http://www1.kaiho.mlit.go.jp/
    KOHO/iloc/docs/iloc-index_e.htm

International Occultation
Timing Association,
7006 Megan Lane,
Greenbelt, MD 20770-3012, U.S.A.
EMAIL: david_dunham@jhuapl.edu
WWW: http://www.occultations.org/

International Meteor Organization.
WWW: http://www.imo.net/

The American Meteor Society.
WWW: http://www.amsmeteors.org/

International Dark-Sky Association,
3225 N. First Avenue,
Tucson, AZ 85719, U.S.A.
EMAIL: ida@darksky.org
WWW: http://www.darksky.org/

International Amateur-Professional
Photoelectric Photometry,
Dyer Observatory, 1000 Oman Drive,
Brentwood, Tennessee 37027-4143, U.S.A.
douglas.s.hall@iappp.vanderbilt.edu/

Webb Society
WWW: http://www.webbsociety.
    freeserve.co.uk/

AGENCIES

International Astronomical Union,
98bis Boulevard Arago,
75014 Paris, France.
WWW: http://www.iau.org/

National Aeronautics and Space
    Administration,
NASA Headquarters,
Washington, DC 20546-0001, U.S.A.
WWW: http://www.nasa.gov/

European Space Agency,
8–10 rue Mario Nikis, 75738 Paris, France.
WWW: http://www.esa.int/

The British National Space Centre,
151 Buckingham Palace Road,
London SW1W 9SS, U.K.
WWW: http://www.bnsc.gov.uk/

Space Telescope Science Institute,
3700 San Martin Drive,
Johns Hopkins University
Homewood Campus,
Baltimore, MD 21218, U.S.A.
WWW: http://www.stsci.edu/

MAGAZINES

Sky & Telescope,
Sky Publishing Corporation,
49 Bay State Road,
Cambridge, MA 02138-1200, U.S.A.
EMAIL: info@skyandtelescope.com
WWW: http://www.skypub.com/

Astronomy,
21027 Crossroads Circle,
P.O. Box 1612,
Waukesha, WI 53187-1612, U.S.A.
WWW: http://www.astronomy.com/

Astronomy Now,
P.O. Box 175,
Tonbridge, TN10 4ZY, U.K.
http://www.astronomynow.com/
    magazine.html

The Astronomer,
6 Chelmerton Avenue, Great Baddow,
Chelmsford, CM2 9RE, U.K.
EMAIL: secretary@theastronomer.org
WWW: http://www.theastronomer.org/

SkyNews,
Box 10, Yarker,
Ontario K0K 3N0, Canada.
EMAIL: skynews@on.aibn.com
WWW: http://
    www.skynewsmagazine.com/

Sky & space,
P.O. Box 1690, Bondi Junction,
New South Wales 1355, Australia.

Glossary

aberration a defect in an optical system. There are six main types: in *chromatic aberration*, which occurs in lenses, coloured fringes appear around objects; *spherical aberration* is a blurring of the image caused when the inner and outer parts of a lens or mirror have different focal lengths; in *astigmatism*, the star image is focused into an ellipse or a cross; *coma* produces elongated images towards the edge of the field of view; *curved field* results when the focal plane of a lens or mirror is not flat; *distortion* is caused by a difference in magnification between the centre and edge of the field, bowing straight lines either outwards (*barrel distortion*) or inwards (*pincushion distortion*).

aberration of starlight a slight displacement in the observed position of a star caused by the motion of the Earth in orbit around the Sun.

absorption lines dark lines crossing a spectrum, caused by absorption of certain wavelengths of light by cooler gas. All stars have absorption lines in their spectra because light leaving their surface passes through cooler gas in their outer layers. Absorption lines can also be produced by gas between us and the stars.

achromatic referring to a lens that has been corrected for chromatic aberration. An achromatic lens actually consists of two separate lenses, called *elements*, that together cancel out the worst effects of chromatic aberration.

airglow a faint background light in the night sky given out by gases in the ionosphere. The sky can therefore never be completely dark as seen from the surface of the Earth.

Airy disk the disk into which the image of a star is spread by diffraction in a telescope. The size of the disk limits the resolution of a telescope: the larger the aperture, the smaller the Airy disk. It is named after the seventh English Astronomer Royal, Sir George Airy, who calculated its size in 1834. In a refracting telescope nearly 84% of the light from a star goes into the Airy disk, the remainder forming faint diffraction rings around the Airy disk. In telescopes with central obstructions, such as the secondary mirrors in reflecting telescopes, more of the light is diverted from the Airy disk into the surrounding diffraction rings.

albedo the proportion of incoming light that is reflected by a surface, such as that of a planet or moon. A dark surface has a low albedo, while a light surface has a high albedo. The albedo of a planet usually differs from place to place, so for practical purposes the mean albedo is used. Planets with rocky surfaces such as Mercury and Mars have low albedos, while those covered with cloud, such as Jupiter and Venus, have high albedos. Albedo can be expressed in two ways: *spherical albedo* assumes that the body is a sphere with a diffuse surface reflecting incoming parallel light in all directions; *geometrical albedo* compares the reflectance of the planet with a flat white surface of the same diameter as the planet in the same place.

alidade a simple instrument for measuring altitudes of celestial bodies above the horizon. In its most basic form the alidade consists of a sighting device attached to a plumb-line that swings freely from the centre of a protractor or similar scale. The altitude of an object can be found from the angle of the plumb-line against the scale.

almanac a book containing timetables of celestial events and predicted positions of celestial objects, usually issued annually.

almucantar a circle on the celestial sphere parallel to the horizon; it is a line of equal altitude, since all objects on an almucantar at a given time are at the same angle above the horizon.

angular diameter the apparent size of a celestial object, usually expressed in degrees, minutes and seconds of arc.

angular distance the apparent distance between two objects on the celestial sphere, such as two stars, usually expressed in degrees, minutes and seconds of arc.

ansae the parts of Saturn's rings that appear like handles on each side of the planet. Singular *ansa*.

apochromat a lens consisting of three or more elements that gives a greater reduction in chromatic aberration than is possible with a normal achromatic (two-element) lens.

apparition the period of time during which a celestial body is well placed for observation, such as an evening apparition of Venus or the apparition of a periodic comet. The word is not used for bodies, such as the Moon, which are continually visible.

appulse the apparent close approach between two celestial bodies, such as two planets or a planet and a star.

apsides the points in an orbit at which two bodies are closest together (*periapsis*) and farthest apart (*apoapsis*). The line joining these points is called the *line of apsides*, and is the major axis of the orbit.

arc (measure of) angles on the celestial sphere are measured in degrees, minutes and seconds of arc. The terms *arc minute* and *arc second* are used to distinguish these measures from units of time. There are 60 arc minutes in a degree, and 60 arc seconds in an arc minute.

asterism a grouping or pattern of stars which does not make a constellation. The stars in an asterism can be a smaller part of one constellation, or members of more than one constellation.

astrometry the branch of astronomy concerned with the precise measurement of the positions of objects on the celestial sphere.

auroral oval either of two rings of permanent, quiet auroral activity that surround the north and south magnetic poles of the Earth. Normally the ovals are fairly narrow and lie about 2000 km from the geomagnetic poles. Under disturbed conditions, though, particularly following solar flares, the ovals expand towards the equator and become broader, most markedly on the side away from the Sun. It is during these expansions that observers at lower latitudes see aurorae.

barycentre the centre of mass, or balance point, of a pair of bodies such as a double star or a moon and planet, around which the two bodies orbit.

Big Bang the hypothetical event that is presumed to have marked the origin of the Universe as we know it. The Universe has been expanding since the Big Bang, which is estimated to have occurred between 10 000 million and 20 000 million years ago.

black body a hypothetical object that is both a perfect absorber of radiation falling on it and a perfect emitter of radiation. Black-body radiation is the spectrum of light and other radiation that would be emitted by a black body at a given temperature.

black hole a volume of space in which gravity is so great that nothing can escape, not even light – hence it is truly black. Black holes are thought to be produced when very massive stars collapse at the end of their life.

Bode's law a series of numbers that roughly describes the average distances of the planets from the Sun in astronomical units, out as far as Uranus. Take the numbers 0, 3, 6, 12, etc., doubling at each step. Add 4 to each number and divide by 10. Table 66 gives the results, compared with the actual mean distances of the planets. The 'law' breaks down for Neptune and Pluto. The German astronomer Johann Bode drew attention to the relationship in 1772, although it had already been pointed out by his countryman Johann Titius; for this reason it is sometimes called the Titius–Bode law.

TABLE 66. *Bode's law.*

| PLANET | DISTANCE (AU) | |
	BODE'S LAW	ACTUAL
Mercury	0.4	0.39
Venus	0.7	0.72
Earth	1.0	1.0
Mars	1.6	1.5
Ceres	2.8	2.8
Jupiter	5.2	5.2
Saturn	10.2	9.5
Uranus	19.6	19.2
Neptune	38.8	30.1
Pluto	77.2	39.5

captured rotation rotation such that a body spins on its axis in the same time as it takes to orbit another body, so that it keeps one face permanently turned towards the object it is orbiting. Our Moon has a captured rotation, as do many moons of other planets. Captured rotation is brought about by tidal forces.

CCD an electronic device used instead of photographic film. It consists of a silicon chip which is sensitive to light, divided into sections known as pixels (picture elements). Light falling on each pixel builds up an electric charge which can be read off to produce an image. CCDs are much more sensitive to light than is photographic film, so exposures can be much shorter. However, they are also smaller in area and have less resolution than film. The initials stand for charge-coupled device.

central meridian (CM) the imaginary north–south line bisecting the disk of a planet, used as a reference for estimating the longitude of planetary features as the planet rotates. The passage of a feature across the central meridian is called a *central meridian transit.*

collimation the act of lining up the optical components of an instrument, such as the mirrors in a reflecting telescope. In spectroscopes the collimator is a lens used to produce a parallel beam of light.

coma (cometary) the cloud of gas and dust, roughly spherical in shape, that makes up the head of a comet. At the centre of the coma is the comet's nucleus, from which the gas and dust escapes. A comet's coma can be between 10 000 and 100 000 km in diameter.

coma (optical) a flaring of star images towards the edge of the field of view.

comes the companion of a double star (plural *comites*).

commensurable an expression used of orbital periods (e.g. of two moons) that are in proportion to one another by exact fractions such as one-half, one-third or three-quarters.

continuous spectrum a spectrum that consists of an unbroken rainbow of colours, as distinct from a spectrum crossed by dark absorption lines or one that consists of emission lines.

continuum a continuous spectrum (q.v.).

coronal hole a cooler and less dense part of the Sun's corona, through which the fastest part of the solar wind flows.

cosmic rays atomic particles that are moving through space at close to the speed of light. They are mostly protons (the nuclei of hydrogen atoms), although the nuclei of most elements are present in small numbers, and also electrons. Some low-energy cosmic rays come from flares on the Sun, but higher-energy cosmic rays are believed to come from outside the Solar System, probably from supernovae and their remnants. The highest-energy cosmic rays of all seem to come from distant galaxies and quasars.

cosmology the study of the origin and evolution of the Universe.

coudé focus a focal point in a reflecting telescope in which the light is reflected out of the telescope tube along the polar axis of the mounting to a fixed observing position. The coudé focus has the advantage that it does not move as the telescope turns, and so heavy equipment such as large spectrographs can be mounted there.

cryogenic   referring to ultra-low temperatures, as needed to liquefy gases. Cryogenic cooling is used to reduce background noise and hence increase the sensitivity of certain instruments. The liquid gases are kept in an insulated flask known as a *cryostat*.

cusp   one of the two 'horns' of the crescent Moon or of a planet in crescent phase.

Cynthian   adjective referring to the Moon.

Cytherean   adjective referring to Venus.

deep sky   that part of space beyond the Solar System. Deep-sky objects include star clusters, nebulae, galaxies, double stars and variable stars.

defect of illumination   the apparent width of the unilluminated section of a planet's disk as seen from the Earth, usually expressed in seconds of arc. For example, if a planet has an apparent diameter of 10 arcsec and a phase of 80%, its defect of illumination would be 2 arcsec.

dichotomy   the moment when the Moon, Mercury or Venus is exactly half-illuminated as seen from the Earth.

differential rotation   the rotation of a body in which different parts spin at different speeds; for example, a gaseous planet or a star spins faster at the equator than at the poles.

diffraction   the slight bending of light around the edge of an object; light of long wavelengths is diffracted more than light of short wavelengths. This effect is utilized in a *diffraction grating*, a series of closely spaced lines (usually thousands per centimetre) ruled on a piece of glass or metal, that spreads light out into a spectrum. Diffraction gratings are commonly used in spectroscopes. See also Airy disk.

disk   the face of a planet, moon or star as seen from the Earth.

dispersion   the spreading out of light into a spectrum, as in a spectrograph. The highest dispersions give the best resolution of features in the spectrum.

diurnal   daily.

Doppler effect   a change in the wavelength of light caused by the motion of the object emitting the light. If the object is moving towards us the wavelengths are shortened, i.e. moved towards the blue end of the spectrum; this is termed a *blueshift*. If the object is receding its light is lengthened in wavelength, i.e. moved towards the red end of the spectrum; this is termed a *redshift* (q.v.). The amount of shift is revealed by the position of lines of known wavelength in the object's spectrum.

doublet   a two-element lens, designed to reduce chromatic aberration.

dwarf star   any star on the main sequence of the Hertzsprung–Russell diagram. The Sun is a dwarf star, but many such stars are actually larger than the Sun. The term is also applied to white dwarfs (q.v.), which are not on the main sequence.

early-type star   a hot star of spectral type O, B or A.

eccentricity (*e*)   a measure of how non-circular an orbit is. The eccentricity of an ellipse ranges between 0 (a circle) and 1 (a parabola). Eccentricity is calculated by dividing the distance between the two foci of the ellipse by the length of the major axis.

element, optical   an optical component, such as a mirror, lens or prism. Usually the term is applied to a lens that makes up part of a more complex lens, e.g. a doublet is a lens with two elements, and a triplet is a three-element lens. The additional elements are introduced to correct the aberrations that are present in a single lens.

elongation   the angle between the Sun and a planet, or between a planet and a satellite, as seen from the Earth. Elongation is measured along the ecliptic in degrees west or east of the Sun.

emersion   the re-emergence of an object after an eclipse or occultation.

emission lines   specific wavelengths of light (or other forms of electromagnetic radiation) given out by atoms of a gas. An *emission spectrum* is a spectrum consisting of bright emission lines, for example as produced by the gas of a nebula. Emission lines can appear as bright lines superimposed on a continuous spectrum if given out by hot gas surrounding a star.

ephemeris   a table of the predicted positions of a celestial object such as the Moon, the Sun or a planet. Plural *ephemerides*.

epoch   an instant in time, such as the beginning or middle of a year, for which positions of stars, orbital elements and other information are given. Since the coordinates of stars are constantly changing because of precession, star positions are referred to a *standard* or *fundamental epoch*. Currently the standard epoch used by astronomers is 2000 January 1, 12h (also written as 2000.0).

equation   in astronomy, either a difference between two values or a correction, as for instance in the *equation of time* (difference between mean and apparent solar time) or a *personal equation* (correction for personal error when measuring or timing something).

escape velocity   the speed at which any object, from a rocket to a gas molecule, must move to break away permanently from the gravitational pull of a body. For the Earth, the escape velocity at the surface is 11.2 km s^{-1}; for the Moon it is 2.4 km s^{-1}.

exit pupil   the image that an eyepiece forms of a telescope's objective lens or mirror; the higher the magnification of the eyepiece, the smaller the diameter of the exit pupil. In order to see the telescope's full field of view, the pupil of the eye must be brought up to the exit pupil of the eyepiece.

extinction   the dimming of starlight by dust in space or by the Earth's atmosphere. Extinction is greater for blue light than it is for red, causing a reddening of starlight. Atmospheric extinction is least at the zenith, where it amounts to a few tenths of a magnitude under clear skies, and increases towards the horizon (see Table 57, on p. 130).

extrapolation   the technique of extending a series of figures in order to estimate an additional value beyond the given range.

field star   a star in the same field of view as an object under study, but which lies at a different distance and hence has no connection with it, for example a foreground star in the same field of view as a distant galaxy.

first contact the beginning of an eclipse, transit or occultation. At a solar eclipse, it is when the Moon starts to move across the face of the Sun; at a lunar eclipse it is when the Moon enters the Earth's umbra.

focal length the distance between a lens or mirror and the point at which it brings parallel light rays to a focus.

focal plane the flat surface at which a lens or mirror forms an image. Some optical systems, notably the Schmidt telescope, form their images on a curved surface known as the *focal surface*.

focal ratio the focal length of a telescope divided by its aperture. For example, a 150-mm telescope of 1200-mm focal length has a focal ratio of $f/8$.

focus (optical) the point at which light rays are concentrated by a lens or mirror to form an image.

focus (of an ellipse) one of the two points whose position determines the eccentricity of an elliptical orbit; plural *foci* (usually pronounced foe-sigh). The two foci lie on the major axis of the ellipse, either side of its centre; the farther apart they are, the greater the eccentricity of the ellipse. The object being orbited lies at one of the foci; the other focus is empty.

following objects move across the sky from east to west because of the rotation of the Earth, so the more easterly of a pair of stars, for example (or the easterly side of a planet), is said to be following. The term is also used of features moving across the face of a body as it rotates, such as sunspots, or spots on Jupiter. The other side is described as preceding (q.v.).

fourth contact the end of an eclipse, transit or occultation. At a solar eclipse it is when the Moon moves completely off the face of the Sun; at a lunar eclipse it is when the Moon leaves the Earth's umbra.

frequency (ν) the number of waves passing a fixed point in a given time, usually one second. Frequency is measured in hertz, and is equal to the speed of the waves divided by their wavelength. Hence the longer the wavelength, the lower the frequency, and the shorter the wavelength, the higher the frequency.

fundamental star a star whose position is determined as precisely as possible, and against which the positions of other stars can be compared. The positions of fundamental stars are published in *fundamental catalogues*.

galactic cluster another name for an open star cluster in our Galaxy, so called because they lie in the spiral arms of the Galaxy rather than in the halo around the Galaxy, where the globular clusters lie.

gamma rays radiation of the shortest wavelengths, 0.01 nanometres and less, shorter even than X-rays.

giant star a star that is swelling up in size as it approaches the end of its life. Giant stars have similar masses to normal stars such as the Sun, but they are larger in diameter and considerably more luminous.

gibbous the phase of the Moon or a planet when it is between half and fully illuminated.

Gould's Belt a band of young, brilliant stars at an angle of between 15° and 20° to the plane of our Galaxy, stretching around the sky from Perseus, Taurus and Orion, via Carina, to Centaurus and Scorpius. Gould's Belt is believed to be a spur on the local spiral arm of our Galaxy.

great circle a circle that divides a sphere into two equal hemispheres. On the celestial sphere, a great circle has the Earth at its centre; examples are the celestial equator, the ecliptic and lines of right ascension. Compare small circle.

green flash an effect caused by atmospheric refraction and absorption in which the last visible segment of the setting Sun turns green, sometimes followed by a green ray like a vertical flame at the instant of setting. The phenomenon lasts for only a few seconds, and is best seen over the sea or a distant horizon when the air is clear (i.e. when there is little reddening of the setting Sun). A similar effect can occasionally be seen as the Sun rises.

greenhouse effect the warming of a planet by the trapping of solar radiation in a planet's atmosphere. The greenhouse effect acts particularly strongly on Venus, raising its temperature to very high levels; it operates to a lesser effect in the atmospheres of other planets.

heavy elements in astronomy, all chemical elements heavier than hydrogen and helium; sometimes termed 'metals'.

heliacal rising the occasion on which a star or planet first appears in the dawn sky, after having been too close to the Sun to be visible.

heliacal setting the last occasion on which a star or planet can be seen in the evening sky before it becomes too close to the Sun to be visible.

immersion the entry of a celestial object into a shadow at an eclipse, or the covering of an object at an occultation.

inclination (i) the angle at which an orbit is tilted with respect to a plane of reference. For objects orbiting the Sun the inclination is given relative to the plane of the Earth's orbit; for objects orbiting the Earth, relative to the Earth's equator; and for double stars, relative to the plane of the sky. The axial inclination of a body is the angle at which its axis of rotation is tilted to the perpendicular to the plane of its own orbit.

infrared radiation with wavelengths longer than visible red light but shorter than radio waves, i.e. between about 700 nanometres and 1 millimetre.

interferometer a device in which radio or optical waves collected by two or more apertures are combined to give improved resolution, such as for separating two closely spaced objects.

interpolation the technique of estimating a value intermediate between two of a range of given values, for instance the position of a planet on a date between two dates tabulated in an ephemeris.

inverse-square law the law which states that the energy received from a source falls off with the inverse square of the distance of the source. For example, a star twice as far away as another identical star appears four times fainter, three times away it appears nine times fainter, and so on. Forces, including gravity, obey the same law.

ion an atom or molecule that has lost one or more electrons (a *positive ion*) or has gained one or more electrons (a *negative ion*).

ionization the process by which electrons are added to or removed from an atom or molecule, so turning it into an ion.

irradiation the optical effect in which a bright object seen against a dark background appears larger or brighter than it actually is.

Kirkwood gaps regions of the asteroid belt, corresponding to particular distances from the Sun, where few asteroids are found. The gaps are caused by Jupiter's gravity, which perturbs asteroids out of orbits whose period is an exact fraction of Jupiter's orbital period.

Lagrangian points five places at which small bodies can exist in stable orbits in the plane of two much larger bodies. Three of the points lie on a line joining the two large bodies (one point between the two bodies, and the other two points on either side of them). The two other Lagrangian points lie 60° ahead of and behind one of the larger bodies in its orbit around the other; it is at these places in the orbit of Jupiter that the Trojan asteroids are found. Objects cannot exist permanently at the three other Lagrangian points of Jupiter's orbit because they would be perturbed by the gravitational pulls of the other planets.

late-type star a cool star of spectral type K, M, C or S.

light curve a graph of the changing brightness of an object such as a variable star, or a planet or moon as it rotates.

light, speed of light travels at 299 792.5 km s^{-1} (often rounded to 300 000 km s^{-1}) in a vacuum; this is the fastest speed in the Universe. All other forms of electromagnetic radiation, from X-rays and gamma rays to radio waves, travel at the same speed.

light-pollution brightening of the night sky caused by artificial sources of illumination such as streetlamps.

light-time the time taken for a beam of light to travel from a celestial body to the Earth. The effect must be taken into account when timing the occurrence of events such as eclipses of the moons of Jupiter, whose times of occurrence are affected by the distance between Jupiter and the Earth.

limb the apparent edge of the disk of a celestial body as seen from the Earth; regions near the visible edge of the Moon are called limb regions. The leading limb of an object crossing the sky as the Earth rotates is called the *preceding limb*; the trailing limb is called the *following limb*.

local standard of rest a volume of space extending out to about 100 parsecs from the Sun in which the velocities of all stars relative to the Sun average out to zero.

lunation the time taken by a complete cycle of phases of the Moon, such as from one full moon to the next. A lunation lasts 29.53 days; it is the same as a synodic month.

magnetosphere the extension of the Earth's magnetic field into space. The magnetosphere is like a magnetic bubble around the Earth. The Van Allen radiation belts lie within the magnetosphere. Other bodies with magnetic fields also have magnetospheres. The boundary of the magnetosphere is called the *magnetopause*.

magnification the amount by which an optical instrument makes an object appear larger. For example, if a line appears ten times longer when viewed through a telescope, the telescope is said to magnify ten times. The magnification of a telescope depends on the instrument's focal length and on the focal length of the eyepiece in use; eyepieces of shorter focal length produce higher magnifications on a given telescope. Magnification can be calculated by dividing the focal length of the telescope by the focal length of the eyepiece. A magnification of, say, ten is written in the form ×10.

major axis the longest diameter of an ellipse, passing through the two foci of the ellipse.

mean the average of a series of values.

meteor the streak of light, lasting no more than a second or so, produced when a speck of dust from space (a meteoroid) burns up in the Earth's atmosphere, usually at a height of about 100 km.

meteorite a chunk of rock or iron from space that reaches the surface of the Earth or of any other body. Large meteorites can produce craters when they hit the ground. Most meteorites are thought to be chips from asteroids, but some fragile stony meteorites called carbonaceous chondrites may come from the nuclei of comets.

meteoroid any small solid object in space. When a meteoroid enters the Earth's atmosphere at high speed it produces a meteor.

Metonic cycle the period of 19 calendar years (6939.6 days) after which the Moon's phases recur on the same day of the year. There are 235 lunations in a Metonic cycle.

minor axis the shortest diameter of an ellipse, at right angles to the major axis.

mock Sun an effect caused by ice crystals in the Earth's atmosphere which refract the Sun's light so that two diffuse areas of light occur either side of the Sun, 22° from it. These mock Suns, also known as *parhelia* or *sundogs*, usually appear on the rim of a halo surrounding the Sun.

neutron star a tiny, very dense star composed of neutrons. Neutron stars have diameters of only about 20 km, but contain the mass of up to three Suns; if the neutron star had a mass greater than three Suns, gravity would cause it to collapse still further into a black hole. Neutron stars are believed to be left behind after massive stars explode as supernovae at the end of their life; in the explosion, the protons and electrons of the star's core are squeezed together to form neutrons.

node the point at which an orbit crosses a given plane, such as the plane of the Earth's orbit or the Earth's equator. There are two nodes: the *ascending node* (☊), when the orbiting body moves from south to north, and the *descending node* (☋), when the body moves from north to south. The *line of nodes* is the straight line joining these two nodes. *Regression of the nodes* is the westward movement of the nodes of an orbit caused by the gravitational pull of other bodies, notably the Sun.

oblateness a measure of the amount by which a rotating object such as a star or planet departs from a perfectly spherical shape. Rotation causes the equatorial regions of a sphere to bulge outwards slightly, so that the sphere appears slightly flattened at the poles; hence oblateness is also known as polar flattening. Oblateness is calculated by taking the difference between the equatorial and polar diameters of the object, and dividing by the equatorial diameter. Saturn has the greatest oblateness of any planet in the Solar System, 0.1.

occulting bar   a bar that may be moved into the focal plane of an eyepiece so as to obscure a bright object and allow a nearby faint object to be observed.

paraboloid   a surface that is curved like a parabola. Main mirrors in telescopes are usually paraboloids, since a paraboloid is free from spherical aberration.

parhelion   a mock Sun (q.v.).

penumbra   the lighter, outer part of a sunspot or shadow. From within the penumbra of the Moon's shadow, a partial eclipse of the Sun is visible. When the Moon is within the penumbra of the Earth's shadow it is said to be *penumbrally eclipsed*; but the Earth's penumbral shadow is so faint that in practice a penumbral eclipse is scarcely noticeable.

period   the interval between the successive occurrences of a cyclical event, such as the time taken for a body to rotate once on its axis or go once around its orbit, or for a variable star to go through one cycle of brightness variations.

perturbation   a slight disturbance of the motion of one body caused by the gravitational pull of other bodies.

phase   the proportion of the sunlit side of the Moon or a planet that is visible from Earth. Mercury and Venus go through a complete cycle of phases similar to those of the Moon. The outer planets show phases only from gibbous to full, being most gibbous at quadrature.

phase angle   the angle between the Sun and the observer as seen from the centre of a given object. When the phase angle is 180° the Sun and the observer lie in opposite directions from the object, and the sunlit side of the object is facing away from the observer. At a phase angle of 0° the Sun and the observer are on the same side, and the object appears fully illuminated.

photometry   the calculation and measurement of the brightness of an object; a device that does this is called a *photometer*. Photometry is often carried out at several wavelengths to determine the colour of a star or other object under study, in order to determine its temperature and to reveal other information about its nature.

photon   the behaviour of light in some situations is best explained by assuming that it is not a wave motion, but a stream of particles. A photon is the name given to such a particle (of light or of other electromagnetic radiation).

planisphere   a circular map with a rotating mask that can be turned to show the stars as they appear from a given latitude at any time on any date.

population index (*r*)   in meteor astronomy, a factor that describes how the number of meteors goes up with decreasing brightness. For example, if there are *n* meteors in the magnitude interval *m* to *m* + 1, there are *rn* between magnitudes *m* + 1 and *m* + 2, *r²n* between *m* + 2 and *m* + 3, and so on. Over the naked-eye magnitude range, *r* is roughly constant. The exact value of *r* depends on the particular shower, but is usually in the range 2.2 to 2.5.

position angle   the relative position of one object with respect to another, such as the two components of a double star or the position of a star around the Moon's limb at an occultation. Position angle is measured in degrees from north via east, south and west. On the celestial sphere, east is the direction towards the eastern horizon.

preceding   term used to describe the side of a planet that leads in its motion across the sky, or of the leading member of a pair of objects such as stars or sunspots. The preceding side can easily be found by watching objects drift through the field of view of a telescope. Compare following.

primary   the larger body of an orbiting pair (e.g. the Earth is the Moon's primary) or the brighter member of a binary star. Compare secondary.

prime focus   the point at which the main mirror or objective lens of a telescope brings light to a focus, without the intervention of other optical components.

pulsar   a star that, every few seconds or less, gives out a rapid flash of energy at radio and other wavelengths. Pulsars are believed to be rapidly rotating neutron stars (q.v.) that flash each time they spin, like a lighthouse beam.

quasar   an object that looks like a star but which emits as much energy as hundreds of normal galaxies. Quasars have high redshifts, and hence must lie far off in the Universe. They are thought to be the bright centres of distant galaxies where matter is falling into a giant central black hole.

radiation belts   belts of atomic particles trapped inside the magnetosphere of a planet. See also Van Allen belts.

radio astronomy   the study of radio waves emitted naturally by objects in space. Radio waves are the longest-wavelength radiation, with wavelengths greater than 1 millimetre.

radius vector   the imaginary line joining an orbiting body and the object it orbits.

red dwarf   a star that is much smaller and cooler than the Sun. Red dwarfs have about one-tenth the mass of the Sun, and are about one-tenth its diameter.

red giant   a large, cool star perhaps ten or more times the diameter of the Sun, produced when a normal star swells up near the end of its life.

redshift   a lengthening in the wavelengths of light from a body, usually caused by the motion of the emitting body away from us (a Doppler shift), although a redshift can also be caused by the presence of strong gravitational fields. The redshift of galaxies is usually regarded as being directly related to their distance from us in the Universe – hence the greater the redshift, the more distant the galaxy.

refraction (atmospheric)   the bending of light by the Earth's atmosphere which increases the apparent altitude of an object above the horizon. It ranges from zero at the zenith to approximately half a degree at the horizon.

residual   the difference between observed and calculated values, such as of the position of a planet in its orbit.

retrograde   motion of a body from east to west, the opposite of the prevailing direction of motion in the Solar System. The term retrograde can apply to either the orbital motion or the direction of spin of a planet or moon.

revolution   the movement of one body in orbit around another, or around a centre of mass.

rotation   the spin of a body on its own axis.

Saros   the length of the cycle of solar and lunar eclipses: the period after which the Sun, the Moon and the nodes of the Moon's orbit return to almost the same relative positions. The Saros lasts 6585.32 days (just over 18 years) and contains 223 lunations.

scintillation   twinkling (q.v.).

second contact   the moment an eclipse becomes total. At a solar eclipse, it is when the Moon completely covers the face of the Sun; at a lunar eclipse, it is when the Moon becomes fully immersed in the Earth's umbra.

secondary   a smaller body that orbits around a larger one (e.g. the Moon is the Earth's secondary) or the fainter member of a binary system. Compare primary.

secondary spectrum   the slight colour fringing around an image in an achromatic lens, resulting from the fact that chromatic aberration cannot be completely eliminated, even by a two-element lens.

semi-major axis   half the longest diameter of an ellipse. The semi-major axis is the average distance of a body, such as a planet, from the object it is in orbit around, such as the Sun.

setting circles   scales marked on the polar and declination axes of an equatorially mounted telescope by which the telescope can be pointed at an object whose coordinates are known.

sidereal   to do with the stars. *Sidereal time* is time based on the rotation of the Earth with respect to the stars rather than with respect to the Sun; the *sidereal period* is the orbital period of a body with reference to a fixed star. Compare synodic.

small circle   a circle that does not divide a sphere into two equal hemispheres, unlike a great circle (q.v.). On the celestial sphere, small circles do not have the Earth at their centre – for example, circles of declination (other than the celestial equator) are small circles.

solar wind   the tenuous stream of atomic particles from the Sun that flows outwards through the Solar System.

spectral lines   narrow lines that cross the spectrum of an object; the lines can be either bright (*emission lines*) or dark (*absorption lines*). Each line in the spectrum corresponds to a particular wavelength at which atoms or molecules absorb or emit light.

spectroscope   a device for taking the spectrum of an object. Spectroscopes use a prism or a diffraction grating to split light into a spectrum; usually the spectrum is then recorded by an electronic detector, in which case the device is known as a *spectrograph*. A spectroscope with good spectral resolution is said to have high dispersion; it spreads out the wavelengths of light more than a low-dispersion device, but the spectrum is fainter and requires a longer exposure time to be recorded.

spectrum, visible   the rainbow-like band of colours that is produced when light is split into its constituent wavelengths. Features in the spectrum, such as bright and dark lines, tell astronomers about the composition and motion of gas in the object under study.

supergiant star   a star many times the mass of the Sun that is swelling up as it ages. Supergiants are the largest and brightest stars known. Many, perhaps all of them, eventually explode as supernovae.

synodic   to do with conjunctions. For example, the *synodic period* of a planet is the time taken for it to return to conjunction with the Sun as seen from the Earth (or conjunction with the Earth as seen from the Sun). The synodic period of a satellite is the mean interval between conjunctions as observed from its parent planet; during this time, the satellite goes through a complete cycle of phases. Compare sidereal.

telluric   to do with the Earth, e.g. telluric lines in a star's spectrum are a result of the passage of the star's light through the Earth's atmosphere.

terminator   the dividing line between the illuminated and dark portions of a planet or satellite, particularly the Moon. The terminator is the sunrise or sunset line, the boundary between day and night.

third contact   the moment when a total eclipse ends. At a solar eclipse, it is when the Sun starts to reappear from behind the Moon; at a lunar eclipse, it is when the Moon starts to emerge from the Earth's umbra.

topocentric   as seen from a point on the surface of the Earth. The *topocentric coordinates* of a nearby body in space, such as the Moon, are slightly different from those that would be measured from the centre of the Earth (geocentric coordinates).

twinkling   the flickering of a star's light caused by air currents in the Earth's atmosphere which distort the path of light rays, causing the star to change in apparent brightness and to flash different colours, particularly when close to the horizon. Planets do not twinkle as much as stars, because they are not point sources, but under bad conditions even planets can twinkle, especially when low down. A large amount of twinkling is a sign of bad seeing.

ultraviolet   radiation with wavelengths shorter than visible violet light but longer than X-rays, from about 10 to 400 nanometres.

umbra   the dark central part of a sunspot or shadow. From within the umbra of the Moon's shadow, a total eclipse of the Sun is visible. The Moon is totally eclipsed when it is completely within the umbra of the Earth's shadow; when it is partly immersed in the Earth's umbra, it is partially eclipsed.

Van Allen belts   two doughnut-shaped zones of atomic particles around the Earth. They consist of electrons and protons trapped inside the Earth's magnetosphere.

wavelength (λ)   the distance between a given point on one wave to the same point on the next wave. The wavelength of light is usually measured in either nanometres or angstroms (an angstrom is one-tenth of a nanometre). Wavelength is equal to the speed of the wave divided by its frequency – hence high-frequency waves have a short wavelength, and vice versa.

white dwarf   a tiny, hot star that is the end-point in the life of stars like the Sun. A typical white dwarf contains as much mass as the Sun compressed into a ball not much larger than the Earth. They cool with age, so the oldest of them are not actually white. The easiest white dwarf to observe is a member of the triple-star system Omicron-2 (o^2) Eridani.

X-rays   radiation with wavelengths shorter than ultraviolet light but longer than gamma rays, between about 0.01 and 10 nanometres.

Zeeman effect   the splitting of spectral lines into two or more parts by a magnetic field.

zodiac   the band of 12 constellations through which the Sun passes each year: Aries, Taurus, Gemini, Cancer, Leo, Virgo, Libra, Scorpius, Sagittarius, Capricornus, Aquarius and Pisces.

Index

A

aberration, optical 23, 29, 30, 37, 185
aberration of starlight 6, 185
   annual 6
   constant of 6
   diurnal 6
absolute magnitude 109
   of a minor planet 88, 89
absorption lines 50, 115, 185
active galaxies 135
airglow 185
Airy disk 28, 185
albedo 185
alidade 99, 185
almanac 185
almucantar 185
altazimuth mounting 32, 33
altitude 4
amplitude, of a variable star 122
Andromeda Galaxy 134
angular diameter 185
angular distance 27, 185
annual aberration 6
annual parallax 7, 113
annular eclipse 95
anomalistic month 21
anomalistic year 21
ansae 185
Antoniadi scale 26
apastron 122
aperture 28, 29
aphelion 66
apochromat 29, 185
apparent magnitude 108–9
apparent place 5
apparent solar day 8
apparent solar time 8
apparition 185
appulse 185
apsides 185
arc, measure of 186
artificial satellites 101–2
   observing 101–2
   predictions 101
ascending node 66, 122, 189
ashen light 74
aspects of the planets 66–7
associations, stellar 130
asterism 186
asteroids – see minor planets
astigmatism 23, 185
astrometry 186
astronomical constants 181
astronomical twilight 8, 13–16
astronomical unit 65
astrophotography 41–6

aurorae 48, 99
   cameras 41, 42, 48
   camera mounts 42–4
   comets 44, 48, 90
   deep-sky objects 48
   eclipses 48
   films 41, 44–6
   filters 46
   focusing 44
   meteors 48, 93
   Moon 48
   noctilucent clouds 100
   planets 48
   stars 42, 48
   star trailing 42
   Sun 48
   zodiacal light 48
   see also imaging
atlases, star 107
atmosphere
   extinction 26, 108, 129–30, 187
   refraction 3, 7, 190
   seeing 26
   transparency 26, 107
atomic time 18
aurorae 98–9
auroral ovals 98
autoguiders 46–7
autumnal equinox 2, 3
averted vision 26
azimuth 4
azimuthal coordinates 4

B

B magnitude 108, 109
Baily's beads 95
Barlow lens 32, 44
Barnard's Star 7, 114
barred spiral galaxies 135
barycentre 5, 18, 186
Barycentric Dynamical Time (TDB) 18
barycentric positions 5
Bayer letters 103
Betelgeuse 114
Big Bang theory 136, 186
binary stars – see double stars
binoculars 28
BL Lac objects 135
black body 114, 115, 186
black hole 118, 135, 186
Bode's law 186
Bok globules 133
bolometric correction 110
bolometric magnitude 110
brown dwarfs 115, 116
butterfly diagram 52

C

Callisto 81
cameras 41, 42, 48
camera mounts 42–4
captured rotation 54, 186
Carrington rotations 51
Cassegrain telescope 29, 30, 36, 38
Cassini Division 83
cataclysmic variables 126, 128
catadioptric telescopes 30, 34, 36
catalogues, star 106
CCDs 40–41, 46–8, 129, 186
CCD autoguiders 46–7
celestial equator 1, 2, 3, 4
celestial latitude 4
celestial longitude 4
celestial poles 1, 2, 4, 5
celestial sphere 1–7, 21
   angular distances on 27
   coordinates on 3–5
central meridian 186
   of Jupiter 80
   of Mars 77
   of Saturn 83
Cepheid variables 113, 124, 128
Ceres 86
Charon 85–6
chromatic aberration 29, 37, 185
chromosphere 50, 53
circumpolar stars 1–2
civil twilight 8
clusters, star 130, 133
   globular 133
   moving 130
   open 130
coatings, mirror 30, 40
collimation 38, 39, 186
colour excess 110
colour index 109, 110, 116, 118
colour temperature 114
colures 3
coma, cometary 88, 186
coma, optical 185, 186
comes 120, 186
comets 88–91
   brightness 89, 91
   coma 88
   imaging 44, 48, 90
   Kuiper Belt 86, 88
   long-period 88
   and meteors 89
   naming 89
   nucleus 88
   observing 89–91
   Oort Cloud 88
   orbits 89

   periodic 88
   tails 89
common proper motion 120
conjunction 66, 67
constants, astronomical 181
constant of aberration 6
constellations 103, 104
   boundaries 103
   names and abbreviations 103, 104
continuous spectrum 186
continuum 186
conversion factors 181
coordinates, celestial 3–5
   ecliptic 4
   equatorial 3–4
   galactic 4
   heliocentric 4–5
   horizontal 4
Coordinated Universal Time (UTC) 18
corona, solar 50, 53, 95
coronal holes 50, 98, 186
cosmic rays 186
cosmology 186
coudé focus 186
counterglow 100–101
Crab Nebula 127, 134
Crab Pulsar 127–8
craters, lunar 55
culmination 2
cusps 73, 187

D

Danjon scale 97
dark adaptation 23, 26
dark nebulae 133, 134
date 7–8
Dawes limit 29, 122
day 7, 8
   apparent solar 8
   mean solar 8
   sidereal 8
declination 1, 3–4, 5
declination axis 33, 39
deep-sky objects xi, 130–38, 187
   observing 136–8
   photographing 48
defect of illumination 187
delta-T (ΔT) 21
descending node 189
dew-caps 35
diamond ring 95
dichotomy 73, 74, 187
differential rotation 187
diffraction 185, 187
diffraction rings 38, 185
direct motion 67, 122

dispersion 187
distance modulus 109
distortion, optical 185
diurnal aberration 6
Dobsonian mounting 32, 33
domes, lunar 64
Doppler effect 187, 190
double stars x, 118–122
    catalogues 106
    eclipsing 120, 123, 126–7
    masses 114
    observing 120–22
    orbits 120–22
    spectroscopic 113, 120
doublet 187
draconic month 21
drive, telescope 33, 34, 42, 43
dwarf nova 126, 128
dwarf star 115, 116, 118, 187
dynamical parallax 113

E

early-type stars 115, 187
earthshine 55
eccentricity, orbital 66, 122, 187
eclipses 21, 94–7
    annular 95
    lunar 96–7
    partial 95
    penumbral 94, 97
    photographing 48
    solar 50, 95–7
    total 95, 97
eclipse year 21
eclipsing binaries 120, 123, 126–7
ecliptic 1, 2–3, 4
    obliquity of 2, 3, 6
ecliptic coordinates 4
ecliptic poles 2
Edgeworth–Kuiper Belt 86
effective temperature 114
electromagnetic spectrum 107, 191
element, optical 29, 187
elements, orbital 66, 89, 120, 122
elliptical galaxies 134, 135
elongation 67, 187
emersion 187
emission lines 187
emission nebulae 133, 138
emulsion, photographic – *see* films
Encke division 83
ephemeris 187
epoch 5, 187
equation of the equinoxes 6
equation of time 3, 17–18
equator, celestial 1, 2, 3, 4
equator, galactic 4
equatorial coordinates 3–4
equatorial mounting 32, 33, 39, 40
equinoctial colure 3
equinoxes 2, 3, 4, 5, 21
    equation of 6
Erfle eyepiece 31
Eros 86, 88
eruptive variables 124, 128
escape velocity 187
Europa 81
exit pupil 31, 38, 187
expansion of the Universe 135–6
extinction 187
    atmospheric 26, 108, 129–30
    interstellar 109–10, 133, 134
extrapolation 187
extrinsic variables 123, 127–8
eye 23
eye relief 31
eyepieces 29, 30–32, 37–8, 43, 44

Erfle 31
Huygenian 31
Kellner 31
Nagler 31
orthoscopic 31
Plössl 31
Ramsden 31
eyepiece projection 44

F

faculae 53
field of view 27, 31, 32, 35, 38, 42, 44
field star 187
filaments 53
films, for astrophotography 41, 44–6
    hypersensitization 45
    pre-flashing 45–6
    reciprocity failure 45
filters
    Hβ 35, 138
    infrared-blocking 46, 48
    light-pollution-reduction (LPR)
        35, 46, 137
    for Mars 77–8
    nebula 35, 137, 138
    O III 35, 138
    photography through 46
    for Saturn 84
    solar 35
    for Venus 74
    Wratten 35
finders 35–6
fireballs 93
first contact 188
first point of Aries 3
Flamsteed numbers 103
flares, solar 53, 98
flash spectrum 50
focal length 29, 32, 37, 42, 44, 188
focal plane 188
focal ratio 29, 188
focus, of an ellipse 66, 188
focus, optical 188
focusing 44
following 27, 188
fortnightly nutation 6
fourth contact 188
frequency 188
fundamental star 188

G

galactic cluster 130, 188
galactic coordinates 4
galactic equator 4
galactic halo 133, 134
galactic latitude 4
galactic longitude 4
galactic poles 4
galaxies 134–6
    active 135
    barred spiral 135
    clusters of 135
    elliptical 134, 135
    irregular 135
    Local Group 135, 136
    observing 138
    recession of 135–6
    Seyfert 135
    spiral 134–5
Galaxy 49, 133, 134, 135
gamma rays 107, 188
Ganymede 81
gegenschein 100–101
general precession 5
geocentric coordinates 3, 4, 5
geometrical albedo 185
giant star 116, 117, 118, 188

gibbous 188
globular clusters 133
    observing 138
GO TO telescopes 34–5
Gould's Belt 188
granulation, solar 49, 53
grazing occultation 97, 98
great circle 188
Great Red Spot 78, 79–80
greatest brilliancy (Venus) 73
greatest elongation 67
Greek alphabet 105
green flash 188
greenhouse effect 188
Greenwich hour angle 18
Greenwich Mean Time (GMT) 17
Greenwich sidereal time 18, 19
guiding 36, 43–4, 46–7

H

heavy elements 188
heliacal rising and setting 188
heliocentric coordinates 4–5
heliocentric latitude 4
heliocentric longitude 4–5
heliographic coordinates 4, 51, 54
Hertzsprung–Russell (HR) diagram
    116, 117, 118, 119
high-velocity stars 114
Hipparcos 109, 113
horizon 1, 4
horizontal coordinates 4
Horsehead Nebula 133
hour angle 4, 18
    Greenwich 18
    local 18
hour circle 3, 4
HR diagram – *see* Hertzsprung–
    Russell diagram
Hubble's constant 136
Hubble's law 135
Huygenian eyepiece 31
Hyades 130
hypersensitization of films 45

I

Iapetus 84
imaging 40–48
    CCDs 40–41, 46–8, 129, 186
    image processing 47, 48
    video cameras 47
    webcams 47
    see also astrophotography
immersion 188
inclination, axial 188
inclination, orbital 66, 122, 188
inferior conjunction 67
infrared 188
infrared-blocking filter 46, 48
integrated magnitude 108
interferometers 114, 188
International Atomic Time (TAI) 18
International Sunspot Number 52
interpolation 188
interstellar extinction 109–10, 133,
    134
intrinsic variables 128–9
invariable plane 5
inverse-square law 188
Io 81
ion 188
ionization 189
irradiation 189
irregular galaxies 135
irregular variables 124, 128

J

Julian date 7–8
    modified 8
Julian year 5
Jupiter 78–81
    belts and zones 79
    features of 79–80
    Great Red Spot 79–80
    observing 80–81
    oppositions 78
    satellites of 81, 98
    Systems I and II 78–9
    white ovals 80

K

Kellner eyepiece 31
Kepler's laws 66
Kirkwood gaps 86, 189
Kuiper Belt 86, 88

L

Lagrangian points 189
late-type stars 115, 189
latitude variation 7
leap second 18
libration 55, 64, 65
light, speed of 112, 189
light curve 122, 123, 189
light-pollution 189
light-pollution-reduction (LPR)
    filters 35, 46, 137, 138
light-time 65, 189
light year 112
limb 189
limb darkening 53
limiting magnitudes 24–5, 28, 29, 94
Local Group 135, 136
local hour angle 18
local sidereal time 18
local standard of rest 113, 189
long-period variables 128
luminosity 109, 110
luminosity class 115–16
lunar eclipses 94–5, 97
    penumbral 94, 97, 190
lunar nutation 6
lunation 21, 54, 65, 189
lunisolar precession 5

M

M numbers 130
Magellanic Clouds 134
magnetosphere 189
magnification 23, 28, 32, 37–8, 44,
    189
magnitude 107–10
    absolute 109
    apparent 108–9
    blue 108, 109
    bolometric 110
    combined 108
    of comets 89
    of eclipse 95
    integrated 108
    limiting 24–5, 28, 29, 94
    of minor planets 88
    photoelectric 108–9
    photographic 108
    photovisual 108
    of planets 67, 70
    visual 108
main sequence 116, 117, 118, 119
major axis 66, 189
Maksutov telescope 29, 30
maria, lunar 55
Mars 74–8

clouds 75
dust storms 75
map of 76
observing 75–8
oppositions 74, 75
polar caps 75–6
polar hood 76
satellites of 75
surface markings 75
mass–luminosity relationship 110
mean equator and equinox 5
mean place 5
mean solar day 8
mean solar time 17
mean Sun 8
Mercury 71–3
motion of 71–2
observing 72–3
transits of 72
meridian 2, 4, 18
Messier objects 130, 131–2
meteors 91–4, 189
fireballs 93
observing 93
photographing 48, 93
population index 94, 190
radiant 91, 94
showers 91–3
sporadic 91, 94
zenithal hourly rate (ZHR) 91, 94
meteorite 93, 189
meteoroid 91, 189
Metonic cycle 189
micrometers 36
Milky Way 134
minor axis 66, 189
minor planets 86–8
Hirayama families 87
observing 88
terminology and naming 87
Trojans 86
Mira stars 125, 128
mirrors 29–30, 37, 38, 40
MKK system 115
mock Sun 189
month
anomalistic 21
draconic 21
sidereal 21
synodic 21
tropical 21
Moon 54–65
craters 55
domes 64
eclipses 94–5, 97
highlands 55
libration 55, 64, 65
maps 56–63
maria 55
motions 54–5
observing 64–5
occultations 97–8
phases 54–5, 65, 189
photographing 42, 47, 48
rays 64
rilles and faults 64
rotation 54
surface features 55–64
terminator 64–5
transient phenomena (TLPs) 64
mountings, camera 42–4
mountings, telescopic 32–5, 39–40
altazimuth 32, 33
Dobsonian 32, 33
equatorial 32, 33, 39, 40
moving clusters 130
multiple stars – see double stars

N

nadir 1
Nagler eyepiece 31
nautical twilight 8
nebulae 133–4
dark 133
emission 133
observing 137–8
planetary 133
reflection 133
supernova remnants 134
nebula filter 35, 137, 138
Neptune 85
Nereid 85
neutron stars 115, 118, 189
Newtonian telescope 29, 20, 32, 34, 36, 38
NGC numbers 130
noctilucent clouds 99–100
node 189
novae 126, 128–9
dwarf 126, 128
recurrent 126, 128
nucleus, cometary 88
nutation 3, 5–6

O

OB associations 130
object glass (objective) 29
oblateness 189
obliquity of ecliptic 2, 3
nutation in 6
observations 23, 26–7
recording 26–7
techniques 23, 26
timing 27
occultations 97–8
occulting bar 190
Omega Centauri 133
Oort Cloud 88
open clusters 130
observing 138
opposition 66
of Jupiter 78
of Mars 74, 75
of Saturn 82
orbits
of artificial satellites 101–2
of binary stars 120–22
of comets 89
elements of 66, 89, 120, 122
laws of 66
of planets 66, 67
Orion Nebula 133
orthoscopic eyepiece 31
osculating elements 66

P

paraboloid 190
parallax 7, 112–13
annual 7, 113
dynamical 113
horizontal 113
planetary 113
secular 113
solar 65
spectroscopic 113
trigonometric 112–13
parhelion 190
parsec 112
penumbra 190
of shadow 93, 97, 190
of sunspot 50, 190
penumbral eclipses 94, 97, 190
periastron 122
perihelion 66

period 190
orbital 67, 122
sidereal 67, 191
synodic 67, 191
period–luminosity law 113
periodic comets 88
perturbation 190
Phaethon 89
phases 190
of Mars 75
of Mercury 72
of Moon 54–5, 65, 189
of planets 67, 71, 187
of Venus 73, 74
phase angle 190
photography – see astrophotography;
see also imaging
photometry 37, 88, 107, 108, 109,
129, 190
photon 190
photosphere 49, 50, 51, 52
plages 53
planes of reference 3, 4, 5
planets 65–86
aspects 66–7
brightnesses 67, 70
data 67–8
occultations by 98
orbits 66, 67
phases 71
photographing 48
rings of 71
satellites 69–70
visibility 67, 70–71
planetary nebulae 133
observing 138
planetary precession 5
planisphere 190
Pleiades 130, 133
Plössl eyepiece 31
Pluto 85–6
polar alignment 39–40, 43
polar axis 33, 39–40
polar distance 4
polar motion 7
Polaris 5
polarization 110
pole
celestial 1, 2, 4, 5
ecliptic 2
galactic 4
Pole Star 5
populations, stellar 118, 134
population index 94, 190
position 3–7
apparent place 5
coordinates 3–5
mean place 5
true place 5
position angle 27, 190
of Moon's axis 65
of Sun's axis 51
powers, magnifying – see
magnification
preceding 27, 190
precession 3, 5, 6, 21
pre-flashing 45–6
primary 190
prime focus 190
probable error 181
prominences 53, 95
proper motion 7, 113–14, 120
Proxima Centauri 112, 113
pulsars 118, 125, 127–8, 190
pulsating variables 124–5, 128
Purkinje effect 26

Q

quadrature 67, 71
quasars 135, 190

R

radial velocity 113
radiant, of meteor shower 91, 94
radiation belts 190
radiation pressure 89, 116
radio astronomy 190
radius vector 66, 190
Ramsden eyepiece 31
rays, lunar 64
reciprocity failure 45
red dwarfs 118, 190
red giants 117–8, 190
redshift 135, 136, 190
Red Spot 78, 79–80
reflecting telescopes 29–30, 32, 33–4,
36, 38, 39
Cassegrain 29, 30, 32, 38
Newtonian 29, 30, 32, 34, 36, 38
reflection nebulae 133
refracting telescopes 29, 32, 33, 36, 38
refraction, atmospheric 3, 7, 190
regression of nodes 189
Relative Sunspot Number 52
residuals 190
resolution 23, 28–9, 120, 122
retrograde motion 67, 122, 190
revolution 190
right ascension 1, 3–4, 5
rilles, lunar 64
rings of planets 71
of Saturn 82–3
rising and setting 1, 2
Roche lobe 123, 127
rotating variables 125
rotation 191
runaway stars 114

S

Saros 21, 191
satellites, artificial 101–2
observing 101–2
predictions 101
satellites, of planets 69–70
of Jupiter 81, 98
of Mars 75
of Neptune 85
of Pluto 85–6
of Saturn 84, 98
of Uranus 85
Saturn 82–4
features of 82
observing 83–4
oppositions 82
passages through ring plane 83
rings of 82–3
satellites of 84, 98
Systems I and II 83–4
Schmidt–Cassegrain telescope 29,
30, 32
Schröter effect 74
Schwarzschild radius 118
Scotch mount 42–3
second 7
leap 18
second contact 191
secondary 191
secondary mirror 29, 30, 38
secondary spectrum 29, 191
secular parallax 113
secular variables 129
seeing 26, 191
semi-diurnal arcs 2

semi-major axis 66, 122, 191
semi-regular variables 125, 128
setting circles 191
Seyfert galaxies 135
shadow bands 95
Shoemaker–Levy 9, Comet 79, 89
short-period comets 88
SI units 179, 180
sidereal day 8
sidereal month 21
sidereal period 67, 191
sidereal time 18, 27, 191
   Greenwich 18, 19
   local 18
sidereal year 21
small circle 191
solar apex 49
solar cycle 51–2
solar day 8
   apparent 8
   mean 8
solar eclipse 50, 94–5
   Baily's beads 95
   diamond ring 95
   shadow bands 95
solar filter 35
solar nutation 6
solar parallax 65
solar time
   apparent 8
   mean 17
solar wind 50, 89, 98, 191
solstices 2, 3
solstitial colure 3
spectra, classification of 115–16
spectral lines 115–16, 191
spectral type 115, 116, 118
spectroscope 191
spectroscopic binary 113, 120
spectroscopic parallax 113
spectrum, electromagnetic 107, 191
spectrum, stellar 115–16, 185, 191
spherical aberration 23, 30, 37, 185
spherical albedo 185
spicules 53
spiral galaxies 134–5
sporadic meteors 91, 94
Spörer's law 52
standard epoch 5, 187
standard error 181
stars 103–130
   atlases 107
   brightest 111
   catalogues 106
   densities 115, 117
   diameters 114–15, 117
   distances 109, 112–13, 130
   dwarf 115, 116, 118, 187
   evolution 116–18
   formation 117
   giants 116, 117, 118, 188
   high-velocity 114
   lifetimes 117
   luminosity 109, 110, 117
   main sequence 116, 117, 118
   mass–luminosity relationship 110
   masses 110, 114, 117
   motions 113–14
   names 103, 105, 122–3
   nearest 111
   occultations 97–8
   photographing 42, 48

populations 118, 134
positions of 5–7
red dwarfs 118, 190
red giants 117–18, 190
runaway 114
spectral classification 115–16
supergiants 115, 116, 117, 118, 191
temperatures 114
white dwarfs 115, 118, 191
   see also double stars; star clusters; variable stars
star atlases and catalogues 106, 107
star clusters 130, 133
   globular 133
   moving 130
   open 130
star diagonals 37
star trailing 42
stationary points 67
Stefan's law 114–15
stepper motors 34
summer solstice 2, 3
Sun 49–54
   active areas 53–4
   chromosphere 50, 53
   corona 50, 53, 95
   eclipses 50, 94–5
   energy 49
   faculae 53
   filaments 53
   flares 53, 98
   flash spectrum 50
   granulation 49, 53
   limb darkening 49
   observing 49, 53–4
   photographing 48
   photosphere 49, 50, 51, 52
   plages 53
   position of 17
   prominences 53
   rotation 51
   solar cycle 51–2
   solar wind 50, 191
   spicules 53
   sunspots 50–4
sundog 189
sunrise and sunset 3, 8, 9–12
sunspots 50–4
   butterfly diagram 52
   numbers 52–3
   penumbra 50, 190
   Spörer's law 52
   umbra 50–51, 191
   Wilson effect 50–51
supergiant star 115, 116, 117, 118, 191
superior conjunction 67
supernovae 118, 123, 126, 128, 129, 134
supernova remnants 134
symbols and abbreviations 181–2
synodic 191
synodic month 21, 54–5
synodic period 67, 191
syzygy 66

T

T associations 130
telecompressor 44
telescopes 23, 26, 27, 28–40
   accessories 35–7
   Cassegrain 29, 30, 36, 38

catadioptric 30, 34, 36
cleaning optics 40
collimation 38, 39
computer-controlled 32, 34–5
directions in 27
eyepieces 30–2
field of view 27, 31, 38, 44
finders 35–6
guiding 36, 43–4, 46–7
imaging through 44, 46–8
limiting magnitude 28, 29
Maksutov 29, 30
mirrors 29–30, 37, 38, 40
mountings 32–5, 39–40
Newtonian 29, 30, 36, 38
reflecting 29–30, 33–4, 36
refracting 29, 32, 33, 36, 38
resolution 23, 28–9, 120, 122
Schmidt–Cassegrain 29, 30
tests and adjustments 37–40
terminator 191
   of Mars 75
   of Moon 64–5
   of Venus 73–4
Terrestrial Time (TT) 18, 21
third contact 191
time 8–21
   apparent solar 8
   Barycentric Dynamical (TDB) 18
   ΔT 21
   equation of 3, 17–18
   Greenwich Mean (GMT) 17
   Greenwich sidereal (GST) 18, 19
   International Atomic (TAI) 18
   local sidereal 18
   mean solar 17
   sidereal 18, 27, 191
   Terrestrial (TT) 18, 21
   Universal (UT) 18, 21
time zones 18, 20
Titan 84
topocentric coordinates 3, 191
total eclipse 94, 95, 97
transient lunar phenomena (TLPs) 64
transit, meridian 2
   of Mercury 72
   of Venus 74
trans-Neptunian objects 86
transparency, of atmosphere 26, 107
Trapezium 120, 133
trigonometric parallax 112–13
Triton 85
Trojan asteroids 86
tropical month 21
tropical year 21
tropics 3
true equator and equinox 5
true place 42
twilight 8, 13–16
   astronomical 8, 13–16
   civil 8
   nautical 8
twinkling 26, 191

U

UBV system 108–9
ultraviolet 191
umbra 191
   of shadow 95, 97, 191
   of sunspot 50–51, 191
units and notation 179–81
unit distance 65

Universal Time (UT) 18, 21
Universe
   expansion of 135–6
   origin of 136
Uranus 85

V

V magnitude 107, 108–9
Van Allen belts 191
variable stars x–xi, 122–130
   cataclysmic 126, 128
   catalogues 106
   Cepheid 113, 124, 128
   dwarf novae 126, 128
   eclipsing 120, 123, 126–7
   eruptive 124, 128
   irregular 124, 128
   light curves 122, 123
   long-period 128
   Mira-type 125, 128
   nomenclature 122–3
   novae 126, 128–9
   observing 129
   pulsating 124–5, 128
   rotating 125
   secular 129
   semi-regular 125, 128
   supernovae 123, 126, 128, 129
   X-ray 127, 129
Veil Nebula 134, 138
Vela Pulsar 127–8
Venus 73–4
   ashen light 74
   cusps 73
   dichotomy 73, 74
   greatest brilliancy 73
   movements 73
   observing 73–4
   phases 73
   transits of 74
vernal equinox 2, 3
Vesta 86
video cameras 47
Virgo cluster 135

W

wavelength 191
webcams 47
white dwarfs 115, 118, 119
Wilson effect 50–51
winter solstice 2, 3
Wratten filters 35

X

X-ray binaries 127, 129
X-rays 107, 191

Y

year 7, 21
   anomalistic 21
   eclipse 21
   Julian 5
   sidereal 21
   tropical 21

Z

Zeeman effect 191
zenith 1, 2, 4
zenith distance 4
zenithal hourly rate (ZHR) 91, 94
zodiac 191
zodiacal light 100–101

The text of this book is set in *Minion*, designed by Robert Slimbach;

The heads are set in *Devinne*, designed by Gustav Schroeder;

and the subheads are set in *Clearview*, designed by James Montalbano.

The book was designed and illustrated by Charles Nix & Associates.